Transformations in

Contributions to Economics

http://www.springer.de/cgi-bin/search_book.pl?series=1262

Sardar M.N. Islam
Mathematical Economics of Multi-Level Optimisation
1998. ISBN 3-7908-1050-9

Sven-Morten Mentzel
Real Exchange Rate Movements
1998. ISBN 3-7908-1081-9

Lei Delsen/Eelke de Jong (Eds.)
The German and Dutch Economies
1998. ISBN 3-7908-1064-9

Mark Weder
Business Cycle Models with Indeterminacy
1998. ISBN 3-7908-1078-9

Tor Rødseth (Ed.)
Models for Multispecies Management
1998. ISBN 3-7908-1001-0

Michael Carlberg
Intertemporal Macroeconomics
1998. ISBN 3-7908-1096-7

Sabine Spangenberg
The Institutionalised Transformation of the East German Economy
1998. ISBN 3-7908-1103-3

Hagen Bobzin
Indivisibilities
1998. ISBN 3-7908-1123-8

Helmut Wagner (Ed.)
Current Issues in Monetary Economics
1998. ISBN 3-7908-1127-0

Peter Michaelis/Frank Stähler (Eds.)
Recent Policy Issues in Environmental and Resource Economics
1998. ISBN 3-7908-1137-8

Jessica de Wolff
The Political Economy of Fiscal Decisions
1998. ISBN 3-7908-1130-0

Georg Bol/Gholamreza Nakhaeizadeh/
Karl-Heinz Vollmer (Eds.)
Risk Measurements, Econometrics and Neural Networks
1998. ISBN 3-7908-1152-1

Joachim Winter
Investment and Exit Decisions at the Plant Level
1998. ISBN 3-7908-1154-8

Bernd Meyer
Intertemporal Asset Pricing
1999. ISBN 3-7908-1159-9

Uwe Walz
Dynamics of Regional Integration
1999. ISBN 3-7908-1185-8

Michael Carlberg
European Monetary Union
1999. ISBN 3-7908-1191-2

Giovanni Galizzi/Luciano Venturini (Eds.)
Vertical Relationships and Coordination in the Food System
1999. ISBN 3-7908-1192-0

Gustav A. Horn/Wolfgang Scheremet/
Rudolf Zwiener
Wages and the Euro
1999. ISBN 3-7908-1199-8

Dirk Willer
The Development of Equity Capital Markets in Transition Economies
1999. ISBN 3-7908-1198-X

Karl Matthias Weber
Innovation Diffusion and Political Control of Energy Technologies
1999. ISBN 3-7908-1205-6

Heike Link et al.
The Costs of Road Infrastructure and Congestion in Europe
1999. ISBN 3-7908-1201-3

Simon Duindam
Military Conscription
1999. ISBN 3-7908-1203-X

Bruno Jeitziner
Political Economy of the Swiss National Bank
1999. ISBN 3-7908-1209-9

Irene Ring et al. (Eds.)
Regional Sustainability
1999. ISBN 3-7908-1233-1

Katharina Müller/Andreas Ryll/
Hans-Jürgen Wagener (Eds.)
Transformation of Social Security: Pensions in Central-Eastern Europe
1999. ISBN 3-7908-1210-2

Stefan Traub
Framing Effects in Taxation
1999. ISBN 3-7908-1240-4

continued on page 384

Peter Meusburger · Heike Jöns
(Editors)

Transformations in Hungary

Essays in Economy and Society

With 97 Figures and 92 Tables

Physica-Verlag

A Springer-Verlag Company

Series Editors
Werner A. Müller
Martina Bihn

Editors
Professor Dr. Peter Meusburger
Dipl.-Geogr. Heike Jöns
University of Heidelberg
Department of Geography
Berliner Straße 48
69120 Heidelberg
Germany
peter.meusburger@urz.uni-heidelberg.de
heike.joens@urz.uni-heidelberg.de

ISSN 1431-1933
ISBN 3-7908-1412-1 Physica-Verlag Heidelberg New York

Cataloging-in-Publication Data applied for
Die Deutsche Bibliothek – CIP-Einheitsaufnahme
Transformations in Hungary: essays in economy and society; with 92 tables / Peter
Meusburger; Heike Jöns (ed.). – Heidelberg: Physica-Verl., 2001
    (Contributions to economics)
    ISBN 3-7908-1412-1

Physica-Verlag Heidelberg New York
a member of BertelsmannSpringer Science+Business Media GmbH

© Physica-Verlag Heidelberg 2001
Printed in Germany

Softcover Design: Erich Kirchner, Heidelberg

SPIN 10841987   88/2202-5 4 3 2 1 0 – Printed on acid-free and non-aging paper

Preface

In the 1990s, Hungary experienced profound changes in economy and society as a result of transforming the socialist one-party state with its centrally planned economy into a multi-party democracy based on a market economy. The aim of this collection of essays is to analyse related dynamics in time and space and thus to contribute to a further understanding of Hungary's ongoing transformation processes and its current situation as one of the leading candidates for EU membership.

This anthology draws together a broad range of topics. It addresses regional development since socialist times (see chapter by Nemes-Nagy), deals with constitutive elements of a modern market economy such as banking (Jöns), foreign direct investment (Hunya), entrepreneurship (Kuczi and Lengyel), knowledge resources (Meusburger I), and the labour market (Dövenyi; Meusburger II), and incorporates essays on people's education and income (Meusburger II), their poverty situation (Spéder), and post-communist voting behaviour (Kovács). Further papers explore urban issues, such as urban development and competitiveness, the housing market or post-socialist residential mobility, either with regard to Budapest (Cséfalvay; Izsák and Probáld) or other cities within the Hungarian urban system (Izsák and Nemes-Nagy; Sailer). Finally, informal as well as institutionalised cross-border co-operations are discussed by looking at the levels of settlements and regions (Rechnitzer).

The majority of the papers explore spatial disparities *within Hungary* by looking at different regions and settlements, individual places, urban structures or the urban hierarchy. They examine the creation of new political, socio-economic and legal frameworks, and institutions, their effects on people living in different places, and the question of how people have dealt with the related, often fundamental changes. In the context of interdisciplinary transformation studies, the main message of such profoundly geographical approaches is that one cannot fully understand Hungary's development and its current situation without taking account of its past and present changing spatial relations, disparities, and local particularities.

A further group of papers concentrates on a comparative perspective *within Europe*, particularly within Central and Eastern Europe, by addressing different national developments (see chapters by Hunya; Kuczi and Lengyel) and the European city system (Cséfalvay). The first and last chapters, on knowledge resources (Meusburger) and cross-border co-operations (Rechnitzer), discuss both regional and international aspects of Hungary's transformation in the 1990s.

An important issue of the international comparisons is European integration at the beginning of the 21st century since Hungary, together with the Czech Republic, Estonia, Poland, and Slovenia, belongs to the first-tier EU accession countries that may join the EU in 2003.[1] In the first decade after the political change in 1989/90, the former socialist countries in Central and Eastern Europe faced serious problems including recession, inflation, budget deficits, unemployment and an increase of poverty. However, macroeconomic analysis shows that Hungary has, in several areas up until the present, been more successful than its neighbouring countries in establishing a market economy of Western European style. Hungary has attracted more foreign direct investment than any other Central and East European country, it has achieved higher growth rates than most EU states, it has reduced its unemployment rates to a level far below that of several EU states, and it has been very successful on export markets. The papers in this book point out that one of Hungary's greatest comparative advantages stemmed from early political and economic reforms as well as from early contacts and experiences with Western Europe. As early as 1968, Hungary launched the New Economic Mechanism. This programme that gradually introduced decentralised management structures, allowed for the influence of market forces, lead to a slow but steady rise in living standards, and made travelling to the west easier. A further decisive move towards market principles came about at the end of the 1970s because the state was heavily in debt. New types of economic organisations appeared during the 1980s, the number of legalised private enterprises increased considerably, and, after a serious financial crisis in 1982, the country joined the International Monetary Fund, the World Bank, and the International Finance Co-operation, promising that new loans would be allocated according to business principles. Some years before the fall of the iron curtain, in 1987, Hungary was also the first amongst the former socialist countries in Central and Eastern Europe to restore the two-tier banking system typical of capitalist market economies. Even longer before the political change, the country had started to open itself to western tourists, to allow students and researchers to study abroad, and to improve bilateral relations with its western neighbours. The strong socio-economic disparities, however, that were identified in several contributions to this book, and particularly the spatial polarisation between Budapest's agglomeration and the country's north-west as prosperous regions on the one hand, and an economically depressed north-east on the other hand, cast a shadow over Hungary's relative success in the 1990s, and gave some contributors cause for one or the other critical remark.

In addition to the analysis of recent socio-economic structures, processes, and relations, some essays offer innovative examples of the application of different theoretical approaches to studies of economy and society in general, and trans-

---

[1]   On March 7, 2001, Sweden's foreign minister Anna Lindh, whose country held the EU's presidency, delivered an optimistic speech about enlargement of the European Union (EU) in Washington D.C. She said that the EU, comprising 15 nations, might be ready to admit new members in 2003. Ten of the twelve countries that are candidates for membership are part of Central and Eastern Europe. In addition to the five first-tier countries, the five other countries from this region are Bulgaria, Latvia, Lithuania, Romania, and Slovakia; Malta and Cyprus are also seeking EU membership (see http://www.rferl.org/nca/features/2001/03/08032001115133.asp).

formation studies in particular. These examples comprise of reflections upon Budapest's competitive position in regard to capital concentration and knowledge transfer (Cséfalvay), the construction and interpretation of regional political profiles of Hungary in the 1990s (Kovács), a comparison of different relative and absolute approaches to poverty (Spéder), the understanding of entrepreneurship in the emerging market economies of former socialist countries (Kuczi and Lengyel), and explorations into the interrelationship between foreign direct investment and international competitiveness in the Central European EU-accession countries (Hunya). In regard to a theory of regional transformation, Meusburger argues for a knowledge-based approach in order to account for Hungary's current regional and international situation, whilst the chapter by Jöns considers her empirical findings on regional transformation in Hungarian banking in the light of Bruno Latour's actor-network theory (ANT) and arrives at rather surprising conclusions for both ANT and a theory of regional transformation.

In regard to theoretical issues, two common themes can be traced through the essays: Firstly, it is clearly shown that, compared with preceding decades, people, institutions and places within Hungary have experienced extreme changes and instabilities in time and space during the first decade of transformation towards a market economy of Western European style. This was the result of the change in ideology that became manifest in 1989/90. By 2001, however, the major changes have been completed in most areas of Hungarian economy and society, and thus a period of stabilisation appears to be on its way. Secondly, there seems to be a core of determinants that have been identified by different authors as being responsible for Hungarian regional transformation in the 1990s, and thus for the country's overall situation at the beginning of the 21st century. These aspects include the mental and material legacies of socialist and pre-socialist times, existing differences in economic development, prevailing contact patterns and potentials for interaction that almost reversed for Hungary after the political change, the availability of human and material resources, the agency and mobility of individual and institutional actors, the accessibility of a place from Budapest or neighbouring countries, and, finally, a fundamental re-evaluation of space after 1989/90.

Altogether, however, this collection of essays comprises of a great variety of viewpoints and theoretical approaches, mirroring to a large extent the authors' different disciplinary and institutional backgrounds. Working in geography, sociology and economics, the fourteen contributors are affiliated with the Hungarian Academy of Sciences (Budapest and Győr), the Eötvös Loránd University (Budapest), the University of Economic Sciences and Public Administration (Budapest), the Kodolányi János University (Székesfehérvár), The Vienna Institute for International Economic Studies (WIIW, Vienna), and the Universities of Heidelberg and Trier. The contributors, therefore, offer mainly Hungarian (ten chapters) but also a few foreign, all German, perspectives (four chapters) on Hungary in the 1990s. In the figurative sense, therefore, this English publication strives to contribute to a multiple transgression of language barriers within the 'scientific community.' It seeks to serve as a gateway to work on Hungary's recent transformation processes done in the Hungarian and German speaking worlds - work that is, however, to a considerable extent embedded within the Anglo-American literature on transformation in Central and Eastern Europe (see lists of references).

What all papers have in common is the analysis and presentation of comprehensive empirical data. The generous funding of Hungarian, German, Austrian and European research foundations, and an excellent co-operation with the Hungarian Central Statistical Office (KSH, Budapest) as well as the Hungarian Ministry of Economic Affairs (Budapest) since the end of the 1980s enabled some of the authors to work with unpublished individual data sets that have never been analysed before (see, for example, chapters by Cséfalvay, Dövenyi and Meusburger). Furthermore, the authors could draw either upon their own knowledge and contacts as Hungarians or upon strong ties with Hungarian researchers and institutions in order to generate new quantitative and qualitative data for their investigations by conducting, for example, comprehensive surveys and qualitative interviews (see chapters by Izsák and Probáld; Izsák and Nemes-Nagy; Jöns; Kuczi and Lengyel; Sailer; Spéder) or by exploring other promising and innovative sources. Displayed in various tables, maps and figures, the empirical results presented in the following essays will inform the reader about recent political, legal, and socio-economic structures and ongoing processes within Hungary, their historical backgrounds, dynamics and interrelations. They will make the reader aware of Hungary's, as well as other Central and Eastern European countries', current problems and future prospects after the first decade of socio-economic transformation to market economies of Western European style. However, the empirical evidence presented in this book does not only emphasise the significance of geographical approaches within transformation studies, but also call for critical reflection, re-interpretation, and further scholarly exchange across national boundaries.

Heike Jöns and Peter Meusburger

Acknowledgements

The authors are grateful to all people who either supported the research that enabled the contributions to this anthology, or helped with preparing the book for publication. Representative of all institutions that provided funding, we should like to thank the Austrian Academy of Sciences (Vienna), the EU Commission (Brussels), the German Research Foundation (DFG, Bonn), the National Scientific Research Fund (OTKA), the Open Society Support Foundation (Soros, Budapest), and the Fritz Thyssen Foundation (Cologne) for having supported several research projects on Hungary in the 1990s. The Hungarian Central Statistical Office (KSH, Budapest) and the Hungarian Ministry of Economic Affairs (Budapest) shall stand in for those institutions and individuals who provided unpublished data for our analysis, and thus deserve our gratitude. As some authors individually express their thankfulness to people who contributed to their research, the editors would like to thank Robert Famulok and Alexander Tonnier for computing some of the data that is presented in this book. Special thanks are due to Verena Reitz for carefully assisting with the editing work, and to Christine Brückner for constructing and reworking most of the illustrations. Similarly, we are indebted to Jan Heine, Martina Hoppe and Yvette Tristram for smoothing out major grammatical flaws from essays that were exclusively written by non-native speakers. Many thanks also to Peter Boda for checking the Hungarian that inevitably underwent changes when the manuscripts were transferred from one PC to another. Mike Heffernan got rather accidentally involved at some stage of this project while being Alexander von Humboldt Research Fellow in Heidelberg. We are especially grateful for his editorial advice. We also owe a great deal of gratitude to Véronique Ragot who was involved in the organisational work and helped to see the project through to its finished form. Finally, we should like to thank Werner A. Müller and Gabriele Keidel at Springer for their assistance and patience.

Contents

Contributors

Zoltán Dövényi, Professor of Human Geography, Geographical Research Institute, Hungarian Academy of Sciences, Népköztársaság utja 62, H-1062 Budapest, Hungary

Zoltán Cséfalvay, Deputy State Secretary for Regional Economic Development, Ministry of Economic Affairs, Szemere utca 6, H-1055 Budapest, Hungary; Research Professor, Kodolányi János University College, Irányi Dániel u. 4, H-8000 Székesfehérvár, Hungary

Gábor Hunya, Senior Research Economist, The Vienna Institute for International Economic Studies, WIIW, Oppolzergasse 6, A-1010 Vienna, Austria

Éva Izsák, Assistant Professor of Human Geography, Department of Regional Geography, Eötvös Loránd University, Ludovika Tér 2, H-1083 Budapest, Hungary

Heike Jöns, Lecturer in Human Geography, Department of Geography, University of Heidelberg, Berliner Straße 48, D-69120 Heidelberg, Germany

Zoltán Kovács, Professor of Human Geography, Geographical Research Institute, Hungarian Academy of Sciences, Népköztársaság utja 62, H-1062 Budapest, Hungary

Tibor Kuczi, Associate Professor of Sociology, Department of Sociology and Social Policy, Budapest University of Economic Sciences and Public Administration, Fővám tér 8, H-1093 Budapest, Hungary

György Lengyel, Professor of Sociology, Department of Sociology and Social Policy, Budapest University of Economic Sciences and Public Administration, Fővám tér 8, H-1093 Budapest, Hungary

Peter Meusburger, Professor of Social and Economic Geography, Department of Geography, University of Heidelberg, Berliner Straße 48, D-69120 Heidelberg, Germany

József Nemes-Nagy, Professor of Geography, Department of Regional Geography, Eötvös Loránd University, Ludovika Tér 2, H-1083 Budapest, Hungary

Ferenc Probáld, Professor of Geography, Department of Regional Geography, Eötvös Loránd University, Ludovika Tér 2, H-1083 Budapest, Hungary

János Rechnitzer, Director of the West Hungarian Research Institute, Centre for Regional Studies, Hungarian Academy of Sciences, Liszt Ferenc u. 10, H-9002 Győr, Hungary

Ulrike Sailer, Professor of Cultural and Regional Geography, Department of Geography, University of Trier, D-54286 Trier, Germany

Zsolt Spéder, Director of the Demographic Research Institute, Hungarian Central Statistical Office, Angol út 77, H-1149 Budapest, Hungary

The Role of Knowledge in the Socio-Economic Transformation of Hungary in the 1990s

Peter Meusburger

1 Knowledge and transformation - some theoretical considerations

At the end of the 1980s, the majority of experts and politicians expected that a transfer of capital, technology, institutions, and management practice from western countries would successfully transform the centrally planned economies of communist states into market economies in a relatively short period of time. In the early discussion, *spatial disparities* and *time lags* in the success of the transformation process did not appear to be an important issue. Ten years later, the lesson has been learned that a transformation process needs, above all, endogenous knowledge resources. Imported capital, technologies, and managerial techniques must be matched with internally available professional skills, experience, creativity, entrepreneurship, as well as attitudes and value systems conducive to a market economy to be effective. Endogenous knowledge resources are the key mechanisms that determine how quickly outside knowledge and technologies are obtained, understood, and incorporated (MALECKI 2000). Even managerial and organisational techniques are culturally determined (SCHREYÖGG 1996, 80-84), and have to be modified and adjusted when they are transferred to a different cultural milieu.

Contrary to the assumptions of many economists, knowledge, technological capabilities and skills are never evenly distributed in space. Various types of knowledge are not a public good and do *not* "travel cheaply and fast" (NAVARETTI and TARR 2000, 2). Many forms of knowledge and cultural traditions are rooted in people and places and cannot easily be transferred from one region to another. Spatial disparities of knowledge and cultural traditions show a remarkable historical persistence. They must not be considered as transitory or short-lived. The speed at which new knowledge and information diffuses over space depends on the type of knowledge, the institution within which the new knowledge is produced, the interest of the producer (inventor) to share his or her knowledge, the *previous* knowledge necessary to understand the contents of new information, the availability of technology necessary for the production and application of knowledge, and the inclination to accept the knowledge.

The first type of knowledge and information, the only one that really travels 'cheaply and fast' is the kind of public news which is easily understood by almost everyone, which does not require previous knowledge or expensive technology, and whose free distribution is in the provider's best interest. This type of information and knowledge is highly *mobile,* can be spread around the globe in seconds and is easily available by telecommunication.

The second type of knowledge and information is more complex and cannot readily be used even by those who have access to it if these constituencies lack the previous knowledge, experience or training necessary to recognise, understand and evaluate the contents of the information. This type of knowledge can only be transferred from one person to another if the receiver's training, skills or capability are broadly equivalent to that of the sender. Previous knowledge necessary to understand an information may consist of a knowledge of a foreign language or of a specialised scientific discipline. The third type of knowledge and information is kept secret by its producer in order to gain an economic, political, scientific or military advantage.

These three forms of knowledge and information vary in the degree of their spatial concentration and in their speed of diffusion. Only the first type of information, which is the least important for the economic competitiveness of Central and East European countries, is (theoretically) ubiquitously available, assuming all places have the technical equipment and all people an equivalent inclination to receive the message. The second type displays a much higher spatial concentration in a few centres or areas, and often circulates only between the upper levels of the urban hierarchy. The third type has the highest degree of spatial concentration and is only exchanged between mother-enterprises and affiliations.

The main hypotheses of this paper are the following:

- The acquisition of new knowledge is the best means to reduce uncertainty. Knowledge increases productivity and enhances competitiveness. Human resources and their value systems are the key factor for transformation, modernisation and international competitiveness. The success of a transformation process depends on knowledge, creativity, experience, absorptive capacities and the ability to benefit from knowledge flows and imported technologies.
- Human resources and value systems necessary for a successful transformation are unevenly distributed in space.
- Regional disparities of knowledge have a cumulative and self-enforcing effect (Matthew effect: 'to he who has, will be given'). For this reason, the starting positions at the beginning of the transformation process were extremely important for subsequent development.

When talking about the importance of knowledge for economic performance, three mistakes should be avoided. The first mistake is to concentrate on easily available indicators such as educational attainment. The level of educational attainment is a very important indicator, but other forms of knowledge are equally significant. This paper defines the term knowledge very broadly as a "capacity for action" (STEHR 1994). Knowledge needed for the transformation process includes educational attainment, professional skills, experience with market economy, embeddedness in important social networks, instrumental knowledge ('Herrschaftswissen', political capital), culture as socially acquired knowledge, social values, the level of information about foreign countries and markets and so on. Identifying which indicators are the most salient for the representation and mapping of knowledge depends on the scale and the topic under scrutiny.

The second mistake is to underestimate the influence of the *historical dimension* of knowledge acquisition. Cultural knowledge, economic traditions, collective memories and historical narratives need a long time to develop and demonstrate a considerable inertia. This kind of knowledge in some places and regions resisted the repression of totalitarian regimes in remarkable ways. The years since 1989 have shown that subjugated knowledge could easily be revived after decades of state propaganda. In other areas, however, the knowledge, skills, and experience needed for the quick adoption of a market economy and for gaining competitiveness in international markets never existed or were completely eradicated by the elimination or expulsion of experts, entrepreneurs, scientists and intellectuals. In these areas, it will take more than one generation to acquire the necessary knowledge, attitudes and experience, and the probability is high that a large proportion of the human resources in these areas will migrate to the already established centres of knowledge. Experience, attitudes and mentalities are not easily mapped. They display themselves, however, in certain indicators, in family histories, biographies, careers, social and cultural activities, and economic performance.

The third mistake is to assume a *linear relationship* between knowledge and economic performance. Knowledge alone does not suffice, it is only one of many factors influencing economic competitiveness. Knowledge enables and empowers, it is a kind of precondition for the effectiveness of other factors. Knowledge is by no means a guarantee of economic success, but a lack of knowledge, information and educational attainment in most cases leads to wrong decisions, impedes access to privileged positions, important networks and scarce resources, and is an obstacle to modernisation. For any social system, the acquisition of new knowledge and the ability to learn, adapt and reorganise are regarded as the best means to cope with the uncertainty of a transformation process. They are preconditions for survival in a continually changing and competitive environment. However, transforming 'new' knowledge into action can be hindered or modified by a number of factors and contexts. The postponement of necessary legal reforms (e.g. the creation of a modern tax and banking system), nepotism, clientelism, the failure to replace incompetent members of the former nomenclature, corruption, organised crime, value systems that are not in favour of entrepreneurship and innovation, the inhibition to take full responsibility and a refusal to deal with the dark sides of the communist past are amongst the most obvious impediments holding back the transformation of a given economy and society. The last point is often neglected in transformation studies although it is very important, when it comes to the establishment of networks and joint ventures with foreigners, refugees and expatriates who are still aware of past atrocities or whose relatives were expelled from these countries or imprisoned and killed during the communist period.

The varying success and the unexpected time-lags and failures of the transfor-mation process have to be studied on various scales. On the macro scale, the question of why some former communist countries performed better in their trans-formation than others is critical. On the meso- and micro-scale, the comparable question is why some regions and places belonged to the winners and others to the losers of the transformation.

2   The importance of a head start in the acquisition of 'new' knowledge

A radical change from a centrally planned economy to a market economy, from an one-party state to a democracy, from a small nomenclature[1] owning almost a monopoly of power and instrumental knowledge to thousands of new decision makers represents a dramatic change of knowledge systems. When establishing new institutions, creating new laws, changing existing structures and finding new markets, tens of thousands of important and far-reaching decisions have to be made within a very short time. The *sudden exposure* to a new economic system, new technologies, new forms of organisation and production, new markets and tough international competition has not left much time to learn and adapt. In thou-sands of institutions the new course had to be set in an extremely short time, and in many cases it was difficult to change the direction once it was set. A piecemeal gradualist approach to the introduction of a market economy or a too long delay in creating new institutions (e.g. banking system), laws (e.g. property rights) and structures necessary for the functioning of market economy caused a delay of for-eign investment and a reduction in competitiveness.

In the transformation process the *time factor* was very crucial for various rea-sons. First, in a competitive market economy, it is not necessarily knowledge or information per se but the *early* acquisition of, and *prior* access to, knowledge that leads to successful action. The earlier the knowledge needed for transformation, adjustment and adaptation to the market economy was acquired, the broader the range of social strata from which successful entrepreneurs, experienced decision makers and new leaders could be recruited in the 1990s, the smaller the proportion of errors and wrong decisions made, the earlier foreign investment was attracted, and the more successful the transformation process. Second, the exchange of thousands of top level decision makers and the recruitment of tens of thousands of new experts need much more time in a rigid, communist one-party state than in a democracy where several elites, parties, institutions and experts continually com-pete for political, economic and cultural power, and are ready to take over new functions after a political change. Third, transformation countries needed urgently

---

[1]   In Hungary, the nomenclature in the late 1980's comprised only 0.4 % of the popula-tion (HARCSA 1995, 271).

experienced managers, scientists and administrators who could compete on the international stage. There was not enough time, to redesign the school system and to wait for new graduates. Fourth, the willingness of people to make a sacrifice for the recovery of their economy was not unlimited. After decades of scarcity, the people had little patience to wait for the fruits of democracy and market economy.

One of the main theses of this paper is that the *point in time* when the communist systems relinquished their rigid ideology, opened themselves for external influences, and changed their method of elite recruitment, was highly significant for an early success of the transformation process. The old elite of Central and Eastern Europe, who had experienced the market economy and democracy that preceded communism, retired in the 1950s and 1960s. Most of the new elites who rose to important leadership roles in the 1990s were educated and gained decisive personal experience in the 1960s and 1970s. In preparation for future leadership roles in a market economy, it made a tremendous difference whether the 15 to 30 year-old age cohort had an opportunity to travel, to study abroad, to establish networks with people in foreign countries in the 1960s and 1970s, or whether they were deliberately cut off from foreign influences and were not even allowed to read books printed in western countries. It mattered a great deal whether small attempts to reform economy and society started in the early 1960s or only at the end of the 1980s. Those countries which had a rigid nomenclature clinging to orthodox Marxism-Leninism and in which the majority of leaders changed their way of thinking only after the political changes that began after 1989, which strictly censured their media and restricted the access to foreign scientific literature, and which did not permit students and scientists to travel to western countries were least prepared for the transformation process because they had delayed the development of new elites and hindered the acquisition of new knowledge. The earlier the knowledge, experience and networks needed in the 1990's for a successful adjustment and adaptation to the market economy had been acquired, the more successful was the transformation process.

Hungary's most important advantage was that its communist system became much less orthodox and repressive from the early 1960s onward. The question why Hungary had these early advantages and why other Central European countries lagged behind is very difficult to answer and will not be elaborated in detail in this paper. Hungary's earlier liberalisation was influenced both by internal and external factors, by historical experience and political traditions. First, Hungary's cultural and political relations to the Soviet Union differed from those of other communist countries. For well known cultural reasons, Hungary was never engaged in a pan-Slavic movement. As Russia had helped to crush the Hungarian revolution of 1848, Russians did not enjoy a similar popularity among Hungarians before the First World War as among Slavic people (e.g. Czechs). Béla Kun's 'revolution' after the First World War, probably contributed to the fact that prior to the Second World War, the proportion of communist voters was much smaller in Hungary than in Czechoslovakia. After the Second World War, the communist hard-liners Rákosi and Gerő had almost no support among the population, and the

Hungarians were the first to start a large-scale revolt against communist repression in 1956. Contrary to the German Democratic Republic (GDR), the Hungarian nomenclature since the mid-1950s had no ambition to be the 'star pupil' of the Soviets, and Hungary was less of a showcase of communism than the GDR.

Even if some important reasons are still unknown, the fact remains that, since the end of the 1950s, Hungary's communist system took a less repressive path. In 1960, internment ended, and in 1961, the last executions of leaders of the 1956 revolution[2] were carried out (ALFÖLDY 1997). In 1963, retaliation[3] ended and a general amnesty was announced for those participating in the 1956 revolution. Kádár tried to integrate the middle classes and old intellectuals with various liberal measures. Restrictions in an access to higher education were reduced, cultural institutions received more freedom, and Hungary allowed its ordinary citizens to travel abroad and to study in western countries in the early 1960s, long before other socialist states[4]. In 1968, the traditional system of central planning ended, and a new economic management gradually evolved. The 1968 Hungarian economic reform expected "to create the framework for a systematically regulated market and for a new management order in line with socialist principles" (TARDOS 1989, 36). Kádár's 'goulash communism'[5] relieved the people's economic burden and political repression. He did not demand agreement with the communist system. As long as the people did not openly question the system, they were not in danger and enjoyed a gradually increasing standard of living. Hungarians could consume goods that were unthinkable in other socialist countries. Finally, Hungary stopped the harassment at its border much earlier than others, which increased the influx of foreign tourists and improved Hungary's image in the west.

Czechoslovakia was on a similar path during the Prague spring in 1968, but this attempt was stopped by the Soviet Union and its allies. There is no sense in speculating whether internal or external factors were more important for the political and economic backlash in Czechoslovakia. It should not be underestimated, however, that Kádár had no opposition from a left wing of the Hungarian communist party. He had convinced the Soviets that the former Stalinist leaders (Rákosi, Gerő) were expelled from Hungarian politics and had to remain in exile

---

[2]   In Hungary, it was forbidden to use the term 'revolution' for the events of 1956 until the end of the 1980s.

[3]   After 1956, more than a hundred thousand people were affected by the repression in Hungary. Tens of thousands were placed in internment, 35 000 were put on trial, among them 25 000-26 000 were imprisoned and at least 239 Hungarians were executed for purely political reasons. Other sources mention up to 350 executions, among them revolutionaries charged with crimes by the communist system (http://www.mfa.gov.hu/sajtonanyag/sajto30.html; ALFÖLDY 1997).

[4]   Another exception was Yugoslavia that was for a long time in opposition to the Soviet Union.

[5]   Kádár's conviction was that the working masses are not interested in politics and that their opinion about the political system is formed by the standard of living and the degree of personal freedom.

in Moscow. Another import difference is that armed Hungarian revolutionaries resisted the Soviet troops for a remarkable period of time (4th to 11th November 1956), whereas the Czechs and Slovaks mainly fought with words and demonstrations and showed no organised military resistance which might have impressed the Soviets.

Most important for the successful transformation was the fact that Hungary changed the principles of recruiting leaders for top positions much earlier than other communist states. In Hungary, the substitution of ideological reliability and party membership as main criteria of promotion with principles of meritocracy, professional skills, educational attainment and experience had already started in the late 1960s and accelerated into the 1980s (LENGYEL 1995). Party membership was still necessary for various top positions, but it was regarded as a formal label. From the 1970s onward, competence, skills and experience gradually became the decisive factors of recruitment for top positions. Many important Hungarian reformers and proponents of democracy and market economy came to power *during* the communist system, years before the market economy was introduced. In Hungary, it were not only members of the opposition, but leading members of the communist party which brought an end to communism. It should not be forgotten that it were prominent Hungarian and Austrian politicians (Gyula Horn and Alois Mock) who symbolically cut the barbed wire at the Austrian-Hungarian border on 27th June 1989 (figure 1). It was Hungary that opened the iron curtain in August 1989 (figure 2), before the fall of the Berlin Wall, and helped thousands of East Germans, pretending to be tourists in western Hungary, to escape to the west.

The acquisition of new knowledge and experience was the most important issue for a succesful transformation process, as the greatest damage that the communist system did was to human capital.[6] The neglect of knowledge and the suppression of intellectuals was not an historical accident, but a fundamental characteristic of the system (MEUSBURGER 1997). Marx underestimated the role of knowledge, the consequences of a vertical division of labour and the insoluble complexity of decision making in a centrally planned economy. Contrary to Bakunin, he did not foresee the huge concentration of power in a centrally planned system. He was not in favour of independent thinking, and he did not agree with the teaching of subjects in school that allowed different interpretations and conclusions (MARX-ENGELS-WERKE vol. 16, 564). For Stalin the best trained part of the old technical intelligence was the main enemy of socialism (LENGYEL 1995, 255). Mao called intellectuals the ninth stinking class, the lowest category of human beings. In Central and Eastern Europe a large proportion of the educated bourgeoisie and upper class was eliminated during the years of terror or escaped to western countries. Among the 200 000 refugees that escaped after the revolution of 1956, more than 25% were intellectuals. About one fourth of the

---

[6]   See KYN, http://econc10.bu.edu/okyn/OKpers/eetrans.html.

Hungarian student population (7 000-8 000 students) sought refuge in western countries (ALFÖLDY 1997).

Figure 1:   The Cutting of the iron curtain by Gyula Horn (on the right) and Alois Mock (on the left)

Source: AP Photo.

In communist systems, education was mainly a means of indoctrination. It is true that some intellectuals played an important role in communist parties and that some fields of science and technology had a high international standard, but vertical social mobility to positions of power was only possible by subordination to the resolutions of the party, and scientists and technical experts were mostly located in the 'administrative and technical staff', not in the 'line staff' where important decisions were taken. Marxism failed first of all because it underestimated the role of knowledge (MEUSBURGER 1998, 127-130), a fact that was already predicted decades ago by MISES (1922) and HAYEK (1944, 1945).

Concerning leadership in the new system, many socialist countries had only two alternatives. First, they could either replace communist hard-liners with experts from universities or with the remnants of the bourgeois opposition. This alternative had the disadvantage that most newcomers lacked the so-called 'instrumental knowledge' (Herrschaftswissen), political capital and networks which are necessary to be successful, but which can only be acquired after years in positions of power. The other alternative was to keep a larger part of the old nomenclature who possessed the necessary political capital and instrumental knowledge but lacked most of the skills needed in a market economy, showed little inclination to change their way of thinking and to give up their privileges.

Figure 2:   The opening of the iron curtain at the Austrian-Hungarian Border in August 1989

Source: AP Photo.

Due to its early introduction of meritocratic principles and its early abandonment of Marxism-Leninism in everyday economic practice, Hungary could more or less evade these traps. It had more competent and internationally experienced leaders in economy and administration and fewer leaders who owed their position predominantly to the communist party. Consequently, there was less of a need to replace a large part of the economic and administrative elite. Due to the early reforms and the lower degrees of corruption in society, members of the Hungarian nomenclature had less opportunities to transform their political capital into economic capital respectively to transfer state resources into private resources.

3   Spatial disparities in the distribution of knowledge and experience

The main argument of this section is that as a result of the spatial distribution of knowledge, experiences, values and traditions, the central region, the axis between Budapest and Székesfehérvár, and the northern part of Transdanubia were much better prepared for structural change and international competition than the eastern and north-eastern areas of Hungary, and thus more attractive to investors and skilled migrants.

3.1   Central-peripheral disparities of knowledge

Two types of spatial disparities of knowledge have to be distinguished. The first type results from vertical division of labour and from co-ordinating social systems in space. It is characterised by a large concentration of power and knowledge in centres at the top of the urban hierarchy and by large central-peripheral disparities of the educational achievement of the work force (MEUSBURGER 2000). In any political system the highest levels of knowledge, skills, competence and power are concentrated in a few centres and urban agglomerations. The centre is the place where the highest levels of authority, power, decision making and knowledge are localised. Centres are the places to which foreign institutions and decision makers establish their first contacts. The centre collects and distributes resources, it sets the rules, norms and standards for the members of the system. Large centres profit from the concentration of power and knowledge and from the access to important networks and are therefore much better prepared to cope with structural change. The centres of power and knowledge benefited first and to a larger extent than other places from the transformation process towards a market economy. Contrary to the centres, most of the work places offered at the periphery are related with lowly skilled and lowly paid routine activities that are co-ordinated and controlled by external decision makers. Thus, the periphery is characterized by less contacts abroad and a lack of the knowledge, information and networks necessary for a successful transformation.

However, the antagonism between centre and periphery was much stronger in centrally planned communist systems than in market economies (e.g. MEUSBURGER 1995a; 1997). At the end of the communist period, the 20 cities at the top of the urban hierarchy (0.64 percent of all Hungarian cities and villages) had a virtual monopoly of decision making, power, and competence. In branches, crucial for government and economy, these 20 cities comprised 80-100% of all Hungarian work places for university graduates.[7] Budapest was a magnet for human resources. People striving for a career often moved to Budapest. According to the Hungarian microcensus of 1996, some 45% of all university graduates living in

---

[7]   For details, see MEUSBURGER (1997).

Budapest came from other areas of Hungary: 11.9% from county capitals, 12.5% from other cities, and 13.4% from villages. Another 7% came from foreign countries. The small cities and villages with less than 5 000 inhabitants that comprised approximately one third of the Hungarian population were deliberately neglected by the central planning authorities. The Hungarian Spatial Plan of 1971, for example, characterised towns with less than 3 000 residents as 'towns without a function'.[8] They almost totally lacked work places for highly qualified, skilled and experienced experts and lost most of their students to the centres.

The spatial distribution of knowledge, skills, experiences with a market economy, technological capabilities, and networks existing at the end of the 1980s determined to a very large extent in which regions and places the necessary structural changes began early, where foreign and domestic financial resources were first invested, where joint ventures were first established, which regions and places were initially successful in changing their internal structures (privatisation, establishment of a banking system) and finding new markets, and which regions failed in doing so. This is the main reason, why central-peripheral disparities of the educational achievement of the work force increased in the 1990s, and why the proportion of work places[9] for male university graduates was almost five times as high in Budapest as in the rural villages (table 1).

Table 1: Educational achievement of the male work force with regard to the size class of their place of work, 1996 (%)

Size of the place of work	Primary school or less	Vocational or Technical school	Secondary School	University or Academy	Sample size (100%)
>= 500	31.0	48.7	14.4	5.9	390
501-1 000	25.2	45.9	19.4	9.5	1 139
1 001-2 000	25.9	48.1	17.3	8.7	2 063
2 001-5 000	27.5	44.9	19.4	8.2	3 653
5 001-10 000	24.1	46.2	20.3	9.4	2 490
10 001-20 000	20.4	44.7	23.9	11.1	4 053
20 001-100 000	17.4	40.9	26.9	14.8	8 578
100 001–1 Million	15.5	37.2	28.0	19.3	5 480
> 1 Million (Budapest)	14.3	28.5	31.2	26.0	7 570
Sample size	6 798	14 014	9 013	5 591	35 416

Source: Hungarian microcensus 1996; author's compilation from individual data sets.

---

[8]   Information of Zoltán Cséfalvay.

[9]   The place of work is a much better indicator for economic attractiveness than the place of residence.

Table 2: Educational achievement of the female work force with regard to the size class of their place of work, 1996 (%)

Size of the place of work	Primary school or less	Vocational or Technical school	Secondary School	University or Academy	Sample size (100%)
>= 500	40.9	23.9	26.5	8.7	264
501-1 000	36.6	20.3	29.3	13.9	815
1 001-2 000	33.0	22.2	30.4	14.4	1 601
2 001-5 000	34.7	21.9	30.8	12.6	2 805
5 001-10 000	31.1	21.6	35.0	12.4	1 980
10 001-20 000	28.5	22.8	35.2	13.4	3 423
20 001-100 000	22.2	12.3	40.3	16.2	7 385
100 001-1 Million	18.9	19.6	41.5	20.0	4 726
> 1 Million (Budapest)	17.8	13.6	44.6	25.1	6 520
Sample size	7 128	5 792	11 398	5 201	29 519

Source: Hungarian microcensus 1996; author's compilation from individual data sets.

3.2   West-east disparities of knowledge

The second type of disparities results mainly from the diffusion paths of various innovations, from different spatial patterns of interaction and contact, and from selective internal migration. Since most of the epoch-making social and economic innovations in Hungary came from the west - beginning with the introduction of Roman law, Christianity, the enlightenment, the abolition of serfdom, the spread of literacy, industrialisation, democracy, and more recent foreign investment since 1989 - the Carpathian basin displayed again and again large socio-economic disparities between the western and eastern parts.

The western regions of Hungary including the later capital of Budapest adopted many important innovations and developments much earlier than the eastern parts, simply because they had early exposure to them and because their fields of interaction were mainly directed towards Vienna and other European centres of innovation. In the 19th century, literacy levels amongst the population of Transdanubia were higher much earlier than among the population in the eastern parts of Hungary. The time lag in this respect between the western and most eastern parts of the Habsburg empire was more than two hundred years at the end of 19th century (MEUSBURGER 1998, 270). Transdanubia industrialised earlier than the east. The western regions invariably enjoyed a head start in acquiring new knowledge, experience and skills, and in participating in networks linked to the cultural and economic centres of Europe.

During the Cold War, the western border areas of Hungary were deliberately discriminated against by the central planning authorities with regard to investments and the distribution of skilled jobs. Nevertheless, the western parts of Hungary had cultural traditions and relations to the west that favoured a quick transition to a market economy after the political change. During the Cold War, the population in western Hungary was much better informed about the west than that in eastern Hungary. In communist countries during the Cold War, all possible means were taken to prevent reception of foreign stations. The Soviet Union forced the other socialist states in Central and Eastern Europe to use technical standards that were not compatible with western technology and would thus prevent the reception of western radio and TV stations until the end of the 1980s. Contrary to other socialist countries, the Hungarian authorities did not intentionally target and block Austrian stations. In Hungary, it was easily possible to buy on the black market radio receivers that transmitted western (Austrian) frequencies and antennas necessary for good reception of Austrian TV. In communist Hungary, listening to foreign radio stations or watching Austrian TV was not as severely restricted and not as strictly controlled as in other communist countries.

Reception of Austrian television was possible as far as to a line east of Győr to the Lake Balaton, given that the technical capabilities existed. It is difficult to determine how many people owned the technical equipment necessary to receive Austrian TV. However, it is well known that persons in western Hungary were able to watch the 1972 Munich Olympics in colour. Austrian radio programs in FM could be heard as far as a line east of Győr to the Lake Balaton, although the reception in the southern areas of Lake Balaton was better than in the more hilly northern areas. Croatian and Slovenian stations could also be received in the south part of Transdanubia. Western television along the border regions became an "uncontrollable stimulant" (PEARSON 1995, 73), and had an important impact on the knowledge, level of information, stereotypes, and attitudes of the population in socialist countries.

In addition to levels of information, cultural traditions also vary in space. The small, self-governing villages in Transdanubia show a stronger social cohesion and more local cultural activities than the large administrative units in the Great Plain. Most villages in Transdanubia had already a higher density of social and cultural clubs at the end of the 19th century (BARTAL 1999). These traditions apparently survived the communist period. Transdanubia at the end of the 1990s had a lower crime rate (figure 3), and a higher density of cultural non-profit organisations per 1 000 inhabitants (figure 4) than the eastern counties (BARTAL 1999). In the western parts of Hungary, especially in the districts Csongrád, Fejér and Komárom, non-profit organisations on average received more financial support from the local communes (BARTAL 1999).

14   P. Meusburger

Figure 3:   The number of violent crimes per 10 000 inhabitants in Hungary, 1997

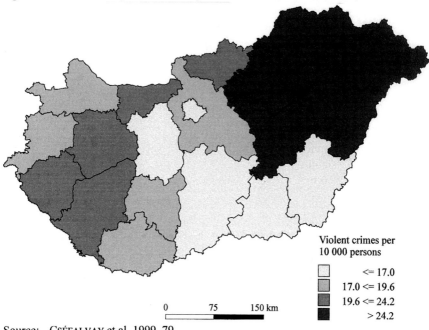

Violent crimes per
10 000 persons

<= 17.0
17.0 <= 19.6
19.6 <= 24.2
> 24.2

0   75   150 km

Source: CSÉFALVAY et al. 1999, 79.

Figure 4:   Proportion of cultural non-profit organisations in Hungary, 1997

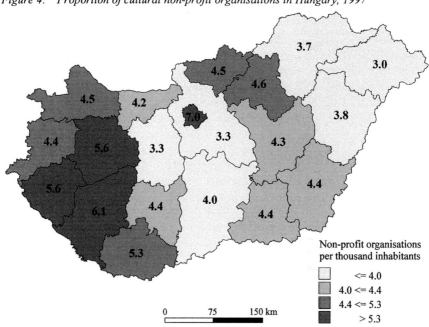

Non-profit organisations
per thousand inhabitants

<= 4.0
4.0 <= 4.4
4.4 <= 5.3
> 5.3

0   75   150 km

Source: BARTAL 1999, 185.

4   The advantage of strong networks to foreign countries

Knowledge first of all is created by discourse, face-to-face contacts, and personal experience. Hungary's most important advantage was, that its new elite, promoted in the 1990s, had studied abroad and gathered international experience much earlier than that of most other communist states. An astonishing proportion of the new Hungarian elite in universities, government and other important institutions of the late 1990s had studied, travelled and gathered personal experience in western countries, had found access to foreign research institutions and foreign companies, and had established networks to important decision makers in foreign countries. From 1960 onward, Hungarian students and scientists were allowed to accept fellowships and to attend scientific congresses in western countries. By comparison, geographers from the German Democratic Republic were not permitted by their political authorities to attend the biannual Congress of German Geographers before 1989. Between 1960 and 1996, approximately 400 outstanding young Hungarian scientists were awarded a German Humboldt Fellowship (the most prestigious German fellowship for foreign scholars below 40 years of age), which gave them an opportunity to work for up to two years at a German university or scientific institution of their choice. A large percentage of these Humboldt fellows later became professors at universities or leading scientists at the Hungarian Academy of Sciences, long before the change of the political system. They informed, educated and trained a large proportion of the new elite who came to power in the 1990s. A large number of Humboldt fellows took leading positions in ministries, in the office of the prime minister, in the Constitutional Court of Justice, in the National Bank of Hungary, in universities, in planning agencies and industrial enterprises. The same applied to DAAD[10] fellows and fellows of other foreign foundations (e.g. Fulbright fellows). These fellows of various prestigious foreign foundations prepared in the 1980s and 1990s important decisions, formulated new laws, introduced new techniques and research methods from the west and were important role models. Consequently, these fellowships can be used as an indicator of when liberalisation started, where the elites received decisive academic orientation and socialisation, and how intensive these relations were to foreign countries.[11]

Many indications reveal that, in the 1970s and 1980s, a large number of Hungarian politicians, scientists, senior administrators and managers of state enterprises had better personal relations with important leaders in Austria and Germany

---

[10]   German Academic Exchange Service (Bonn).

[11]   With regard to the number of Humboldt fellows, it is well known that Czechoslovakia was on a similar path in the late 1960s until the 'Prague spring' was smashed by the tanks of communist brother-countries in 1968. Czechoslovakia led the socialist countries in Humboldt fellowship recipients until 1968, but then fell back and lagged far behind Hungary until 1990.

than their counterparts in other socialist countries. Hungarian universities started a number of partnerships and exchange programs with universities in western countries in the early 1980s that became very successful. One of many examples may illustrate the importance of these contacts. In 1982 the first working program of the co-operation between the Faculties of Law of Eötvös Loránd University in Budapest and the University of Heidelberg was established.[12] Since 1983, both faculties organised each year one to three joint seminars that took place alternatively in Budapest and Heidelberg. Long before the introduction of a market economy could be foreseen, they organised symposia on civil law, on legal forms of enterprises, on international contract law, copyright law, and many other topics that became important after the political change. After 1989, four of the Hungarian proponents of this co-operation were promoted to important positions. Lajos Vékás became rector of the Eötvös Loránd University and later rector of the Collegium Budapest. Ferenc Mádl, a Hungarian-German, first became head of the state property agency, in 1993, he was appointed Minister of Cultural Affairs, and in August 2000 he became Hungary's President. László Sólyom became the first president of Hungary's Constitutional Court of Law after 1989, while János Németh is the acting president of this institution. Numerous other examples could be found in regard to similar relations with the US, Great Britain, France or other countries.

Figure 5:   The proportion of Humboldt Fellows in Hungary and Czechoslovakia

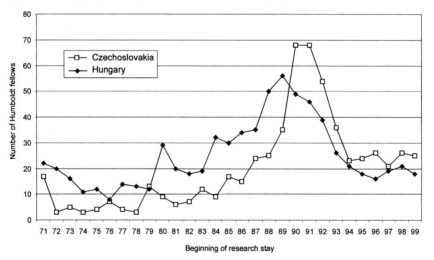

Source: Alexander von Humboldt Foundation (Bonn); data compilation and design by
           Heike Jöns.

---

[12]   I am very thankful to Gert Reinhart (professor at the University of Heidelberg and one of the German proponents of this cooperation) for providing me with information about this cooperation.

These strong ties to the west were not restricted to Hungary's political and economic elite but concerned also the middle and lower classes. A large number of Hungarian towns and villages in Transdanubia have had a German minority (some even a majority) since the end of Turkish occupation in the late 17th century when large devastated areas were repopulated by Germans, Slovaks, Croats and other ethnic groups. Before the Second World War, approximately 620 000 Germans lived in Hungary. Between 1946 and 1948, 163 000 were forced to leave Hungary, most of whom (125 000) settled in the so-called 'American occupation zones' in the Austrian provinces of Salzburg and Upper Austria, and in the German provinces of Bavaria and Baden-Württemberg. After the Hungarian revolution of 1956, almost 200 000 Hungarian refugees stayed for a certain time in Austria before migrating to other countries.[13] About 80 000 Hungarian refugees finally settled in United States, 22 000 in Great Britain, 16 000 in the Federal Republic of Germany, 14 000 in Switzerland and 13 000 in France.[14] At the end of the 20th century, approximately 240 000 Germans lived in Hungary. Hundreds of thousands of Hungarians, predominantly those living in Transdanubia and Budapest, have at least one grandparent of German origin.[15]

When new opportunities opened after the fall of the iron curtain, many Hungarians used their personal connections to relatives and former neighbours now living in Germany, Austria, the US and other countries to get advice during their adjustment to the free market. Many Hungarian expatriates came back or intensified their relations with Hungary. Numerous joint ventures of foreign entrepreneurs in Hungary and many Hungarian sales offices abroad were based on such personal relations. This advantage of good personal relations cannot be overestimated. It contributed a great deal to Hungary's success story in developing German and Austrian markets and to the role of Austrian and German firms as investors in Hungary. German and Austrian decision-makers also influenced investments by multinationals from the US and other countries. When IBM was searching world-wide for a production site of hard disks, it were German managers who gave Székesfehérvár in Hungary the advantage to other locations in Central and Eastern Europe as well as in East Asia.

In the 1990s, German and Austrian capital was the most important source with regard to the *number* of joint ventures in Hungary. While the US, Japan, France and other countries invested predominantly in a few large enterprises, Germany and Austria, apart from a few large companies, invested in thousands of small enterprises. Between 1990 and 1999, 20.7 billion USD were invested in Hungary, more than 50% of foreign investment coming from Germany and Austria. Hungary received in the 1990s an average of 1 900 USD foreign direct investment per

---

[13] Some refugees escaped to Yugoslavia and came later to western countries.
[14] See http://www.mfa.gov.hu/sajtonanyag/sajto30.html.
[15] Many Germans changed their names already before or after the First World War and became assimilated as Hungarians in order to have better employment opportunities in the civil service.

capita (the average of the EU is 2 500 USD), followed by the Czech Republic with 1 600 USD per capita (CSÉFALVAY 2000, 5).

In the 1990s, Hungary was the most important country for Austrian investment. More than 5 600 Hungarian enterprises received investments from Austria. After Germany and the US, Austria is the third largest investor in Hungary. At the end of the 1990s, approximately 6 000 Hungarian firms exported goods to Austria, and 14 000 Hungarian firms imported goods from Austria. Between 1991 and 1998, Hungarian exports to Germany increased by 307% from 2.7 billion USD to 8.4 billion USD (see Ungarische Wirtschaft 4 (2), 1999). In 1998, 36.5% of total Hungarian exports went to Germany and 10.1% to Austria. More than two-thirds of all Hungarian exports to Germany went to Bavaria, Baden-Württemberg and Nordrhein-Westfalen. The Bavarian share of Germans exports to Hungary was almost 30% in 1998 (see Ungarische Wirtschaft 4 (2), 1999).

If trade relations and foreign investment were simply a function of spatial distance - as the geographic location theory suggests -, then the Czech Republic should have received a much larger trade volume with Austria and Germany, and should have attracted a larger number of investors from Germany and Austria than it did. There is no doubt, that the attitude of the Czech government to the issue of ethnic cleansing and post-war atrocities against Sudeten Germans, the delayed reform of the Czech banking system and the different way of privatisation impeded a similar boom of Austrian and German investment.

Figure 6:   Foreign capital investments in Hungary with regard to population size, 1997

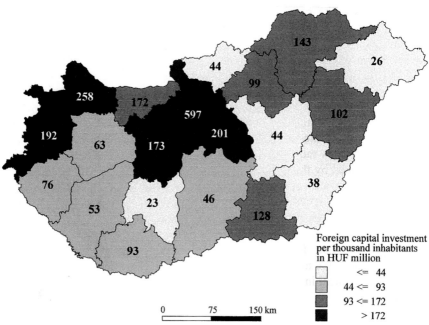

Source: CSÉFALVAY et al. 1999, 56.

The extent and spatial distribution of direct foreign investment and co-operation with foreign firms was not only important for an early beginning of the transformation process, but also played a crucial role for export and technology transfers. First, the most important channels of knowledge accumulation and transmission of ideas for enterprises are co-operation with foreign firms, competition on an international market, and "learning by exporting" (NAVARETTI and TARR 2000, 2). It is well documented that multinational enterprises are the most important channels for transferring technologies and that the vast majority of royalty payments for international technology transfers are made from subsidiaries to their parent firm (NAVARETTI and TARR 2000, 8). Second, in countries with a high proportion of direct foreign investment, economic performance is less sensitive to fluctuations of domestic demand. Due to intra-firm trade links, these affiliations are integrated in international production and sales networks.

Countries sceptical about foreign ownership (e.g. Poland and Czech Republic) had to pay the prize at the end of the 1990s. Whereas the economic situation in most transformation countries started to deteriorate between 1996 and 1998, Hungary alone managed to speed up its export-led GDP growth after 1997, largely thanks to foreign-owned companies and their rapidly expanding activities (see chapter by Gábor Hunya). In 1999, most Central and East European countries plunged into recession, Hungary being an exception once again (HAVLIK et al. 1999). Hungary's export boom was to a large extent fuelled by the affiliations of foreign firms. At the end of the 1990s, almost three quarters of total exports were being produced by foreign multinationals, and 73% of Hungary's export were directed towards the European Union (CSÉFALVAY 2000, 5). This means that Hungary was economically more integrated into the EU than many member states of the EU; among the EU members the average share of total exports to other EU members was only 66% (CSÉFALVAY 2000, 5). Also the efficiency of foreign affiliates started to improve somewhat earlier than that of domestic companies after the transformation shock (SZANYI 1998). The per capita 'value added' profit rates of many multinational corporations in Hungary are as high as 55-60% of the EU average, whereas small to medium sized domestic enterprises show per capita 'value added' profit rates of 38-40% (CSÉFALVAY 2000, 7).

Immediately after the change of the political system, a kind of self-enforcing effect - influenced by the availability of highly skilled human resources, foreign investment, the revival of cross-border relations, and the re-emergence of economic and cultural traditions - provided the central region of Hungary and western Transdanubia again with higher salaries, an in-migration of talent, better infrastructure, better opportunities to be mobile and with more frequent face-to-face contacts to foreigners from western countries. The number of phone lines per 1 000 inhabitants, the number of passenger cars per 1 000 inhabitants, the spatial pattern of investment in industrial parks, the number of nights spent by foreigners in the relevant regions, are only a small sample of indicators hinting at these west-east disparities in economic performance. With regard to GDP per capita, the Central Budapest Region at the end of 20th century was again the most highly developed region, followed by western Transdanubia, which reached three quarters of the figure estimated for the central Budapest region. GDP per capita in the northern Great Plain was a mere two thirds, while the per capita Foreign Direct Investment was only a third of that found in the western Transdanubian region (CSÉFALVAY 2000, 8).

Figure 7:   Number of phone lines per 1000 inhabitants in the regions of Hungary, 1998

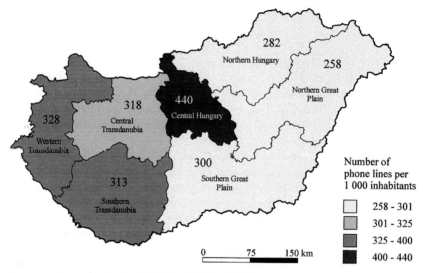

Source: Hungarian Central Statistical Office (KSH).

Figure 8:   Number of passenger cars per 1000 inhabitants, 1998

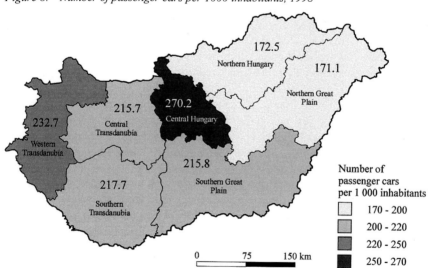

Source: Hungarian Central Statistical Office (KSH).

Figure 9:   Nights spent by foreigners in the regions of Hungary, 1998

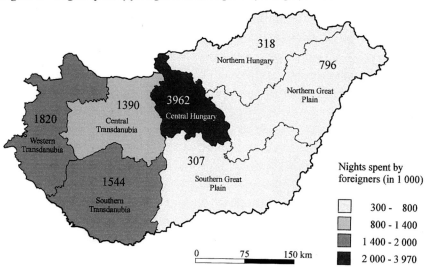

Source: Hungarian Central Statistical Office (KSH).

Figure 10: The distribution of industrial parks in Hungary by investment, 2000

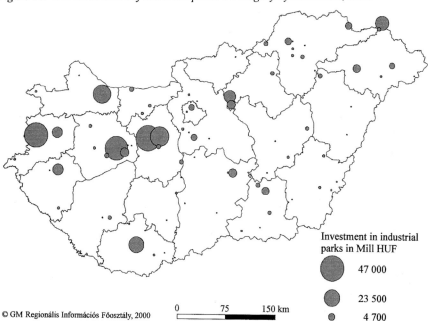

Source: Hungarian Ministry of Economic Affairs; made available by Zoltan Cséfalvay.

Figure 11: Regional disparities of economic performance, 1998 (per capita GDP)

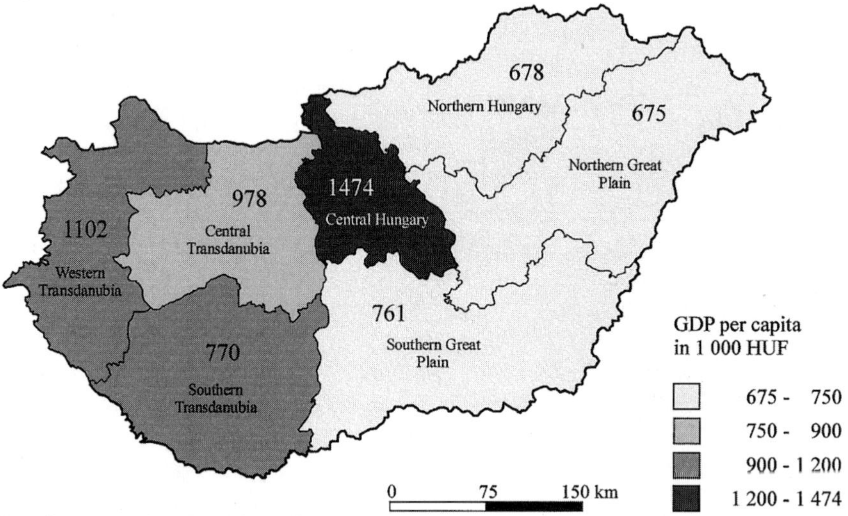

Source: CSÉFALVAY et al. 1999, 56.

5   Skilled leaders and entrepreneurs as a motor of transformation

Unlike Russia, where the new entrepreneurs came predominantly from the political, economic and military nomenclature and where the chances of becoming an entrepreneur were principally determined by the size of political capital (instrumental knowledge) and not by entrepreneurial experience or knowledge of the market economy, in Hungary, the recruitment of entrepreneurs was primarily influenced by their occupation, their level of education and by demographic factors such as age and gender (see chapter by Tibor Kuczi and György Lengyel). A large part of Hungarian entrepreneurs were recruited from the middle classes, the main reason being that in Hungary people could gather experience with private enterprises and elements of market economy since the early 1980s, much earlier than in the other communist countries.

The private sector was allowed to grow, when by the end of 1978 the extent of indebtedness had become excessive and more market elements were necessary to improve economic performance. After 1981, the private sector could develop with high speed. In 1982, the so-called 'intra-firm voluntary working units' were allowed to produce goods and offer services outside the central planning system. These voluntary working units consisted of particularly hard-working, skilled and motivated workers. They had the chance to work and do business in a kind of proto free market economy and worked for their own profit. This contributed a great deal to the earlier and quicker development of the private sector. In spite of these reforms and first careful steps of private enterprises, the party and state apparatus controlling the economy had no intention of giving up the authority to define the objectives of the economy and to control the state-enterprises (VOSZKA 1989, 110). It was only in the years 1985-1986 that most state-owned enterprises were released from the supervision of the state administration (VOSZKA 1989, 111). In 1980, the proportion of the active population working in the private sector was 3.4%. According to official statistics, the share of private activities was already 7% in 1982. "Taking small agrarian enterprises into account, this value is more than 20%, [...]. If non-registered activities are included as well, the share of the private sector may easily be over one third of the national income" (TARDOS 1989, 38). While the share of the private sector was 11% in 1990, it rose to 20% in 1992.

In a period of transformation, statistics about entrepreneurs have to be interpreted very carefully. Many people only became independent entrepreneurs (with no employees) in order to avoid unemployment. While in Hungary the ratio of enterprises per 1 000 inhabitants is 90, in the Netherlands, for example, this number is only 28, and even in southern Europe, where the number of self-employed is traditionally high, this ratio was considerably lower (ÉKES and KUPA/KREKÓ 2000, 39). However, many self employed Hungarians lacked most of the skills and experience necessary to survive the first years after establishing a new business, although the new entrepreneurs incorporated a large proportion of university and secondary school graduates. The percentage of university graduates among entrepreneurs correlated positively to the size of their business, while the size of the business and basic schooling correlated negatively (table 3).

Table 3: Educational achievement of Hungarian entrepreneurs with regard to the size of their enterprise, 1996 (%)

Business or enterprise size	Primary school or less	Vocational school or technical school	Secondary school	University or Academy	Sample size (100%)
No employees	19.7	36.1	32.9	1.4	5410
1-2 employees	11.9	32.9	38.7	16.5	1 216
3-20 employees	9.8	25.1	41.2	23.9	502
> 20 employees	7.7	29.2	29.2	33.9	65
Sample size	1 263	2 499	2 475	956	7 193

Source: Hungarian microcensus 1996; author's compilation from individual data sets.

Since the late 1990s, multinational companies increasingly tend to delegate 'high value-added' activities of their production chain (research, marketing) to their Hungarian subsidiaries (CSÉFALVAY 2000, 8). Small and medium-sized companies, managed by highly skilled and experienced entrepreneurs, more and more play the role of suppliers of high value and innovative products.

6   The lack of formal education as an obstacle for employment and endogenous development

The kind of knowledge needed for international competitiveness is unevenly distributed in space, so are the perspectives of catching up with booming areas, and the spatial contexts impeding the politically aspired goal to even out imbalances in regional development. One of the most serious bottlenecks postponing development is a lack of basic education. Already in the socialist period, the proportion of the population lacking the basic eight grades of primary school is an excellent indicator of underdeveloped areas (in socialist times they were called 'problem areas'). In the 1990s, this indicator predicted very accurately which areas would lag behind in the transformation process. Those persons who did not attend or complete eight grades of primary school[16] were the least prepared for a market economy, they were the most vulnerable in a system of meritocracy, and the first to be pushed out of the labour market.

Lowly educated persons in young age groups continually showed the highest unemployment rates and became the main losers of the transformation process. Already in 1990, when unemployment rates were still very low in Hungary (DÖVÉNYI 1995; see also chapter by Zoltán Dövényi), 15 to 25 year old persons who did not complete compulsory school had unemployment rates of 20 to 30% (figure 12). Also the increase of unemployment in the following years was highest amongst the lowly educated (table 4).

The proportion of the population without eight grades of basic schooling was relatively high in the eastern regions of Hungary and in the inner periphery of southern Transdanubia (figure 13). The proportion of the population with less than eight grades of basic education decreased very rapidly in the 1990s. It is, however, not so much the *absolute* level of educational achievement that matters for the labour market, but rather the ranking on an ordinal scale of skills and educational achievement. Absolving eight grades of primary school does not mean to leave the category of the lowly skilled or to improve significantly one's chances on the labour market. Therefore, figure 13 shows relative regional disparities by using the location quotient.

---

[16]   According to the 1961 school reform that was acknowledged in 1985, children are required to attend school until age 14 in Hungary. If a child does not complete school by age 14, he or she is required to attend school until age 16. If a child does not complete school by age 16, he or she can leave school without a graduation certificate.

Figure 12: Unemployment rates according to age and educational achievement, 1990

Unemployment rate in %

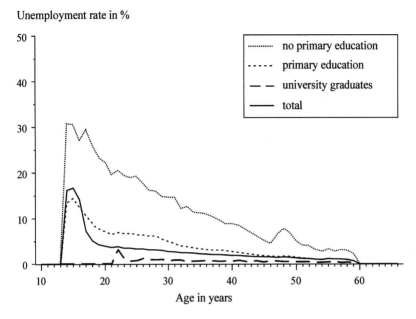

Source: Hungarian census 1990; author's compilation from individual data sets.

Table 4: The development of unemployment rates between 1990 and 1994 with regard to gender and level of educational achievement (%)

	Men 15-60 years old Rates of unemployment		Women 15-55 years old Rates of unemployment	
Educational achievement	1990	1994	1990	1994
Did not attend or complete primary school[17]	7.7	15.2	3.5	10.0
Primary school	3.1	9.5	1.5	6.6
Vocational or Technical school	2.5	11.0	1.6	7.2
Secondary school	1.7	6.7	1.1	6.0
University or academy degree	0.9	2.6	0.5	1.2

Source: Hungarian census 1990 and Hungarian microcensus 1994; author's compilation from individual data sets.

---

[17]   The population group 'who have not completed primary school' includes illiterates, persons who never attended school, those who did not complete eight years of schooling because they dropped out of school, and those who attended only an old type of elementary school or frequently had to repeat a grade.

Figure 13: Proportion of Hungarians above 15 years of age who had not completed eight grades of primary school, 1990 and 1996

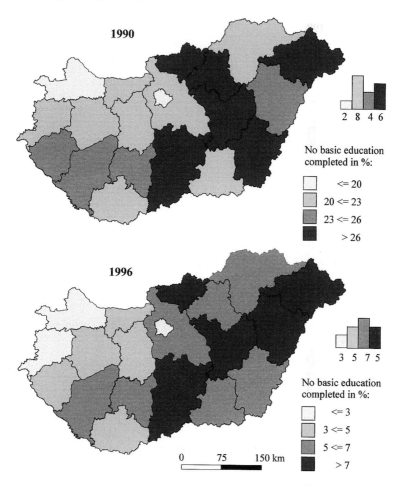

Source: Hungarian census 1990 and Hungarian microcensus 1996; author's compilation from individual data sets.

*Figure 14: Proportion of Hungarians above 15 years of age who had not completed eight
grades of primary school, 1980 and 1990 (location quotient)*

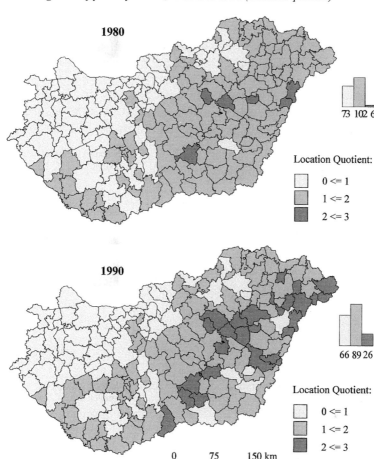

Source: Hungarian census 1980 and 1990; author's compilation from individual data sets.

Among those having not attended or completed eight grades of elementary education, gypsies[18] are the most vulnerable and disadvantaged. Although the group of

---

[18]   In this case, it would be scientifically incorrect to use the terms Roma and Sinti instead
of gypsy. In the census of 1990, these people identified themselves as gypsies and not
as Roma or Sinti. Even official documents of the Hungarian government and organizations of gypsies did not use the terms Roma and Sinti until the early 1990s. In 1985,
the Advisory Board for Hungarian Gypsies was founded. The 1993 Hungarian law
about the rights of national and ethnic minorities (chapter 6, § 42) indicates the language of these minorities as 'gypsy' (ZUR LAGE DER ZIGEUNER 1994, 49). In 1989, the
'Democratic Union of Hungarian Gypsies' and the 'Social Democratic Party of Gyp-

gypsies is estimated to be not more than 5-6% of the Hungarian population, a focus on their level of education, their unemployment rates and their places of residence is very revealing. At the end of 19th century, approximately 80% of the gypsies in Hungary were illiterate. In 1961, out of the then 202 000 registered gypsies, 40% could not read or write (GRONEMEYER 1981, 206). Until the early 1960s, 15-20% of Hungarian gypsy children did not attend elementary school (SEEWANN 1987, 22). At the end of the 1960s, every third Hungarian gypsy child completed primary school; this share only rose to 50% by 1985 (ZUR LAGE DER ZIGEUNER 1994, 51).[19] During the 20th century, gypsies have improved their level of education significantly, but most of them still belong to the category of the lowest skilled.

Table 5: Unemployment rates of gypsies and total population in Hungary according to age, 1990 (%)

Age	Male gypsies	Total male population	Female gypsies	Total female population
15 Years	14.1	1.9	13.4	2.0
16 Years	24.3	3.3	16.0	2.7
17 Years	39.2	4.6	17.4	3.1
18 Years	25.9	5.6	12.9	4.0
19 Years	23.5	5.0	11.2	3.4
20 Years	20.6	4.4	9.4	2.4
25 Years	16.9	3.6	7.2	1.7
30 Years	15.0	3.2	5.5	1.5
35 Years	13.7	2.8	4.6	1.3

Source: Hungarian census 1990; author's compilation from individual data sets.

Out of the persons of 15 years and older who declared themselves as gypsies in the 1990 census, 55.2% (45.9% of male gypsies and 64.0% of female gypsies) did not complete eight grades of primary school (the respective proportion for the

sies in Hungary' were founded. It was not until several years after the fall of the iron curtain that the influence of western European Roma and Sinti organizations caused a change in self-identification. In 1995, the 'Hungarian National Council of Romany Representatives' was founded, in the following years Roma magazines and various Roma organizations came into being. The methodological difficulties of defining ethnic categories are well-known. Therefore, in this paper, only those 142 683 Hungarians are considered as gypsies that identified themselves as gypsies under the census question 'nationality'. The actual number of gypsies is much higher and varies between 380 000 and 1 000 000 depending on the source and the interests of those who are counting. A number of 600 000 seems to be realistic (KOCSIS and KOVÁCS 1991, 78; DARÓCZI and BÁRSONY 1991, 8; KOCSIS 2000, 129). Despite of some methodological problems, these census statistics currently provide the best data available on gypsies.

[19]   According to SEEWANN (1987, 22), only 21% of all Hungarian gypsy children and only 7% of all gypsy children who consider Rom as their native language completed the eight-year mandatory school requirement between 1962 and 1970.

overall population was 8%). Only 39.7% of gypsies completed the eight-grade compulsory school and only 5.1% completed a level of education beyond primary school (table 6). At the end of the 1990s, approximately 70% of the Roma children completed primary schooling, but only one third (compared with 90% of non-Roma families) continue schooling by attending vocational, technical or secondary schools.[20]

Table 6:   Educational achievement of Hungarian gypsies over 14 years old, 1990 (%)

Educational level	Male gypsies	Female gypsies	Gypsies total
Did not attend or complete 8 grades of primary school	45.86	64.01	54.97
Primary school	46.17	33.29	39.71
Technical school or vocational school	6.94	2.00	4.46
Secondary school	0.86	0.62	0.74
University or Academy	0.15	0.08	0.11
Total	100.00	100.00	100.00

Source: Hungarian census 1990; author's compilation from individual data sets.

Table 5 shows that already in 1990 the unemployment rates of male gypsies between 15 and 35 years of age were four to seven times as high as that of the total population in the same age groups. These data prove that unemployment of gypsies was extremely high before 1989. In the 1980s, they were the first to be dismissed by state-firms and had not yet found alternative employment niches. Also at the end of the 1990s, unemployment rates of Roma were four to five times higher than the national average. There are villages, where 90-100% of the Roma population is unemployed. Long term and youth unemployment are also far more common in the Roma community.[21]

A minority group who for various reasons displays a social distance to the school system of the majority, and, as a result, has a significantly lower educational achievement than the general population, is rarely able to escape from a marginalised and economically disadvantaged position. In an information and knowledge society, illiterates or persons without a completed primary school education are partly excluded from the labour market or forced into less attractive and unstable segments of the workforce for functional reasons, since the modern economy requires skilled workers. On the other hand, a missing primary school certificate is often used as pre-text to justify a deeper prejudice against gypsies.

Gypsies have consistently demonstrated high intelligence and creativity in their struggle for economic survival. It is part of their survival strategy to find niches, tricks, and legal loopholes before the *gadje* (non-gypsy) discovers something peculiar. However, a formal education cannot be substituted by such a kind of

---

[20]   See http://www.mfa.gov.hu/sajtonanyag/Nat%20and%20Eth.html.
[21]   See http://www.mfa.gov.hu/sajtonanyag/Nat%20and%20Eth.html.

knowledge. Most employers, especially foreign companies and large enterprises, require the completion of *formal* education (primary school) even for lowest paid routine jobs.

Table 7:   Educational level of Hungarian gypsies over 14 years old according to their community size, 1990 (%)

Number of Inhabitants in place of residence	Did not attend or complete primary school	Primary school	Vocational school	Technical School	Second-ary school	University or Academy
up to 500	63.26	32.68	3.78	0.04	0.22	0.02
501-1 000	59.67	35.82	4.07	0.07	0.36	0.02
1 001-2 000	56.80	38.73	4.06	0.05	0.33	0.09
2 001-5 000	57.92	37.97	3.65	0.03	0.39	0.04
5 001-10 000	55.31	39.87	4.08	0.04	0.63	0.07
10 001-20 000	57.43	37.96	3.82	0.09	0.60	0.11
20 001 – 100 000	52.25	41.22	5.45	0.07	0.86	0.16
100 001 - 1 Million	44.15	47.52	6.18	0.15	1.75	0.24
Budapest	33.54	54.86	7.05	0.28	3.55	0.71
Total	54.97	39.71	4.39	0.07	0.74	0.11

Source: Hungarian census 1990; author's compilation from individual data sets.

Table 8:   Number of Hungarian gypsies over 14 years old who did not complete primary school according to gender and size of their place of residence, 1990 (%)

Number of inhabitants in place of residence	Total	Men	Women
up to 500	63.3	53.6	72.7
501-1 000	59.7	50.4	68.7
1 001-2 000	56.8	48.5	64.8
2 001-5 000	57.9	49.2	66.3
5 001-10 000	55.3	45.9	64.7
10 001-20 000	57.4	48.7	66.1
20 001 – 100 000	52.3	42.8	62.0
100 001 - 1 Million	44.2	33.4	54.7
Budapest	33.5	26.3	42.4
Total	55.0	45.9	64.0

Source: Hungarian census 1990; author's compilation from individual data sets.

The gypsies' educational achievement varies with the size of their place of residence. The larger the size of their place of residence, the more the proportion of gypsies without a basic education decreases, and the more the share of those with a primary school education or further education increases. In 1990, 63.3% of gypsies who lived in towns with less than 500 inhabitants had not completed primary school, but only 33.5% of gypsies who lived in Budapest had not completed primary school. Out of the gypsies over 15 years old who did not complete primary school 19.1% lived in towns with less than 1 000 inhabitants and 59.7% lived in towns with less than 5 000 inhabitants.

It is worth noting that there are large gender-specific educational disparities among gypsies. Whereas in Budapest, 'only' 42.4% of female gypsies over 15 years old did not complete primary school, this percentage rose to 72.7% in small villages with less than 500 inhabitants. Such large gender-specific disparities are only seen in the poorest developing countries today. Considering the fact that educational achievement of children is mainly affected by the maternal level of education, opportunities of young gypsies growing up in the rural periphery are not very promising.

As lowly skilled gypsies often retreated in areas (and houses) abandoned by others and as social marginalisation often corresponds to spatial exclusion, the majority of Hungarian gypsies until the early 1990s lived in economically underdeveloped peripheral areas with poor infrastructure and in substandard residential areas of Budapest.

Figure 15: The distribution of gypsies in Hungary, 1990 (location quotient)

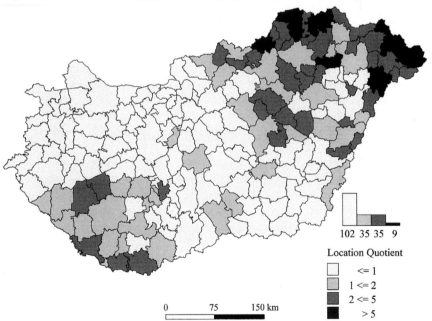

Location Quotient

□ <= 1
▨ 1 <= 2
▩ 2 <= 5
■ > 5

0   75   150 km

Source: Hungarian census 1990; author's compilation from individual data sets.

According to the definition of 'underdeveloped regions' given by the Hungarian Central Statistical Office in 1985, the share of gypsies living in economically underdeveloped regions (14.5%) was four times higher than the Hungarian aver-

age (3.8%). Large concentrations of gypsies can be found in places along the
north-eastern border of Slovakia, including Encs (location quotient 9.08), Záhony
(6.80), Edelény (6.43), Vásárosnamény (6.41), Nyírbátor (6.02), Gönc (5.84), Ózd
(5.82) and Fehérgyarmat (5.44). These areas have five to nine times more gypsies
than the national average.[22] The second region in which gypsies are strongly concentrated includes the so-called 'inner periphery' in the Baranya and the Somogy
region.

7   Conclusion

During the past centuries the Carpathian basin displayed large west-east disparities in socio-economic structures, literacy, technological standards, and economic
performance. In certain periods of transition and modernisation, these spatial patterns of knowledge systems became more distinct and played a crucial role in the
creation and adoption of innovation. One of these decisive turning points in recent
history was the change from a centrally planned communist system to a democracy and market economy. Forty years of communism did not lead to more spatial
equality. On the contrary, the disparities between the large centres and the underdeveloped rural periphery increased during the period of central planning. This
historical legacy was projected into the 1990's and partly increased. After the
change of the political and economic system in 1989, the expansion of the tertiary
sector, foreign investment and many other factors favoured the capital region.
Nevertheless, the market economy and cross-border relations to Austria initiated
substantial economic growth in the western areas of Hungary. Only when the supply of highly-skilled, low cost labour was becoming scarce in central and western
Hungary in the late 1990s, foreign direct investment moved increasingly towards
the east of Hungary.

In the 1990s, Hungary has been the most successful Central and East European
country to attract foreign direct investment and to achieve a stable growth after
1996. In the first period of transformation, the role of educational achievement,
technological standards, and professional skills for economic performance were of
great importance. Foreign investment, economic performance and success on
foreign markets depend, however, not only on technical and professional
knowledge, but also on value systems, attitudes, identities, stereotypes, cultural
traditions and social relations with decision makers in foreign countries. An
explanation of the spatially varying success of the transformation process needs
further research on so-called soft indicators describing the persistence or change
of attitudes, of shared orientation and behaviour patterns, identities and knowledge systems. Such indicators could concentrate on the degree of corruption,
nepotism, and organised crime, on the willingness to take individual responsibility

---

[22]   These figures comprise only those who declared themselves as gypies.

and risks, on the attitude of state bureaucracy towards the citizen, on media policy, on the readiness to deal with the dark sides of communist past and on the language used to describe expropriation, ethnic cleansing,[23] concentration camps, and resistance against communist repression. Hungary seems to be in the lead in abandoning behaviour patterns and 'old mentalities' acquired in the communist period of central planning. The experience of other Central and East European countries demonstrate how long it can take to change such knowledge and value systems.

Some of the comparative advantages Hungary had in the 1990s are gradually fading. A well-skilled - yet low cost - labour force is becoming scarce. The benefits of early liberalisation and of a head start in establishing good economic and political relations to the west are being compromised. Influx of foreign capital could also slow down. The proportion of companies where the majority of assets are of high technical standard is still rather low (see HARSÁNY and NYERS 2000). Continuity or acceleration of the modernisation of the Hungarian economy needs new comparative advantages, such as larger investments of companies in research and development, better funding of universities and other research institutions, the strengthening of entrepreneurship, a further increase of technological standards, a higher proportion of domestically produced technological innovations, an internationally competitive level of financial and business related services, and many other programmes. The newly released Széchenyi plan of the Hungarian Government (CSÉFALVAY 2000) is, in this context, a step to the right direction.

References

ALFÖLDY, G. (1997): Ungarn 1956: Aufstand, Revolution, Freiheitskampf. Schriften der Philososphisch-historischen Klasse der Heidelberger Akademie der Wissenschaften 2, Heidelberg

ASCHAUER, W. (1995): Bedeutung und regionale Verteilung von Joint Ventures in Ungarn. In: MEUSBURGER P., KLINGER A. (eds): Vom Plan zum Markt: Eine Untersuchung am Beispiel Ungarns. Physica Verlag, Heidelberg, 62-79

BÁRÁNY, Z. (1994): Minderheiten, Ethnopolitik und die osteuropäischen Roma. Ethnos-Nation: Eine europäische Zeitschrift 2, 6-17

BARTAL, A. M. (1999): Nonprofit alapismeretek kézikönyve. Ligatura, Budapest

BERÉNYI, I. (1992): The Socio-Economic Transformation and the Consequences of the Liberalisation of Borders in Hungary. In: KERTÉSZ, A., KOVÁCS, Z. (eds): New Perspectives in Hungarian Geography. Geographical Research

---

[23] Media and tourist guides in some areas of Central and Eastern Europe in the year 2000 still preferred the term 'the property was transferred to the responsibility of the state' to the term 'expropriation', or rather use the term 'transfer of people' instead of 'expulsion' (Vertreibung) or 'ethnic cleansing'. Such a use of words reveals much about attitudes, identities and 'old thinking'.

Institute, Hungarian Academy of Sciences, Budapest (= Studies in Geography in Hungary 27), 143-157

BERÉNYI, I. (1995): Einige Probleme des wirtschaftlichen Strukturwandels in Ungarn. In: MEUSBURGER, P., KLINGER, A. (eds): Von der Planwirtschaft zur Marktwirtschaft: Eine Untersuchung am Beispiel Ungarns. Physica Verlag, Heidelberg, 99-113

BERÉNYI, I. (2000): Hungarian Agriculture in Transformation: Spatial Aspects. In: KOVÁCS, Z. (ed.): Hungary Towards the 21st Century: The Human Geography of Transition. Geographical Research Institute, Hungarian Academy of Sciences, Budapest, (= Studies in Geography of Hungary 31), 207-225

BEREY, K. (1991): A cigánytelepek felszámolása és újratermelődése. In: MTA Politikai Tudományok Intézete (ed.): Cigányélet. Budapest, 106-144

CROWE, D. (1991): The Gypsies in Hungary. In: CROWE, D. (ed..): The Gypsies of Eastern Europe. Sharpe, Armonk NY, 117-132

CSÉFALVAY, Z. (1994): The Regional Differentiation of the Hungarian Economy in Transition. GeoJournal 32, 351-362

CSÉFALVAY, Z. (1995a): Raum und Gesellschaft Ungarns in der Übergangsphase zur Marktwirtschaft. In: MEUSBURGER, P., KLINGER, A. (eds): Vom Plan zum Markt. Eine Untersuchung am Beispiel Ungarns. Physica Verlag, Heidelberg, 80-98

CSÉFALVAY, Z. (1995b): Die Dualität des ungarischen Arbeitsmarktes. In: FASSMANN, H., LICHTENBERGER, E. (eds): Märkte in Bewegung: Metropolen und Regionen in Ostmitteleuropa. Böhlau Verlag, Wien, Köln, Weimar, 113-129

CSÉFALVAY, Z. (1997): Aufholen durch regionale Differenzierung? Von der Planwirtschaft zur Marktwirtschaft: Ostdeutschland und Ungarn im Vergleich. Erdkundliches Wissen Bd. 122, Steiner Verlag, Stuttgart

CSÉFALVAY, Z. (2000): Széchenyi Plan: The Background, Objectives and Funding. In: Economic Trends in Hungary 2, Ecostat, Budapest, 4-12

CSÉFALVAY, Z., LANDESMANN, M., MATOLCSY, G. (eds) (1999): Hungary's Accession to the EU: The Impact on Selected Areas of Hungarian-Austrian relations. Bilaterial Synthetic Report. Budapest, Vienna

DARÓCZI, A., BÁRSONY, J. (1991): Roma in Ungarn. In: Roma in Ost- und Südost-Europa, Grundsatztagung für Verantwortliche in Politik, Rechtsprechung und Verwaltung, für in der Flüchtlingsarbeit Verantwortliche und Tätige bei Städten, Landkreisen, Gemeinden, Kirchen, Wohlfahrtsverbänden und Initiativen (15.-17.5.1991), Bad Boll-Tagungsbericht, 8-15

DÖVÉNYI, Z. (1993): Unemployment as a New Phenomenon of the Transition. In: HAJDÚ, Z. (ed.): Hungary: Society, State, Economy and Regional Structure in Transition. Centre for Regional Studies, Pécs, 185-208

DÖVÉNYI, Z. (1995): Die strukturellen und territorialen Besonderheiten der Arbeitslosigkeit in Ungarn. In: MEUSBURGER, P., KLINGER, A. (eds): Vom Plan zum Markt. Eine Untersuchung am Beispiel Ungarns. Physica Verlag, Heidelberg, 114-129

ÉKES, I., KUPA/KREKÓ, I. (2000): On Key Social Trends. In: Economic Trends in Hungary 2, Ecostat, Budapest, 36-42

ENYEDI, G. (1998): Transformation in Central European Postsocialist Cities. Centre for Regional Studies of Hungarian Academy of Sciences, Pécs (= Discussion Papers 21)

FASSMANN, H. (1995): Wegbegleiter nach Europa. Ökonomische Krisen und wachsende Arbeitslosigkeit in Ost-Mitteleuropa. In: MEUSBURGER, P., KLINGER, A. (eds): Vom Plan zum Markt. Eine Untersuchung am Beispiel Ungarns. Physica Verlag, Heidelberg, 1-18

FASSMANN, H., LICHTENBERGER, E. (eds) (1995): Märkte in Bewegung. Metropolen und Regionen in Ostmitteleuropa. Böhlau Verlag, Wien, Köln, Weimar

FEHÉR, E. (2000): Subsidies for Research and Development (R&D) in Hungary. In: Economic Trends in Hungary 2, Ecostat, Budapest, 69-70

GORZELAK, G. (1996): The Regional Dimension of Transformation in Central Europe. Regional Studies Association, London (= Regional Policy and Development 10)

GRONEMEYER, R. (1981): Unaufgeräumte Hinterzimmer: Ordnungsabsichten sozialistischer Zigeunerpolitik am Beispiel Ungarn. In: MÜNZEL, M., STRECK, B. (eds): Kumpania und Kontrolle. Focus-Verlag Giessen, 193-224

GRONEMEYER, R. (1983): Zigeunerpolitik in sozialistischen Ländern Osteuropas. In: GRONEMEYER, R. (ed.): Eigensinn und Hilfe: Zigeuner in der Sozialpolitik heutiger Leistungsgesellschaften. Focus-Verlag, Giessen, 43-183

HAMILTON, I. F. E. (1998): Re-evaluating Space: Locational Change and Adjustment in Central and Eastern Europe. Geographische Zeitschrift 83, 67-86

HARCSA, I. (1995): Ungarische Kader in den Achtziger Jahren. In: MEUSBURGER, P., KLINGER, A. (eds): Vom Plan zum Markt: Eine Untersuchung am Beispiel Ungarns. Physica Verlag, Heidelberg, 270-284

HARSÁNY, L., NYERS, J. (2000): Technical Standards and State-Of-The-Art Technologies in Company Practices (1995-1998). In: Economic Trends in Hungary 2, Ecostat, Budapest, 27-35

HAVLIK P. et al. (1999): The Transition Countries in 1999: A Further Weakening of Growth and Some Hopes for Later Recovery. The Vienna Institute for International Economic Studies, Vienna (= Research Reports 257)

HAYEK, F. A. (1944): The Road to Serfdom. University of Chicago Press, Chicago

HAYEK, F. A. (1945): The Use of Knowledge in Society. American Economic Review 35, 519-530

HOÓZ, I. (1991): A társadalmi folyamatok és a cigánynépesség. In: MTA Politikai Tudományok Intézete (ed.): Cigányélet. Budapest, 54-77

KISS, É. (2000): The Hungarian Industry in the Period of Transition. In: KOVÁCS, Z. (ed.): Hungary Towards the 21st Century: The Human Geography of Transition. Geographical Research Institute, Hungarian Academy of Sciences, Budapest (= Studies in Geography in Hungary 31), 227-240

KOCSIS, K. (1994): Contribution to the Background of the Ethnic Conflicts in the Carpathian Basin. GeoJournal 32, 425-433

KOCSIS, K. (2000): The Roma (Gypsy) Question in the Carpatho-Pannonian Region. In: KOVÁCS, Z. (ed.): Hungary Towards the 21st Century: The Human Geography of Transition. Geographical Research Institute, Hungarian Academy of Sciences, Budapest (= Studies in Geography of Hungary 31), 227-240

KOCSIS, K., KOVÁCS, Z. (1991): A magyarországi cigánynépesség társadalomföldrajza. In: MTA Politikai Tudományok Intézete (ed.): Cigánylet, Budapest, 78-105

KOVÁCS, Z. (2000): Hungary at the Threshold of the New Millennium: The Human Geography of Transition. In: KOVÁCS, Z. (ed.): Hungary Towards the 21st Century: The Human Geography of Transition. Geographical Research Institute, Hungarian Academy of Sciences, Budapest (= Studies in Geography of Hungary 31), 11-27

KYN, O.: Eastern Europe in Transition. http://econc10.bu.edu/okyn/OKpers/eetrans.html

LENGYEL, GY. (1995): Kader und Manager: Unterschiedliche Muster der Rekrutierung von Führungskräften in der Planwirtschaft. In: MEUSBURGER, P., KLINGER, A. (eds): Vom Plan zum Markt: Eine Untersuchung am Beispiel Ungarns. Physica Verlag, Heidelberg, 249-269

MACHOWSKI, H., WILKENS, H. (1999): Wirtschaftslage und Reformprozesse in Mittel- und Osteuropa. Deutsches Institut für Wirtschaftsforschung Berlin, Bonn

MALECKI, E. (2000): Knowledge and Regional Competitiveness. Erdkunde 54, 334-351

MARX-ENGELS-WERKE (1962). Edited by the Institut für Marxismus-Leninismus beim ZK der SED. Volume 16, Dietz, Berlin

MEUSBURGER, P. (1995a): Spatial Disparities of Labour Markets in Centrally Planned and Free Market Economies: A Comparison Between Hungary and Austria in the Early 1980's. In FLÜCHTER, W. (ed.): Japan and Central Europe Restructuring: Geographical Aspects of Socio-Economic, Urban, and Regional Development. Wiesbaden, 67-82

MEUSBURGER, P. (1995b): Zur Veränderung der Frauenerwerbstätigkeit in Ungarn beim Übergang von der sozialistischen Planwirtschaft zur Marktwirtschaft. In: MEUSBURGER, P., KLINGER, A. (eds): Vom Plan zum Markt: Eine Untersuchung am Beispiel Ungarns. Physica Verlag, Heidelberg, 130-181

MEUSBURGER, P. (1997): Spatial and Social Inequality in Communist Countries and in the First Period of the Transformation Process to a Market Economy: The Example of Hungary. Geographical Review of Japan 70 (Ser. B.), 126-143

MEUSBURGER, P. (1998): Bildungsgeographie: Wissen und Ausbildung in der räumlichen Dimension. Spektrum Akademischer Verlag, Heidelberg

MEUSBURGER, P. (2000): The Spatial Concentration of Knowledge: Some Theoretical Considerations. Erdkunde 54, 352-364

MISES, L. v. (1922): Die Gemeinwirtschaft: Untersuchungen über den Sozialismus. Fischer Verlag, Jena

NAVARETTI, G. B., TARR, D. G. (2000): International Knowledge Flows and Economic Performance: A Review of the Evidence. The World Bank Economic Review 14 (1), 1-15

NEMES-NAGY, J. (1994): Regional Disparities in Hungary during the Period of Transition to a Market Economy. GeoJournal 32, 363-368

NEMES-NAGY, J. (2000): Regional Inequalities in Hungary at the End of the Socio-Economic Transition. In: KOVÁCS, Z. (ed.): Hungary Towards the 21st Century: The Human Geography of Transition. Geographical Research Institute, Hungarian Academy of Sciences, Budapest (= Studies in Geography of Hungary 31), 87-98

OESTERER, M., MEUSBURGER, P. (1998): Die Situation der ungarischen Zigeuner auf dem Arbeitsmarkt beim Übergang von der sozialistischen Planwirtschaft zur Marktwirtschaft. In: Berliner Geographische Arbeiten 86, 141-158

PEARSON, R. (1995): The Making of '89: Nationalism and the Dissolution of Communist Eastern Europe. Nations & Nationalism 1 (1), 69-79

PICKLES, J., SMITH, A. (eds) (1998): Theorising Transition: The Political Economy of Post-Communist Transformations. Routledge, London

RECHNITZER, J. (1998): Allocation of Foreign Investments in the Transdanubian Region. In: BASSA, L.; DERTÉSZ, A. (eds): Windows on Hungarian Geography. Geographical Research Institute, Hungarian Academy of Sciences, Budapest (= Studies in Geography in Hungary 29), 185-197

RECHNITZER, J. (2000): The Features of the Transition of Hungary's Regional System. Centre for Regional Studies of Hungarian Academy of Sciences, Pécs (= Discussion Papers 32)

RYLL, A. (1994): Transition to a Market Economy as the Transformation of Coordination. In: WAGENER, H.-J. (ed.): The Political Economy of Transformation. Physica Verlag, Heidelberg, 45-59

SACHS, J. (1993): Poland's Jump to the Market Economy. MIT Press, Cambridge

SCHREYÖGG, G. (1996): Transferring Managerial and Organizational Methods to Eastern Europe. In: HAX, H. et al. (eds): Economic Transformation in Eastern Europe and East Asia: A Challenge for Japan and Germany. Springer Verlag, Berlin, Heidelberg, 79-90

SEEWANN, G. (1987): Zigeuner in Ungarn. Südosteuropa 36, 19-32

SIKLÓS, L. (1970): The Gypsies. The New Hungarian Quarterly 11, 150-162

STEHR, N. (1994): Knowledge Societies. Sage, London

SZABÓ, G. (1991): Die Roma in Ungarn: Ein Beitrag zur Sozialgeschichte einer Minderheit in Ost- und Mitteleuropa. Lang, Frankfurt a.M.

SZANYI, M. (1998): The Role of Foreign Direct Investment in Restructuring and Modernizing Transition Economies: An Overview of Literature on Hungary.

The Vienna Institute for International Economic Studies, Vienna (= Research Reports 244)

TARDOS, M. (1989): The Blueprint of Economic Reforms in Hungary. In: UNITED NATIONS ECONOMIC COMMISSION FOR EUROPE (ed.): Economic Reforms in the European Centrally Planned Economies. New York, 34-41

THE 1956 HUNGARIAN REVOLUTION. http://www.mfa.gov.hu/sajtonanyag/sajto30. html

THE NATIONAL AND ETHNIC MINORITIES IN HUNGARY. http://www.mfa.gov.hu/ sajtonanyag/Nat%20and%20Eth.html

THE WORLD BANK (ed.) (1999): Hungary: On the Road to the European Union. A World Bank Country Study. Washington

VOSZKA, É. (1989): Role and Functioning of the Enterprise in Post-Reform Hungary. In: UNITED NATIONS ECONOMIC COMMISSION FOR EUROPE (ed.): Economic Reforms in the European Centrally Planned Economies. New York, 109-116

ZUR LAGE DER ZIGEUNER IN UNGARN (1994). Ein Untersuchungsbericht. Ethnos-Nation: Eine europäische Zeitschrift 2, 49-60

New Regional Patterns in Hungary

József Nemes-Nagy

1 Introduction

In Hungary, most of the basic socio-economic changes launched by the political transition of 1989 were completed by the end of the 1990s. The institutions of political democracy and market economy were established.

As a result of these processes, a new regional pattern emerged which is substantially different from that of the socialist period. This study presents the most important features of this altered regional pattern at the level of the most significant administrative units, the county level, and also addresses socio-economic disparities characteristic at the microregional and settlement levels. Altogether, the following spatial levels and units are considered:

- 7 statistical-planning regions (the proposed NUTS 2 units);
- 19 counties and the capital, Budapest (NUTS 3 units, actual regional authorities);
- 150 statistical microregions (without administrative institutions);
- 3 100 settlements (local authorities).

The most significant quantitative features of regional development and spatial inequalities are surveyed at these spatial levels (e.g. the variation in employment structure, industrial production, income, and foreign capital investment).[1] In order to characterise the new spatial pattern, some complex mathematical-statistical methods and indices have also been used in the analysis.

Before examining the new regional pattern, a brief summary of its historical background is given that presents the specific features of the country's development. Unlike most Central Eastern European countries, Hungary was characterised by socio-economic factors which prepared for and forecasted the processes that took place after the 1989-90 political transition. Important economic reforms happened in the five to ten-year period after 1968 and more evidently in the 1980s. It is important to note that in Hungary, the economic transition started earlier than the radical transformation of political institutions.

---

[1] Since the regional characteristics of unemployment are dealt with by other studies under the research project, they are not discussed here in much detail. Unemployment, however, as a regionally characteristic indicator of economic crisis, was included in the complex mathematical-statistical analysis. For details on the development and spatial disparities of unemployment, see the paper of Zoltán Dövényi in this volume.

40   J. Nemes-Nagy

Figure 1:   Statistical-planning regions in Hungary

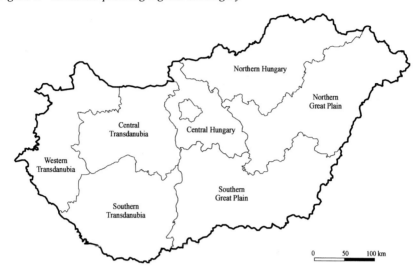

Source: Hungarian Central Statistical Office (KSH).

Figure 2:   Hungarian counties

Source: Hungarian Central Statistical Office (KSH).

2   The relevant historical background: spatial processes during the period of socialism

Following World War II, Hungary came under the Soviet sphere of interest. This determined Hungary's political system, the mobility of the citizens, foreign affairs, the economic structure, the functional apparatus, and thus indirectly, the regional processes and settlement development for over forty years. Considering regional and settlement development, we can define three periods:

- 1945-1968: the dominance of the 'Soviet model'
- 1968-1982: the emergence of a 'Hungarian model'
- 1982-1989: a deepening economic and political crisis.

2.1   The dominance of the 'Soviet Model'

Following the relatively quick restoration of war damages throughout the country, the Stalinist political and economic model had become totally dominant by 1949. The primary ideological purpose of a centrally planned economy was to increase material production with heavy industry by raising it to a privileged position. For three decades, industry was the driving force behind the economy, putting an end to the predominance of agriculture with respect to both production and employment. The industrialisation of the 1950s and 1960s was also the motor of regional economic development and the key factor for regional development policy, even if the latter did not yet exist on its own.

The one-sided orientation towards heavy industry in the 1950s represented a grave, uncalled-for aberration in a country lacking raw materials and resulted in a less efficient economy as well as economic dependence on the Soviet Union. Parallel to the development of heavy industry in the 1950s the construction of so-called 'socialist cities' took place. The majority of these cities were built alongside the central mountain range. Only one of them, the metallurgical centre of Sztálin-város (known today as Dunaújváros), probably the most typical example of all socialist cities, was built on the banks of the Danube, south of Budapest. Today most of these cities suffer from the difficulties of transformation and environmental pollution. At the same time, industry in Budapest became more and more overdeveloped due to the expansion and reconstruction of the old plants. The government did not recognise these problems until the mid-1960s, at which time it enacted regulations to curb industrialisation in the capital city and even to relocate some industries.

The development of the transportation and communication infrastructure (particularly the road and telephone networks) was neglected and these systems were left permanently inadequate and unbalanced. The greatest achievement of the early socialist period in infrastructure development was the introduction of electricity to the villages (completed by the early 1960s). The peasantry, pressured by ideological policies, was forced to join co-operatives. Starting in the 1960s, they were allowed to keep small household farms and thereby slightly improve their living conditions. Until the 1980s, the village infrastructure developed slowly and inconsistently.

Up to the mid-1970s, the labour force was the most important source of regional economic growth. This (excess) labour supply consisted mainly of women and was the most attractive factor in the regional industrial decentralisation. Both industrialisation and collectivisation in agriculture set off large-scale migration and daily commuting into Budapest and other industrial centres. As a result, villages weakened demographically and cities became socially unstable. The system of full employment and the related, artificially restricted salaries were without doubt the main factors causing the regional equalisation tendencies which began in the 1960s.

2.2   The 'Hungarian Model'

There was a shift in Hungarian internal politics and economic policy in the period from the mid- to late 1960s which was also noticeable in regional and settlement processes. The plan-directive system was replaced by a state system of indirect economic regulations. This was accompanied by a civil liberalisation process (e.g. travel abroad became relatively easy). The most fundamental socio-political (ideological) change was that while previously only production mattered, the new social image drawn by the reform emphasised economic growth, and therefore, consumption and higher standards of living became accepted values.

The emphasis gradually shifted from production and branch economics (which used to be the central factors in regional development) to management, infrastructure development and the objective of improving the population's living conditions. This change terminated the close links between material production and community infrastructure development. The last 'socialist' city was built in the late 1960s (Leninváros, today's Tiszaújváros, at the River Tisza, in the county of Borsod-Abaúj-Zemplén) in order to serve a huge new petrochemical plant.

Slow levelling processes were typical of regional development in the 1970s, during the period of the so-called 'developed socialism' (NEMES-NAGY 1990). The spatial economic structure that evolved at this time consisted of a heavily industrialised and overdeveloped capital city with a growing industrial axis and lagging agrarian areas. However, the diversification in production on large scale collective farms, and the development of their subsidiary activities (industry, building, service) facilitated the expansion of agricultural areas in the 1970s.

In the mid-1970s, the limited availability of extensive types of development resources slowed down the large macroregional movements and allowed microregional differentiation to become more conspicuous. The focus was on quality rather than quantity. Nevertheless, the general trend of the 1960s and 1970s was the levelling of economic development and social conditions among the large regions, counties and main settlement types. The most important element in this process was levelling off the regional differences in personal income.

The 1970s were a decade of dynamic development for provincial cities, especially for county seats. It was a period of mass housing production (mainly housing projects) and the basic infrastructural developments which were connected to housing production. It was during this time that the largest housing projects were built in Budapest. All of these improvements are good examples of the development goals. This increase in the population's material welfare enabled masses of people to begin using private automobiles around this time.

2.3   Period of the 'Open Crisis'

By the end of the 1970s, the nation was at the edge of a debt crisis, which was due to the wasteful economic policy and a standard of living growing at a higher rate than economic results as measured by increasing productivity could justify. As a result, economic growth came to a halt and began to decrease with the standard of living. The 1980s brought about a twofold change in the economy:

- the expansion of the main industrial branches came to a halt;
- new types of economic organisations appeared.

The decline of traditional heavy industry and the regional depression associated with it caused a pronounced regional realignment. This had a 'downward' pulling effect on the proportion of spatial development, and affected those industrial areas previously considered developed. Dependence upon external markets and resources in the 1980s characteristically affected cities and sub-regions organised around large industrial plants (metallurgy, chemistry) and areas specialising in a single industrial branch. In the early 1980s many large scale industrial plants formerly considered flagships of the state economy simply lost their value. Consequently, one of the ideological cornerstones of the system, the 'workers ideology', was on the verge of crisis, too.

Following the economic recession, the basic socialist values of production (labour) and consumption were joined by a new element of the 1980s, enterprise. This was acknowledged as a social and political value, although in a forced way. A multitude of new 'semi-private' small enterprises were formed following comprehensive governmental decrees. In the mid-1980s, almost one million people joined the urban secondary economy, but four-fifths of them only on a part-time basis.[2] The thousands of new small organisations which were connected to the socialist sector were not capable of deterring depression in the production sector. They had a large service profile, and appeared in Budapest, in larger cities and in special recreation areas (e.g. at Lake Balaton). The increasing differences between

---

[2]   For further details about regional processes in this important period, see SILINCE 1987; NEMES-NAGY and RUTTKAY 1991; RECHNITZER 1993; HORVÁTH 1993; SIK 1994.

the 'old' and the 'new' economies, which were considered permanent during these historical times, on the regional and settlement levels preceded the dual tendency that manifested itself after the economic transition. In this latter period, the entire country was in deep economic depression and only a few regional settlement concentrations were able to recover from the crisis. The political transformation started in this unstable and unsound economic environment.

3   Spatial processes and changes after the political transformation

Building a multi-party democratic constitutional state and a market economy in the 1990s had far-reaching consequences for all levels of society. Because these comprehensive changes happened over such a short period of time, radical regroupings within the socio-economic spatial structure have taken place as well (ENYEDI 1996a; TURNOCK 1998).

The most fundamental change has taken place in the structure of ownership. Following privatisation (RECHNITZER 1998), land ownership compensation, new capital investments (especially the foreign capital amounting to 16 billion USD by 1997), and private ownership became dominant by the mid-1990s. In 1996, more than half of all wage earners in every county worked for private or at least partly-private enterprises. Today there are approximately 700 000 individual entrepreneurs and more than 300 000 business enterprises in Hungary. The role of the state is now dominant only in the fields of education, health services, and public administration. At the same time, the number of actual wage earners decreased from 4.8 to 3.6 million between 1990 and 1995. Unemployment figures peaked at over 700 000 in 1993 but dropped to less than 400 000 by 1998.

Hungary was forced completely to reorient its external economic ties. This led to a subsequent deepening of the regional crisis in manufacturing and agricultural industries which had been producing for the Soviet market. Economic decline was at its peak in 1993.

The complete opening of borders served as a catalyst for changes in the spatial structure. The system of state-monopolised and centrally organised international relations began to allow cross-border co-operation. Previously closed border regions soon became zones of dynamic activity, although not all sectors of the border were affected in the same way (see the chapter by János Rechnitzer). At the Austrian border, there were large scale investments while at the southern border, there was controversy surrounding a business boom fuelled by the Balkan crisis, and based on activities that were only partly legitimate. On the eastern borders a network of illegal businesses sprang up. Therefore the country's geographical periphery cannot be regarded as part of an economic and social entity. On the contrary, the western border region can be considered a dynamic edge.

3.1   Macroregional pattern of development

The transformation of the spatial development pattern can be shown by the per capita GDP values and by the so-called 'economic health' indicator, which is an index obtained by factor analysis. With the help of these two kinds of indices, it is possible to distinguish between both long- and short-term changes. With the use of GDP per capita data, which has been estimated on the county level since 1994 by the Hungarian Central Statistical Office (KSH), it is possible to make comparisons with the proportions existing in 1975. These indicate the typical spatial pattern of the socialist period (the regional GDP data for 1975 has been estimated by the author based on official statistical information and data from BARTA 1977).

Indexes of economic health were calculated for 1990 and 1996. These reveal the crisis symptom characteristic of this decade. Entrepreneurial activity measured by firm-density, income level of employees based on taxable incomes, unemployment rate and the proportion of joint ventures to the total number of firms were merged and result in a single factor with a common high (greater than 0.8) factor, and with 72% variation. High positive figures mean relatively high income levels, low unemployment rates, a great number of business ventures, and a large volume of foreign capital investment. In contrast, high negative figures indicate counties where the signs of crisis are dominant, while figures close to 0 represent the average.

The GDP data reflecting the 1975 development level and the new pattern indicate extremely intense changes in the relative positions of the counties (table 1). From among the positive shifts, the position of the capital city with its extremely high level of development deserves particular attention. The dominance of Budapest over the rest of the country has increased from 1975 to 1996. The economic health indicator demonstrates that in the transition process the initial position of the capital was already very favourable compared to the other country's regions. The factor for 1990 was prominently high, and although this structural advantage decreased slightly by the middle of the decade, it is still significant today.[3]

Outside the capital, counties with steadily improving positions can be found only in the western, Transdanubian (Dunántúl) part of the country. The advancement of Vas county is most notable. Győr-Moson-Sopron also has a development level clearly above the average, and it is still improving its position. These two

---

[3]   However, without intending to underestimate the economic and social potential of Budapest, it should be noted that the Hungarian Central Statistical Office has estimated that the per capita GDP in Hungary for 1994 was only 37% of the European Union average. This means that the high figure calculated for Budapest is not even half that of the EU average. Therefore, if measured according to 'European' criteria (the famous 75% GDP per capita 'underdevelopment limit' in the EU), the countryside and the capital city are in need of support because they fall far below the 75% of the EU per capita GDP level.

western counties are, in an economic sense, the winners of the economic-political transition.

Table 1: The dynamics of county level differences in economic development

Regions, counties	GDP per capita (average=100)			'Economic health' (factor values)		
	1975	1996	Change	1990	1996	Change
Central Hungary						
Budapest	139	186	47	3.22	2.80	-0.42
Pest	61	74	13	0.72	0.48	-0.23
C. Transdanubia						
Fejér	106	102	-4	0.24	0.40	0.15
Komárom-E.	131	89	-42	0.53	0.11	-0.42
Veszprém	116	80	-36	0.23	0.39	0.16
W. Transdanubia						
Győr-M.-S.	111	110	-1	0.92	1.01	0.09
Vas	82	109	27	0.46	0.89	0.43
Zala	88	93	5	0.09	0.73	0.63
S. Transdanubia						
Baranya	108	77	-31	0.27	0.01	-0.26
Somogy	71	75	4	-0.37	-0.22	0.15
Tolna	77	90	13	-0.24	-0.43	-0.20
N. Hungary						
Borsod-A.-Z.	111	70	-41	-1.01	-1.29	-0.29
Heves	100	73	-27	-0.82	-0.50	0.33
Nógrád	77	57	-20	-0.95	-1.01	-0.06
N. Great Plain						
Hajdú-B.	83	78	-5	-0.44	-0.87	-0.42
Jász-N.-Sz.	93	76	-17	-0.69	-0.80	-0.12
Szabolcs-Sz.-B.	59	59	0	-1.65	-1.56	0.08
S. Great Plain						
Bács-K.	79	76	-3	0.00	-0.15	-0.15
Békés	89	76	-13	-0.65	-0.77	-0.12
Csongrád	109	93	-16	0.13	0.78	0.66
Hungary	100	100		0.00	0.00	

Source: GDP 1975: Author's estimation based on KSH data and BARTA 1977; GDP 1996: Hungarian Central Statistical Office (KSH).

The present dynamics of the western regions are primarily due to their favourable geographical location. A previous study that accounted for the economic influence of regions in neighbouring countries determined the so-called 'economic geographical potential' of the Hungarian regions by using the 'potential model' analogous with the one used in physics (NEMES-NAGY 1997b). In this study, it was shown that the present spatial pattern is strongly influenced by the proximity of the economically powerful regions in Austria, southern Germany and northern

Italy. The prime beneficiaries of this external impact are the three counties mentioned above that benefit from large-scale daily contact with adjacent Austrian regions. Another unique advantage of this area is that the western border zone (for political-military reasons) was neglected during the heavy industrial stage of socialist industrialisation and had a less obsolete and more flexible economic structure and professional culture at the beginning of the transition. These factors strengthen the region's advantages because a common mentality has been built upon socio-psychological heritage.

While there has been significant development in the two western counties, we are witnessing a radical decline in the northern Hungarian region. Borsod-Abaúj-Zemplén county has gone through a long period of depression, and Nógrád has fallen to the bottom in terms of economic development. Heves is the only county which shows a positive result in the first part of this decade. Northern Hungary is a typical example of a formerly developed region now in a crisis due to its outdated heavy industrial bases. The development level in northern Hungary does not differ much from that in the Great Plain (Alföld) because of its economic demise.

Neither the development indicators of the traditionally less developed macroregion nor those of the agrarian Great Plain give any reason for optimism. Csongrád is the only county whose development level exceeds that in the Great Plain. One of the main reasons for Csongrád's better position is the attractiveness of Szeged city, and the high density of small scale entrepreneurship. The city has evolved due to immigration and an inflow of capital from Yugoslavia, which was heavily interwoven by 'black and grey' elements.

The crisis (sharp decrease in production, high unemployment rates) in the first half of the nineties in the eastern counties also had many causes. It was due to local problems and the peculiar mechanism of the more developed regions' economy (the capital in particular) to react to the difficulties by passing most of the burden to the peripheries. The commuters were the first in the labour force to be dismissed and the small plants in the countryside were the first to be closed down.[4] The other cause of depression in the east was that the north-eastern counties were those hit most severely by the industrial crisis and the crises of agricultural mass production oriented towards the collapsing Soviet market. Finally, as a consequence of the insufficient macroregional infrastructure, the incoming foreign capital remained in the western part of the country and in Budapest. In the east, through privatisation, foreign capital selected and acquired only those companies with the most promising markets.

The central part of the country was also strongly divided on the county level because of its rather unstable character during the transition period. Komárom-Esztergom county, for example, has suffered the most during the last twenty

---

[4]   The high unemployment rate in Szabolcs-Szatmár-Bereg county was actually the result of the capital's ability to react quickly.

years. Veszprém county faces similar problems. Fejér county directly adjoins Veszprém and has undergone a spectacular rate of development since 1990. The reason for the present high GDP in Tolna (the county that has ranked second after Vas in achieving lasting development) is primarily due to the great production potential of the Paks Nuclear Power Plant completed in the early 1980s. Tolna's relatively favourable position has nothing to do with the post-transition processes. This is also evident because Tolna is not among the highest ranked counties according to the index of economic health. Improvement with regard to the GDP indicator is also remarkable in Pest county, which is closely linked to the administratively separated capital city. However, the GDP per capita allocated according to the actual place of production certainly underestimates Pest's position. According to the economic health indicator, Pest ranks average among the counties. Bács-Kiskun county is also part of this heterogeneous and unstable middle zone. Although it is located east of the Danube, Bács-Kiskun does not share the sluggish character of most of the Great Plain, but boasts rigorous small enterprise activities.

3.2   The collapse and restoration of industry

The transformation in the spatial development pattern was accompanied by sharp changes in the counties' economic structure in the 1990s (table 2). Although the most significant trend in the economic structure transformation in the nineties was the increase in the proportion of the service sector, this only caused a noticeable shift between the capital and the countryside. Budapest's outstanding economic performance was due primarily to the spectacular growth in commerce, business and financial activities, and resulted in high incomes (PROBÁLD 1995). In 1990, 46.6% of Hungary's active wage earners were employed in the tertiary economic sector. The share increased to 59.3% by 1996. At present, the majority of the labour force is employed in the service sector in all Hungarian counties. In the capital city, this sector has an especially important role, shown by a substantial increase from 62.5% to 75% between 1990 and 1996.

The declining agricultural sector had not contributed significantly to regional changes: the counties that were characteristically agrarian in the early 1990s have remained agrarian. 15.5% of the total Hungarian working population was engaged in agriculture in 1990, while by 1996 that figure had dropped to 8.0%. Among the counties, only Bács-Kiskun and Békés have more than 15% of wage earners in the agricultural sector. It is rather unfortunate that economic crisis and underdevelopment today is strongly connected to the agrarian character and has led to much social and political tension. The shares of the three GDP sectors in the counties are very similar to those in employment structure.

In contrast to agriculture, the industrial sector has become the most important reorganising factor in the regional economic pattern described above.[5] This is due to its profound structural change and adaptation processes, although its relative significance is diminishing compared to the tertiary activities.

Table 2: The sectoral economic structure in the counties in 1996

Regions, counties	Employment			GDP		
			Distribution (%)			
	I.	II.	III.	I.	II.	III.
Central Hungary						
Budapest	0.7	23.7	75.7	0.5	21.8	77.7
Pest	6.0	31.9	62.1	6.1	38.3	55.5
Central Transdanubia						
Fejér	10.4	41.1	48.5	7.5	47.0	45.5
Komárom-Esztergom	7.0	40.6	52.4	7.0	43.4	49.6
Veszprém	8.2	36.9	54.8	6.3	38.7	55.0
Western Transdanubia						
Győr-Moson-Sopron	7.8	37.7	54.5	7.8	38.0	54.2
Vas	8.1	42.2	49.8	6.4	48.5	45.1
Zala	6.3	39.9	53.9	7.5	38.4	54.1
Southern Transdanubia						
Baranya	11.4	33.1	55.4	9.3	26.2	64.6
Somogy	12.5	25.4	62.1	13.2	26.7	60.1
Tolna	14.2	36.0	49.9	14.1	39.1	46.8
Northern Hungary						
Borsod-Abaúj-Zemplén	5.7	37.3	57.0	6.3	39.3	54.4
Heves	6.9	38.4	54.7	9.2	32.3	58.4
Nógrád	5.6	39.8	54.6	7.2	32.2	60.6
Northern Great Plain						
Hajdú-Bihar	12.3	31.6	56.0	13.6	27.7	58,7
Jász-Nagykun-Szolnok	12.3	34.2	53.5	12.6	34.3	53.1
Szabolcs-Szatmár-Bereg	10.0	34.8	55.2	12.0	28.8	59.3
Southern Great Plain						
Bács-Kiskun	15.8	31.5	52.6	15.6	27.8	56.6
Békés	16.2	32.5	51.3	17.4	30.4	52.3
Csongrád	15.3	28.9	55.7	12.0	29.7	58.3
Hungary	8.0	32.7	59.3	6.6	30.7	62.7

Source: Hungarian Central Statistical Office (KSH); I: Agriculture, II: Mining, manufacturing, building industries, III. Tertiary and Quaternary branches.

---

[5]   For a long-range (1965-95) regional analysis of industrial production, see NEMES-NAGY (1997a).

The rearrangement of manufacturing has taken place in two distinct phases during the nineties. The first phase can be characterised by a comprehensive, uniform, and radical decline (table 3).

Table 3: County-level characteristics of industrial production in the 1990s

Counties	Year of least industrial production	Minimum level of industrial production	Industrial production in 1997
			(1989=100)
Vas (W. Tr.)	1991	64.4	285.9
Fejér (C. Tr.)	1992	66.8	244.2
Győr-Moson-Sopron (W. Tr.)	1992	55.6	202.3
Komárom-Esztergom (C. Tr.)	1992	67.9	100.1
Somogy (SD. Tr.)	1993	53.0	97.1
Hajdú-Bihar (N. G. P.)	1992	69.3	90.0
Pest (C. H.)	1992	59.5	88.5
Zala (W. Tr.)	1992	82.4	82.4
Heves (N. H.)	1993	61.8	79.9
Tolna (S. Tr.)	1997	74.7	74.7
Csongrád (S. G. P.)	1993	62.5	65.9
Jász-Nagykun-Szolnok (N. G. P.)	1993	57.8	63.1
Budapest (C. H.)	1993	56.0	62.9
Békés (S. G. P.)	1993	57.5	61.8
Nógrád (N. H.)	1994	47.9	60.6
Baranya (S. Tr.)	1995	50.6	59.9
Bács-Kiskun (S. G. P.)	1997	59.6	59.6
Szabolcs-Szatmár-Bereg (N. G. P.)	1996	50.5	55.1
Borsod-Abaúj-Zemplén (N. H.)	1992	45.3	52.3
Veszprém (C. Tr.)	1993	48.4	51.4

Source: Author's calculation based on regional data of the Hungarian Central Statistical Office (KSH).

In comparison to 1989, industrial production has dropped considerably in all counties (by more than 50% in Nógrád, Borsod, and Veszprém; by 30-40% in the majority of the counties and in Budapest; Zala was an exception with a decrease of less than 20%). The industrial crisis first affected those counties with a heavy industrial character before 1989, and was at its worst point in 1992-93 in most regions. Consequently, by this time the industrial development gap between the counties narrowed due to the sharp decline of the top-ranking areas. The real transition, however, came later.

Despite the fact that industrial production began to grow again in all counties in the mid 1990s (usually accompanied by a further decrease in employment), in the traditional industrial northern Hungarian counties and in the less industrialised plains areas, a slow recovery was underway. Production could not even reach its 1989 level until 1997.

In Vas, Győr-Moson-Sopron, and Fejér counties, however, huge direct capital investments included some large greenfield projects and enabled some multinational firms to settle there. The manufacturing industry there has almost doubled its production since 1989. Komárom-Esztergom is also on its way towards permanent recovery from economic depression. As a result of these recent developments, the most important Hungarian industrial zone is the western border area. The capital's role as an 'industrial base' has vanished and the northern industrial axis has completely disappeared.

Table 4: Enterprises with foreign direct investment in Hungary

Regions, counties	Number of organisations		Subscribed capital (billion HUF)	
	1993	1996	1993	1996
Central Hungary	12 265	14 560	784.6	1 402.8
Budapest	10 953	12 921	725.1	1 237.1
Pest	1 312	1 639	59.5	165.7
C. Transdanubia	1 470	1 815	90.3	150.4
Fejér	439	529	40.8	60.6
Komárom-E.	446	528	35.7	60.3
Veszprém	585	758	13.8	29.5
W.-Transdanubia	2 006	2 955	74.2	197.0
Győr-M.-S.	930	1 274	37.4	120.0
Vas	525	678	22.0	46.8
Zala	551	1 003	14.8	30.2
S.-Transdanubia	1 355	1 943	35.0	87.2
Baranya	715	994	20.6	62.1
Somogy	417	679	8.9	17.0
Tolna	223	270	5.5	8.1
N.-Hungary	681	787	44.9	206.1
Borsod-A.-Z.	358	352	23.1	143.1
Heves	171	270	12.7	51.7
Nógrád	152	165	9.1	11.3
N. Great Plain	832	1 018	40.8	122.4
Hajdú-B.	351	407	20.8	83.9
Jász-N-Sz.	214	272	11.5	22.5
Szabolcs-Sz.-B.	267	339	8.5	16.0
S. Great Plain	2 390	3 052	43.4	114.2
Bács-K.	1 003	1 075	17.9	25.4
Békés	282	273	14.6	17.7
Csongrád	1 105	1 704	10.9	71.1
Hungary	20 999	26 130	1 113.2	2 280.1

Source: HUNGARIAN CENTRAL STATISTICAL OFFICE 1996.

Foreign direct investment played a decisive role in the radical transformation of the regional industrial pattern by giving preference to the western region and to the capital (CSÉFALVAY 1995; ENYEDI 1996b; HASTENBERG 1996; SWAIN 1998; see also table 4). The first serious evidence of the foreign incentives to cross the Danube River came as late as 1998 when multinational companies launched some new industrial and commercial business projects in eastern and northern towns. The impact of these projects, however, has not yet been felt in the general regional development pattern.

3.3   Regional disparities in personal income

While analysing the effects of economic-political transition, the population's income is as important as GDP, capital investment and sectoral change. The need to include this aspect can be illustrated by the fact that Hungarians, facing serious social consequences of massive unemployment, feel economic changes mostly in their personal income levels.[6]

In Hungary, there is no reliable information about the population's actual total and real personal income. This is due to the inadequacy of data collection and the widespread practice of keeping income information concealed. The representative data collection by the Hungarian Central Statistical Office (KSH 1998) and still unpublished income tax returns, however, allow the assessment of changes in the period between 1988 and 1996. Total income consists of the official wages or salaries received by employees, income from business ventures or small-scale agricultural production, social welfare benefits, and other complementary income sources. The most important social benefits are pensions, family allowances and (in the nineties) those relating to unemployment. In contrast, taxable incomes, because of the well-known characteristics of the Hungarian tax system, reflect the spatial differentiation between wages and salaries, while the incomes from entrepreneurship and capital have little influence in determining their level.[7]

Real personal incomes decreased between 1987 and 1995 in all regions (according to the Hungarian Central Statistical Office, they decreased by 37% in the whole country, by 30% in Budapest, by 38% in other towns, and by 40% in villages). Therefore, in the short assessment below, the 'improvement' mentioned relative to the capital and the western counties does not mean more than a smaller than average decrease in real incomes. According to the relative income levels (table 5) in 1987 and 1995, it is evident that all four figures exceed the national average in the capital city.

---

[6]   It is worth mentioning that the county-level inequalities measured by personal income tend to be much smaller than those of per capita GDP.

[7]   Data for taxable income has been made available to the author by the Ministry of Finance on a settlement-basis.

Table 5: Changes in counties' income-positions in the early 1990s

Counties, types	Income per capita					
	Total income			Taxable income		
	Country average=100					
	1987	1995	Change	1988	1995	Change
Improving position						
in both income types						
Budapest (C. H.)	113.8	125.6	11.8	137.7	152.1	14.4
Vas (W. Tr.)	95.3	103.8	8.5	90.4	105.4	15.0
Tolna (S. Tr.)	98.4	104.0	5.6	89.6	90.3	0.7
Zala (W. Tr.)	98.7	100.2	1.5	87.3	97.7	10.4
Somogy (S. TR.)	92.0	93.5	1.5	81.4	82.0	0.6
Győr-Moson-Sopron (W. Tr.)	100.1	100.2	0.1	97.8	106.9	9.1
Opposite movement						
in the two income types						
Baranya (S. Tr.)	100.3	102.1	1.8	97.6	89.3	-8.3
Fejér (C. Tr.)	100.7	99.9	-0.8	102.3	104.2	1.9
Csongrád (S. G. P.)	103.1	101.7	-1.4	88.8	94.4	5.6
Veszprém (C. Tr.)	102.2	93.0	-9.2	96.2	98.0	1.8
Worsening position						
in both income types						
Pest (C. H.)	98.1	97.0	-1.1	102.5	98.4	-4.1
Borsod-Abaúj-Zemplén (N.H.)	89.6	88.0	-1.6	89.6	81.0	-8.6
Bács-Kiskun (S. G. P.)	96.9	93.3	-3.6	82.2	76.6	-5.6
Hajdú-Bihar (N. G. P.)	91.2	86.3	-4.9	83.6	80.4	-3.2
Békés (S. G. P.)	98.4	92.8	-5.6	82.5	80.8	-1.7
Jász-Nagykun-Szolnok (N. G. P.)	98.8	91.6	-7.2	84.2	81.4	-2.8
Heves (N. H.)	98.0	90.8	-7.2	91.2	86.8	-4.4
Strong deterioration						
in both income types						
Nógrád (N. H.)	93.4	87.6	-5.8	94.0	78.0	-16.0
Komárom-Esztergom (C. Tr.)	103.9	96.2	-7.7	109.1	98.0	-11.1
Szabolcs-Szatmár-Bereg (N.G.P.)	91.5	81.2	-10.3	70.4	64.3	-6.1
Hungary(%)	100.0	100.0		100.0	100.0	
Hungary (thousand HUF)	66	212		50	156	

Source: Total Income: KSH 1998; taxable income: MINISTRY OF FINANCE (unpublished).

The gap between the capital's income levels and the rest of the country has increased noticeably. The countryside has also become more divided due to the changing performance of the various regions. The western counties have achieved higher positions, but the relative income level of the northern regions continues to drop. Even the best situated regions in the countryside (this includes Győr-Moson-Sopron and Fejér counties) have at least one income indicator which, sometimes only to a small extent, shows a lower than average figure. These incomes, however, still exceed the average countryside figures. In all, there were eight Hungarian counties where an income indicator reached the national average in some years. Yet, in eleven regions, personal income showed a below average level all of the time.

These movements have not resulted in a complete 'inversion' of the spatial pattern of personal incomes, which is indicated by the Spearman rank correlation coefficient of 0.61 for total county income data in the years from 1987 through 1995. This figure suggests that although there were some changes in the counties' relative positions, the regions which had higher incomes originally are still leading the list, and the traditionally less developed counties have not improved considerably. Four large groups can be distinguished when concentrating on changes in income positions.

The first group consists of five counties in Transdanubia and the capital city; their ranking in both income types has improved. Besides Budapest, Vas' improvement has also been spectacular. The next group consists of four counties showing income levels moving in opposite directions. While total income has stayed relatively stable in Baranya county, the taxable income level has worsened. In Fejér, Csongrád, and Veszprém, the situation has improved. The rest of the counties show a decline in both income types. Nógrád, Komárom-Esztergom, and Szabolcs-Szatmár-Bereg can be separated from the other counties of the country's northern region and the Great Plain by their changing income level. These three counties are characterised by a particularly strong deterioration in income positions.

Despite the strong correlation between the two different income ranks, the actual changes in the total income positions are less than those of taxable incomes. For example, Budapest strengthened its leading position by 11.8% thereby showing the greatest advancement, while Szabolcs suffered a decrease by 10.3%. With regard to taxable incomes, relative changes in position were greatest at the top and bottom, with extremes of 15.0% in Vas, and –11.6% in Nógrád. An important reason for this discrepancy is that in the industrial regions most seriously affected by depression (Nógrád, Borsod, Komárom), the drop in (taxable) wages was partially offset by the relatively high tax-free pensions, a reflection of the brighter days of the past.

Data for settlements is available for taxable incomes between 1988 and 1996. This database is appropriate for examining income variations for different spatial scales (table 6). The calculations provide clear evidence of the general regional development trend in the 1990s: income disparities increased at all aggregation levels between 1988 and 1994. In the last two years, the figures show relative stability in the income inequalities albeit on a much higher level.

It is possible to do a more thorough evaluation of the changing regional patterns on the basis of figures 3 and 4, which display income data on the settlement level. The maps showing spatial income patterns help to clarify the most important elements in the spatial structure. The settlements ranking among the first 1 000 in the hierarchy of approximately 3 100 settlements are set up by per capita taxable income levels and are indicated in black on the maps (hence the title 'Income for the top 1 000 settlements').

Figure 3:   Income for the top 1 000 settlements (1988)

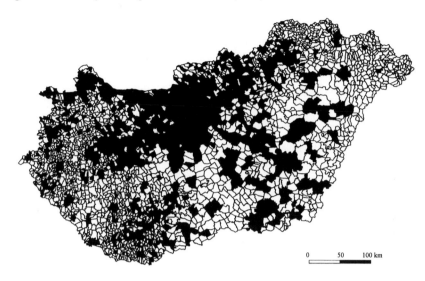

Source: Ministry of Finance.

Figure 4:   Income for the top 1 000 settlements (1996)

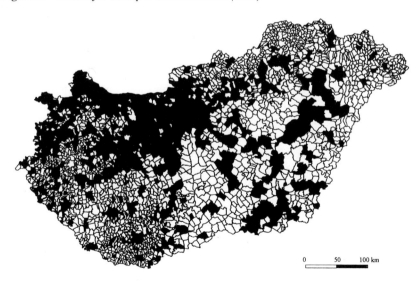

Source: Ministry of Finance.

Table 6: Disparities in per capita taxable personal income by various scales

Years	Weighted standard deviation (%)			
	Settlements	Microregions	Counties	Regions
1988	25.4	22.2	19.6	17.3
1989	27.6	24.0	21.2	18.5
1990	30.2	26.1	23.1	19.9
1991	30.8	25.5	21.2	17.9
1992	34.6	29.8	26.4	22.1
1993	35.6	30.8	27.1	22.7
1994	36.5	31.5	27.7	23.5
1995	35.9	30.7	26.7	22.6
1996	35.7	30.5	26.4	22.3

Source: MAJOR and NEMES-NAGY 1999.

How can the situation in 1988 be characterised? The region with a high-income population consists of the capital and its surroundings (all of Pest county) with a continuous long band stretching in a north-east – south-west direction (the traditional 'industrial axis'). There was another smaller area emerging from this central region to the west, the industrialised zone consisting of Komárom and Győr-Sopron along the Danube. As early as 1988, there were bigger 'patches' of relatively high income along the Austrian border, and also in Baranya and Tolna counties. In the eastern part of Hungary only the largest counties reached a higher position. The north-eastern region near the Romanian border region is 'white' and so is the great 'blank' area of the Middle Tisza region (Közép-Tiszavidék) consisting of Heves and Szolnok counties. Similar patterns exist in Transdanubia, where there are extensive 'inner peripheries' in parts of Vas and Zala with the smallest villages, or along the internal border zone of Somogy, Fejér, and Tolna counties.

What are the most important changes revealed by the 1996 map? The favourable position of the two regions has definitely vanished. First, the high income ring surrounding the capital city has narrowed relative to the Great Plain. This is probably due to the shrinking commuter belt around Budapest. Second, the zone of the well-off north-eastern industrial 'half-axis' has been confined to the line of towns at the foothills. The income drop is conspicuous in the broad eastern Slovakian border zone. On the other hand, the Budapest-Balaton axis and the western regions along the Danube have retained their favourable position. The southern strip along the Danube, where stable, strategic industrial plants such as Százhalombatta, Dunaújváros and Paks are located, has also maintained its position. At the same time, the western border region has developed into a high-income zone, where both towns and smaller settlements are enjoying the benefits of spectacular growth. There are some signs of similar changes in Zala too. In Baranya, however, mining activity has ceased and the high income area is restricted to Pécs and a few other towns. The income levels of the towns along the most important highways on the Great Plain still stand out. However, the pattern revealed by the spatial variation of taxable income on the Great Plain indicates a much more polarised situation than is suggested by real income proportions. This is because only a small part of the incomes from small-scale farming appear in the tax revenues, although this sector is of key importance in providing stability for this region. If the income data were corrected to incorporate actual living conditions and the basic infrastructure elements (housing, cars, etc.), it would indicate a much better financial situation, especially in Bács-Kiskun county.

3.4   Microregional inequalities: 'winners' and 'losers'

The two income maps (figures 3 and 4) provide insight to the inequalities on the microregional level, i.e. below county level. This picture is a more complex mosaic. Since there is no microregional GDP data available, regions can only be evaluated based on accumulated index numbers.[8] The classification is based on 7 indicators of the 150 microregions (1995-96):

1. taxable income per capita
2. unemployment rate (June 1996)
3. sole proprietors per 1 000 inhabitants
4. personal vehicles per 1 000 inhabitants
5. telephone lines per 1 000 inhabitants
6. foreign capital per inhabitant
7. share of joint ventures within the total number of enterprises

On the basis of these indicators, four aggregated categories of regions can be distinguished:

*   *'Winner'* regions are those where the values of at least 6 indicators are higher than the countryside average (i.e. the average without the capital city).
*   *'Emerging'* regions are those where 4-5 indicators result in figures above the countryside average.
*   *'Stagnating'* regions are those with only 2-3 indicators above the countryside average
*   *'Losers'* are regions where the value of no more than one indicator is higher than the countryside average.

It is necessary to use the countryside average as the basis for comparisons because of the capital's great economic weight and high development level. The capital raises the national averages to such a degree that except for really dynamic re-

---

[8]   The methodology described in this section is only one of numerous possibilities for microregional analysis. Several authors have applied other methods and indicators for multivariate analysis of the transition, but the main research results are very similar (see, for example, CSATÁRI 1996; KSH 1995; NEMES-NAGY 1995b).

gions or towns such as Győr, Sopron, Székesfehérvár and a few towns in the capital's agglomeration, almost all of the other regions are below average.

A map of the calculation results on the microregional level (150 spatial units) clearly shows the most evident change in the spatial structure: the collapse of the north-east – south-west oriented heavy industrial axis located along the mountain region (figure 5). The chemical industry was the only industrial branch in this zone that was able to revive after the shock caused by the transition to a market economy. A number of industrial centres in this region are currently facing environmental and social disasters. Ózd, for example, a former iron and steel centre in the north-eastern Slovakian border zone, is one of the towns suffering the most. At the same time, there are some cities in the industrial axis (Székesfehérvár, and more recently, Tatabánya) where there is dynamic renewal due to recent developments in high-tech manufacturing branches.

Figure 5:   Microregional disparities in Hungary 1995

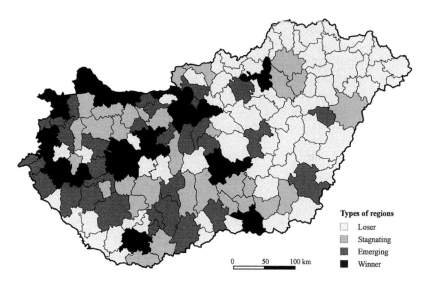

Source: Hungarian Central Statistical Office (KSH).

The more stable western and north-western regions and the depressed eastern and north-eastern parts of the country are also well separated in the new pattern created on the smaller spatial unit level. The dynamic axes in the western part of the country are easily recognisable: the Budapest - Győr - Mosonmagyaróvár - Sopron (Bratislava-Vienna) axis, and the Budapest - Székesfehérvár - Balaton axis. These two axes stretch out of the capital city and are linked by the 'emerging' zone along the Austrian border, where both towns and centres are part of one of the highest performing microregions. The key development factors in this border region are the external impacts, especially the effects of the relatively close and economically powerful southern German, northern Italian and Austrian regions. The Balaton region has grown primarily because of its concentration of small entrepreneurship which depends heavily on the vicissitudes of tourism.

There are some characteristically agrarian 'shadow regions' and formerly prospering, but now depressed smaller industrial areas between the Transdanubian development axes. The most problematic contiguous area is the southern border zone, since it was partially affected by the Croatian war in the early 1990s, which led to a collapse in foreign investments.

In both the southern and eastern parts of Hungary, almost all counties have emerged and expanded from their local dynamic centres. These cities attract wholesale and retail commercial functions and advanced financial and economic service activities.

Pest and Bács-Kiskun form a transitional zone on the eastern bank of the Danube. This region is close to the capital and benefits from it economically. It also has relatively favourable transportation connections and small-scale agricultural production as a stabilising element. Foreign capital has reached all but the largest cities such as Kecskemét and Szeged. The more distant northern and eastern regions form an almost contiguous 'loser-zone', hit by massive and permanent unemployment. Positive economic features only exist in the county centres and in a few stable, strategically important industrial cities such as Tiszaújváros.

Settlements have been classified using the same indicators as for microregions, distinguished by 'winner', 'emerging', 'stagnating', and 'loser' categories. This further division within the microregions revealed that cities throughout the country were doing much better than villages. The analysis also shows the number of persons living in the four settlement types in each region and in the whole country (table 7).

In the whole country, the share of the two more favourable settlement types is about 60%. This is a high proportion due to the capital city's presence in this category. About 40% of the population falls into the other 'second class' groups. 50% of the population falls into the more favourable settlement type. It is evident that larger and medium sized towns contribute disproportionately to the two preferable groups, while the villages usually dominate the unfavourable categories. Regional inequalities are quite evident here: in the central region and in western Hungary, development reaches outside the cities, while the proportion of persons living in the 'loser' settlements is highest in the north-eastern regions.[9]

---

[9]   Of course, there are 'loser' social groups living in 'winner' settlements, and 'winner' groups in 'loser' villages. The population figures in the table describe the settlement level only.

Table 7: The 1996 population distribution for the characteristic transition types

Spatial units, regions	Population distribution living in different settlement types				
	Total	Winners	Emerging	Stagnating	Losers
			Population (in thousands)		
Hungary (total)	10 174	4 076	1 964	1 812	2 320
Countryside (total)	8 287	2 190	1 964	1 812	2 320
C.-Hungary	2 880	2 225	298	285	71
C.-Transdanubia	1 113	318	365	276	153
W.-Transdanubia	994	522	169	225	77
S.-Transdanubia	991	253	325	178	233
N.-Hungary	1 290	239	191	273	585
N. Great Plain	1 539	208	296	203	830
S. Great Plain	1 364	309	318	368	368
Hungary (%)	100.0	40.1	19.3	17.8	22.8
Countryside (%)	100.0	26.4	23.7	21.9	28.0
C.-Hungary	100.0	77.2	10.4	9.9	2.5
C.-Transdanubia	100.0	28.6	32.8	24.8	13.8
W.-Transdanubia	100.0	52.5	17.0	22.6	7.8
S.-Transdanubia	100.0	25.6	32.8	18.0	23.5
N.- Hungary	100.0	18.6	14.8	21.2	45.4
N. Great Plain	100.0	13.5	19.3	13.2	54.0
S. Great Plain	100.0	22.7	23.3	27.0	27.0
Hungary (%)	100.0	100.0	100.0	100.0	100.0
Countryside (%)	81.5	53.8	100.0	100.0	100.0
C.-Hungary	28.3	54.5	15.2	15.8	3.1
C.-Transdanubia	11.0	7.8	18.6	15.2	6.6
W.-Transdanubia	9.8	12.8	8.6	12.4	3.3
S.-Transdanubia	9.7	6.3	16.5	9.9	10.1
N.-Hungary	12.7	5.9	9.7	15.1	25.2
N. Great Plain	15.1	5.1	15.1	11.2	35.8
S. Great Plain	13.4	7.6	16.2	20.4	15.9

Source: Author's calculation based on regional data of the Hungarian Central Statistical Office (KSH).

4   The future of regional development in Hungary

Is the spatial pattern that has evolved in Hungary at the end of the 20th century transitional? Considering the most significant regional types found in the new spatial structure and on the basis of the above analyses, the present situation and future development trends in the coming decade will most likely be variable (ENYEDI 1994; EHRLICH and RÉVÉSZ 1995).

The dynamic regions of the country, the capital, its surroundings and the western counties, and some larger and medium size cities (NEMES-NAGY 1995a) have created the economic and social structural basis which will enable them to utilise the increased potential of these zones, the locational advantages, this attractiveness of the market, and their opportunities for innovations. Budapest, ranked on the second or third level in the European 'city-competition' (BARTA and CONTI 1994), is undergoing an extremely dynamic transformation.[10] The 'central core' formed by the capital and the 'dynamic edge' formed by the Austrian border over the rest of the country seem to be stable, although the positive prospects in some smaller towns are uncertain. In the smaller towns, sudden development was due primarily to large foreign capital investments and thus lacked domestic linkages. The multinational companies may change their profiles, close down plants or move further east. The massive migration into the dynamic regions which was expected by several analysts has not taken place because of the spread of capital-intensive improvements and the low mobility of the Hungarian labour force. This has resulted in a relative labour shortage in regions such as Győr city, but it has also allowed stable social structures to be maintained.

Extensive regions of the country (especially the regions east of the Tisza River, northern Hungary, and the border zone of southern Transdanubia) are still struggling with both deeply-rooted and new difficulties. They are the so-called 'problematic regions'. The regional development policy of decentralisation, though reformed in its institutions, has not yet been able to improve the situation (HORVÁTH and ILLÉS 1997).

Will these less developed regions 'catch up'? Besides the internal factors, the development of the neighbouring countries (Slovakia, Ukraina, Rumania, and the Balkan states) is crucially important. The great cities near the borders (Pécs, Szeged, Debrecen and Nyíregyháza) may act as 'gates' via Southern and Eastern Europe and their logistic importance may grow, but their role is uncertain due to the instability on the other side of the border. A more favourable situation would evolve if the economies of the neighbouring countries stabilise and co-operation further extends (SÜLI-ZAKAR 1997). The inflow of regional development funds expected from Hungary's European Union membership should contribute a great deal to the improvement of these regions and is very much anticipated (HORVÁTH 1997).

There are millions of people living in regions with unstable transitional economic and social conditions, particularly in the south-central part of the country. Mobilised socio-economic resources provided relative stability but were insufficient for further advancement. Local incentives are particularly important in re-

---

[10] See also the chapter by Zoltán Cséfalvay.

gional development, since a spontaneous and rapid increase cannot be expected from international sources or from central government funds. Foreign capital still prefers the western part of the country and the national regional development grants aim to improve the least developed eastern regions.

The lasting character of the fundamental elements of today's spatial pattern (the capital-countryside dualism, the west-east socio-economic gradient and the urban-rural economic disparities) is confirmed by the low intensity of interactive links within the country. The positive poles are not considerably attractive to the other regions since they utilise their own resources and receive other resources from abroad. However, the increasing speed of economic growth since 1996-97 will hopefully diminish regional disparities, and its positive effects will also reach the stagnating and less developed regions. The evolution of 'another new spatial pattern', more stable than the present one, is only probable in the long run, even if a consistently high (3-5%) economic growth rate is achieved.

References

BARTA, GY. (1977): A területi gazdasági különbségek változása 1960-1975 között. (Changes of Regional Economic Inequalities between 1960-1975). Területi Statisztika 2, 522-537

BARTA, GY., CONTI, S. (1994): Budapest's Changing Position in Europe and Hungary. In: HAJDÚ, Z., HORVÁTH, GY. (eds): European Challenges and Hungarian Responses in Regional Policy. MTA RKK, Pécs, 170-182

CSATÁRI, B. (1996): A magyarországi kistérségek néhány jellegzetessége. (Some Characteristic Features of Hungarian Microregions). MTA RKK-KTM PHARE Iroda, Kecskemét

CSÉFALVAY, Z. (1995): Raum und Gesellschaft Ungarns in der Übergangsphase zur Marktwirtschaft. In: MEUSBURGER, P., KLINGER, A. (eds): Vom Plan zum Markt: Eine Untersuchung am Beispiel Ungarns. Physica-Verlag, Heidelberg, 80-98

EHRLICH, É., RÉVÉSZ, G. (1995): Hungary and its Future Prospects 1985-2005. Akadémiai K., Budapest

ENYEDI, GY. (1994): Regional and Urban Development in Hungary until 2005. In: HAJDÚ, Z., HORVÁTH, GY. (eds): European Challenges and Hungarian Responses in Regional Policy. MTA RKK, Pécs, 239-253

ENYEDI, GY. (1996a): Regionális folyamatok Magyarországon az átmenet időszakában. (Regional Processes in Hungary in the Transition Period). Hilscher Rezső Szociálpolitikai Egyesület, Ember-település-régió sorozat, Budapest

ENYEDI, GY. (1996b): Külföldi müködőtőke befektetések hatása a regionális fejlődésre Magyarországon. (Foreign Direct Investment and Regional Development in Hungary). In: DÖVÉNYI Z. (ed.): Tér - Gazdaság - Társadalom. MTA FKI, 247-256

HASTENBERG, H. VAN (1996): Regional and Sectoral Characteristics of Foreign Direct Investment in Hungary. In: Workshop Transition Processes in Eastern Europe. ESR, The Hague, 121-136

HUNGARIAN CENTRAL STATISTICAL OFFICE (ed.) (1996): Regional Statistical Yearbook. Budapest.

HORVÁTH, GY. (1993): Entrepreneurship and Regional Policy in Hungary. In: HAJDÚ, Z. (ed.): Hungary. CRS, Pécs, 263-277

HORVÁTH, GY. (1997): Európai integráció, keleti bővítés és a magyar regionális politika. (European Integration, Enlargement and the Hungarian Regional Policy). Tér és Társadalom 3, 17-56

HORVÁTH, GY., ILLÉS, I. (1997): Regionális fejlődés és politika. (Regional Development and Policy). Európai Tükör 16. ISM, Budapest

KSH (ed.) (1995): Kistérségi vonzáskörzetek: A regionális térszerkezet jellemzői az átmenet éveiben. (Microregions: The Regional Structure in the Years of Transition). Hungarian Central Statistical Office, Budapest

KSH (ed.) (1998): Jövedelemeloszlás Magyarországon 1995. (Income Distribution in Hungary 1995). Hungarian Central Statistical Office, Budapest

MAJOR, K., NEMES-NAGY, J. (1999): Területi jövedelemegyenlőtlenségek a kilencvenes években. (Regional Income Inequalitites in the 1990s). Statisztikai Szemle 5, 397-421

NEMES-NAGY, J. (1990): Területi kiegyenlítődés és differenciálódás Magyarországon. (Spatial Convergence and Divergence in Hungary). Földrajzi Értesítő 1-4, 133-149

NEMES-NAGY, J. (1995a): Soprontól Nyíradonyig (városok a piacgazdasági átmenetben). (The Hungarian Cities in Economic Transition), Comitatus 8-9, 15-22

NEMES-NAGY, J. (1995b): A piacgazdasági átmenet terei ('Spaces' of Hungarian Transition). Falu - Város - Régió 7-8, 6-11

NEMES-NAGY, J. (1997a): Radikális változások a magyar ipar térszerkezetében. (Radical Changes in the Regional Structure of Hungarian Industry). In: Földrajz - Hagyomány és Jövő: A 125 éves MFT jubileumi konferenciája. Budapest, 24

NEMES-NAGY, J. (1997b): A fekvés szerepe a regionális tagoltságban. (Importance of Location in Regional Disparities). Munkaerőpiac és Regionalitás: MTA Konferencia, Szirák, 1997, október (A konferenciakötetet kiadta MTA KKI), Budapest 1998, 147-165

NEMES-NAGY, J., RUTTKAY, É. (1991): Rural and Urban Forms of Private Enterprises in Hungary. In: HORVÁTH, GY. (ed.): Regional Policy and Local Government. CRS, Pécs, 163-173

PROBÁLD, F. (1995): Regionale Strukturen des Arbeitplatzangebotes in der Agglomeration von Budapest. In: MEUSBURGER, P., KLINGER, A. (eds): Vom

Plan zum Markt: Eine Untersuchung am Beispiel Ungarns. Physica-Verlag, Heidelberg, 182-208

RECHNITZER, J. (1993): Szétszakadás vagy felzárkózás (A térszerkezetet alakító innovációk). (Division or cohesion: Innovation in regional development). MTA RKK, Győr

RECHNITZER, J. (1998): A privatizáció regionális összefüggései. (Privatisation in regional context). AVÜ, Budapest

SIK, E. (1994): From the Multicolored to the Black and White Economy: The Hungarian Secondary Economy and the Transformation. International Journal of Urban and Regional Research 18 (1), 46-70

SILINCE, J. A. (1987): Regional Policy in Hungary: Objectives and Achievements. Transaction 12, 451-464

SÜLI-ZAKAR, I. (1997): A Kárpátok Eurorégió a régiók Európájában. (The Carpathian Euroregion in the Europe of the Regions). Educatio 3, 438-452

SWAIN, A. (1998): Governing the Workplace: The Workplace and Regional Development Implications of Automotive Foreign Direct Investment in Hungary. Regional Studies 7, 653-671

TURNOCK, D. (1998): Socio-economic Stress, Spatial Imbalance and Policy Responses in the New Democracies. In: PINDER, D. (ed.): The New Europe. J. Wiley & Sons, Chichester, 323-339

Foreign Banks Are Branching Out: Changing Geographies of Hungarian Banking, 1987-1999

Heike Jöns

1 Introduction[1]

Walking through the streets of Budapest in spring 1999 could have given you the following impression: the supermarkets (Spar), the milk products sold there (Danone, Müller), and the property markets (OBI) come from different Western European countries such as the Netherlands, France and Germany. Almost all fast food restaurants (McDonalds, Pizza Hut, KFC) and many hotels (Hilton, Mariott) have their origins in the US; shoes and clothes offered in downtown are designed in Italy or France (Benetton, Marco Polo); medicine is predominantly produced in Switzerland (Novartis, Roche) and the banks as well as the car dealerships have their roots everywhere in the so-called Western world - usually including Japan and other Asian countries with major (car) companies - but not in Hungary itself.

In 1998, the ownership structure of the commercial banks, that are the main focus of this article, revealed a foreign share of almost 80%.[2] Thus, the high amount of foreign direct investment (FDI) in the Hungarian banking sector fits the picture of a great openness towards foreign influences during the political-economic shift towards a market economy in Hungary.[3] In 1999, about two years after the privatisation of the large state-owned commercial banks was completed, there were 25 commercial banks with FDI of more than 10% and only two commercial banks with FDI of less than 10%. Just one commercial bank was fully privately Hungarian-owned. These commercial banks were all fiercely competing for corporate and private customers. Considering major concentration processes in Western European bank branch networks during the 1990s,[4] it seems to be a paradox that these market-experienced and often polyglot banks with representations

---

[1] This study was partly sponsored by the German Research Foundation (DFG) as part of Project ME 807/9-2. Regarding the process of writing this article, I would like to express my sincere gratitude to Janet Bojan for reading an earlier version of this text. Many thanks also to Aaron Steele for ironing out some English flaws at the beginning of this undertaking, and to Yvette Tristram for doing the same towards the end. I am responsible for any remaining ones. The conclusion owes much to discussions with Tim Freytag and Mike Heffernan. I am very grateful for their constructive and encouraging comments.
[2] See section 2 and figure 2.
[3] See HAMILTON 1995, 81; JONES 1999.
[4] See LEYSHON and THRIFT 1995; EUROPÄISCHE KOMMISSION 1999; and section 3.1.

in major global cities would have to spread their branch networks into small Hungarian towns in order to succeed in the face of tough competition. Surprisingly, this situation arose in Hungary at the end of the 1990s.

Why has this situation evolved? Who were the actors in the Hungarian banking market during the 1990s? Which business and branch network strategies have they followed? Which regions and which kinds of settlements were attractive for the banks? In which places have people or small and medium-sized enterprises (SME) benefited from access to the full range of banking services, and which places were neglected by the banks? What role have the banks played in the formation of a market economy in Hungary during the 1990s? And, finally, what can be concluded from the banks' network strategies and the theoretical framework of actor-network theory in regard to a theory of regional transformation? The aim of this article is to answer these questions and thus to contribute to a better understanding of the interrelation between banking and regional development during the first decade of political-economic transition from a centrally planned to a market economy in Hungary. In order to do so, this article will briefly recall the main steps of legal changes and examine three institutional developments during the 1990s in more detail:

- the number and types of financial institutions in operation;
- the ownership structure and business strategies of these institutions;
- the development of their branch networks.

In the first section, this study draws upon a broad range of literature concerned with the legal and institutional transformation of the Hungarian banking sector, and its competitive situation.[5] The role of foreign-owned banks has been given special attention,[6] including comprehensive analyses on the basis of geographical perspectives.[7] Nevertheless, most studies focus on certain aspects of the legal and institutional transformation rather than relating different processes to each other. This is done to a certain extent by SZELÉNYI and URSPRUNG (1998) for the first decade of legal transformation of banking, but their main focus is on a yearly account of events. They do not analyse the dynamics of different interrelated processes over the whole period of time in detail, and do not include the spatial organisation of branch networks. Thus, different political, legal, institutional *and* spatial dynamics are woven together in the following and interpreted in regard to a theory of regional transformation in Hungary.

So far, only a few studies have paid closer attention to the role of *bank branch networks* for economic transformation in Hungary.[8] In several aspects this study

---

[5] See, for example, LENGYEL 1994, 1995; BALASSA 1996; BOTOS 1998.
[6] See, for example, BUCH 1997; BUSCH and WEISIGK 1997, JONES 1999.
[7] See KLAGGE 1997a, 1997b.
[8] See, for example, CSÉFALVAY 1994; ILLÉS 1994; LENGYEL 1994, 1995; JÖNS 1996, JÖNS and KLAGGE 1997, KLAGGE 1997b.

continues and extends these works, of which the last three include more detailed reviews of geographical research on banking in Hungary, elsewhere, and in general. For the most recent discussion of research on 'Transition and transformation' in post-socialist Europe conducted by geographers, see DINGSDALE (1999), FAßMANN (1999), HAMILTON (1999), and WARDENGA (2000).

The empirical work of this article is based on a database containing the branches of all credit institutions operating in Hungary between 1990 and 1999,[9] and 24 semi-structured interviews conducted by the author in different banking authorities and banks (headquarters and branches) in Budapest, Győr, Székésfehervár and Békéscaba in spring 1996 (18) and spring 1999 (6).

2   Legal framework and institutional development in the Hungarian banking sector since the 1980s

The creation of a well functioning banking system is essential for the successful conversion of a centrally planned into a market economy. In Hungary, this was recognised very early. On January 1, 1987, Hungary was the first among the former socialist countries in Eastern Europe to restore the two-tier banking system typical of capitalist market economies. Within the socialist economic order, a Soviet-type one-tier system existed after a major restructuring of the banking system had taken place between 1947 and 1950. The dominant bank in this one-tier banking system, which was steered by the Ministry of Finance, was the National Bank of Hungary (NBH). It combined the functions of the former independent central bank and formerly competing commercial banks, keeping the accounts of all corporations in Hungary. In addition, there were four specialised financial institutions operating under the control of the NBH. Each financial institution held a monopoly for certain banking activities.[10] A third element introduced in 1957 was

---

[9]   The database consists of information mainly drawn from the Hungarian Financial and Stock Exchange Almanac (see KEREKES 1990 etc.). This data was checked, completed and expanded by referring to Annual Reports of the National Bank of Hungary (NBH), the Hungarian Banking and Capital Market Supervision (HBCMS; including the former State Banking Supervision) and individual banks. The data for the period 1990 to 1996 was build up in collaboration with Britta Klagge (Hamburg).

[10]   See LENGYEL 1994, 382ff. The Országos Takarékpénztár (OTP) had the monopoly for dealing with private savings, loans and credits, and from 1972 onwards, with local governments as well. The Magyar Külkereskedelmi Bank (MKB) was responsible for financial transactions related to foreign trade, the State Bank for Development (SBD) for financing and supervising large investment projects arranged by the state. Before 1972 it dealt with building companies as well. The small Általános Értékforgalmi Bank (ÁÉB) was concerned with foreign accounts, claims and international property issues.

constituted by small savings co-operatives, 263 of which were operating in 1980.[11] While they were very active collecting private savings and granting household loans in rural areas, these co-operatives were bound by strong restrictions regarding their size, activities and networks.[12]

Competition between the financial institutions was completely prevented by a division of labour, regional restrictions and pre-set interest rates for loans and savings deposits. Thus, the activities of financial institutions were to a large extent mechanical. Due to the central control of the financial system, the supply of services was extremely limited as well.[13] Generally speaking, banking was a medium for distributing state funds and redistributing private and corporate funds.[14] Compared with market economies, it played a completely different role within the framework of the centrally managed economy: Banking had almost no momentum and lacked significance in everyday life.

The move towards market principles came about at the end of the 1970s because the state was heavily in debt. Not only was Hungary in need of new capital for corporations, but, when the country joined the International Monetary Fund, the World Bank, and the International Finance Co-operation after a serious financial crises in 1982, it had to present new concepts on how to finance the deficit of the national budget. Moreover, Hungary had to promise that new loans would be allocated according to business principles. Finally, the growing number of legalised private enterprises during the 1980s led to a tremendous expansion of account management for corporate customers and made a decentralisation of the financial institutions necessary.[15]

As early as 1979, the Central-European International Bank (CIB) had been founded by the NBH and six foreign investors (major European and Japanese banks). CIB operated as a dollar-based bank with offshore status from January 1980 onwards, concentrating on foreign trade and currency transactions.[16] Following this first step of institutional decentralisation, an average of two to three smaller specialised financial institutions were founded each year during the 1980s. These development funds and limited partnerships dealt mostly with small

---

[11]   See LENGYEL 1994, 383. After World War II about 1 000 co-operative banks operated in Hungary. They were liquidated in 1952 and were allowed again in the form of savings co-operatives in 1957 (see SZÖKE 1998, 584).

[12]   There was, for example, no competition between OTP, which was responsible for retail banking, and the savings co-operatives because of regional restrictions. OTP could maintain branches only in settlements with more than 5 000 inhabitants, the savings co-operatives were only allowed to establish branches in settlements with less than 5 000 inhabitants (see WASS VON CZEGE 1987, 411).

[13]   See BOGNÁR and FORGÁCS 1994, 53; ILLÉS 1994, 169. Securities were first introduced in 1983. Credit cards or check books were not used either (LENGYEL 1994, 383).

[14]   See, for example, WASS VON CZEGE 1987, 408f. However, in this respect it was "the most important tool in the indirect control of the economy" (LENGYEL 1994, 381).

[15]   See LENGYEL 1994, 382, 385; LENGYEL 1995, 112.

[16]   See KEREKES 1998, 181; KLAGGE 1997b, 158, 160.

Hungarian enterprises, mainly financing the - often risky - development and launching of new products, and co-operation with foreign companies.[17] In addition to the NBH, which was reorganised for the first time in 1985, when the central and commercial bank functions were separated, 19 financial institutions were operating in Hungary before the banking reform took place in 1987.[18] The first regular bank with foreign direct investment (FDI) started operation in 1986 (Citibank), the second followed in 1987 (Unicbank).

During the 1980s, not only the number of financial institutions, but also the range of banking services increased.[19] Nevertheless, it was already recognised at the beginning of the 1980s that the pressure on the financial system could not be remedied through a liberalisation of the one-tier banking system. Therefore, the government decided in favour of a market-orientated monetary policy and the establishment of state-independent relations between banks and enterprises.[20]

On January 1, 1987, the re-established two-tier banking system was organised as follows: The National Bank of Hungary (NBH) became the independent central bank, concentrating on monetary policy. The customers and branches of the former NBH lending department and parts of one specialised financial institution (SBD) were divided among three newly created institutions that became major non specialised commercial banks in the new system: Kereskedelmi Bank (K&H), Magyar Hitel Bank (MHB) and Budapest Bank (BB). MKB and ÁÉB got commercial banking licenses in 1987.[21] OTP received a commercial banking license in 1989 when its monopoly for retail banking was removed. The small financial institutions founded during the 1980s received either banking licenses, were liquidated or integrated into larger banks. The foundation of new credit institutions with or without FDI was welcomed, but, it was not until 1998 that the opening of *foreign bank branches as legally independent institutions* was allowed. The commercial banks operated as independent joint-stock companies and were allowed to act without regional or sectoral restrictions from 1989 onwards. The savings co-operatives continued to operate - like the already existing banks with FDI - and benefited from the revocation of the restrictions as well. The supervision of the banking market has been carried out by the State Banking Supervision since 1987.

---

[17] See LENGYEL 1995, 112; JÖNS and KLAGGE 1997, 15. Bad debts of small enterprises are a main reason why small Hungarian-owned banks which had started as specialised financial institutions in the 1980s got major liquidity problems during the 1990s, and often had to be liquidated (see footnote 26).

[18] These institutions were 14 small specialised financial institutions (LENGYEL 1995, 112) plus OTP, MKB, SBD, ÁÉB and Citibank (for the meaning of the abbreviations, see footnote 10).

[19] See LENGYEL 1994, 385, and footnote 13.

[20] See BOGNÁR and FORGÁCS 1994, 63; LENGYEL 1994, 385.

[21] In order to support the newly founded banks, the former customers of the NBH were not allowed to change banks for the first six months in 1987, and only one bank had got a branch in each town (JÖNS and KLAGGE 1997, 16; LENGYEL 1994, 386).

70   H. Jöns

Figure 1:   Institutional development in the Hungarian banking sector, 1987-1999

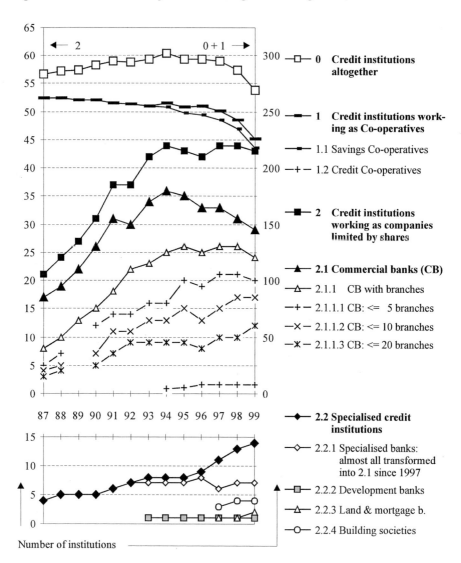

Source: Author's compilation after Annual Reports of the National Bank of Hungary, the
          Hungarian Banking and Capital Market Supervision (including the State Banking
          Supervision), individual banks and KEREKES 1990, 1991, 1992, 1994, 1995,
          1996, 1998.

Along with the banking reform the first boom of new banks entering the market occurred in 1987.[22] This boom marked the shift from a steady increase of financial institutions during the 1980s to an expansive rise (figure 1). By 1994 the number of credit institutions had increased to 44. This expansive period was based on slowly growing trust of foreign investors in the Hungarian banking sector due to the political transformation in 1989/90 and new legislation that aimed at fulfilling the conditions of a market economy of Western European style. The most important laws came into force in 1991 and 1992:[23]

- the Act on Credit Institutions outlining the operation of different types of credit institutions, safety regulations and banking supervision;[24]
- the Central Bank Act defining the tasks of the National Bank of Hungary;
- the Accounting Act primarily ruling banking activities;
- the Bankruptcy Act and the Act on Investment Funds.

Accordingly, 1991 showed the second boom in new banks entering the market. Another significant rise in the number of banks followed in 1993 when the bankruptcy, banking and accounting laws were amended for the first time.[25]

Compared with 1987, the banks starting operation in 1991 (7) and 1993 (5) were mainly foreign-owned and joint-venture banks. At the same time it became normal that an average of two to three banks per year had to be liquidated or to merge with other institutions.[26] The reasons for the failure of these, mainly state-owned banks were combinations of bad assets inherited from socialist times, misconceptions of managers lacking market experience, negligent lending behaviour and obscure transactions, and the new regulations - above all the asset valuation system introduced in the Act on Credit Institutions.[27] The latter had not only revealed the banks' bad asset portfolios but asked the banks to build up risk reserves for these. According to several doubtful and non-performing assets that were often related to state-owned enterprises and inherited from socialist times, many credit institutions had to set aside very high reserves and, thus, got serious capital, funding and portfolio problems. In order to remedy the financial problems of the

---

[22] Six banks started operation in 1987.
[23] See SZELÉNYI and URSPRUNG 1998.
[24] See STATE BANKING SUPERVISION 1991a, 1991b.
[25] See NATIONAL BANK OF HUNGARY 1993, 159ff. Several adjustments were undertaken in the following years with regard to EU regulations (see footnote 33).
[26] 1991 saw the first bankruptcy of a *specialised* state-owned bank. In 1992, the first state-owned *commercial* bank had to be liquidated because the safeguarding criteria were no more fulfilled, while another bank of this type was taken over when WestLB entered the market.
[27] In few cases the management had intended to make money themselves. SZELÉNYI and URSPRUNG (1998, 9) point out that an important phenomenon during the early years of the two-tier banking system was "the banks' and companies' concern to avoid initiating bankruptcy and liquidation proceedings against their debtors." This led to an accumulation of doubtful assets.

majority of banks the government conducted consolidation programs for the banks and their debtors in 1992-93 and 1993-94.[28] Although some smaller banks had to be liquidated and some mergers occurred as well, SZELÉNYI and URSPRUNG (1998, 8-9) regarded the consolidation process as being instrumental in ensuing adequate overall capitalisation of the banks in 1997 and thus having been "successful in solving the problems derived from the way in which the system was created and eliminating the consequences of the transformation crises."[29]

After the expansive period until 1994 the number of credit institutions fluctuated (figure 1): it decreased in 1995-1996, rose in 1997-1998 and declined again in 1999. In order to understand this development one has to take into account the different types of credit institutions that differ in organisational forms,[30] the minimum amount of subscribed capital[31] and their scope of activities:[32]

- (commercial) banks;
- specialised credit institutions;
- co-operative credit institutions (savings or credit co-operatives).[33]

---

[28] See BALASSA (1996) and SZELÉNYI and URSPRUNG (1998, 9-10). The loan consolidation of 1992-93 included a bank-orientated loan consolidation and a enterprise-oriented loan consolidation at the banks. The bank consolidation of 1993-94 aimed at a recapitalisation of the banks.

[29] At that time all the banks who survived were able to meet the required 8% minimum capital adequacy ratio (SZELÉNYI and URSPRUNG 1998, 8).

[30] Banks and specialised credit institutions operate exclusively as public companies limited by shares (See STATE BANKING SUPERVISION 1997, 21).

[31] Banks have to be established with a subscribed capital amounting at least HUF two billion and co-operative credit institutions need at least HUF hundred billion. The minimum amount of subscribed capital for specialised credit institutions depends on their activities (See STATE BANKING SUPERVISION 1997, 21).

[32] Banks are permitted to pursue the full range of activities allowed for credit institutions; specialised credit institutions have the right to pursue some activities (the actual scope of activities depends on the individual license, but in total it is less than commercial banks can do). Co-operative credit institutions can do almost all the activities of the banks, but the actual scope of activities depends on the amount of the subscribed capital. In contrast to savings co-operatives, credit co-operatives are only allowed to deal - except for currency conversion - with their own members (see STATE BANKING SUPERVISION 1997, 19).

[33] These types are defined in Act No. CXII. of 1996 on the credit institutions and the financial undertakings (see STATE BANKING SUPERVISION 1997). The Act on Credit Institutions of 1991 defined slightly different categories. At that time a specialised banking system of Anglo-American type was established. Commercial banks were not allowed to pursue certain financial institutional and capital market activities, such as offering and trading of publicly traded securities, investment trust management or insurance activities, in order to guarantee their prudential operation and to protect deposits (STATE BANKING SUPERVISION 1991, 5f.). For meeting the requirements of OECD membership and EU harmonisation, a transition towards a universal financial system became necessary. Continuous modifications of the Act on Credit Institutions and the

Responsible for reaching the 1994 maximum of 44 credit institutions again at the end of 1997 and 1998 was the slow but continuous increase of specialised financial institutions. Their growth in number accelerated in 1997 when two new types of financial intermediaries entered the market: building societies and mortgage credit institutions. The fact that these institutions, which provide long-term funds for households and businesses, were legally instituted ten years after the banking reform took place very well illustrates the time-consuming and step-by-step character of such a transformation process.[34] The group of specialised credit institutions also includes one development bank and several specialised banks working in areas such as car purchase financing, consumer financing, export financing, investment facilities, reorganisation projects and regional development. This specialisation makes it more or less possible to do business without too much overlapping - although car purchase financing was already very competitive in 1998. Therefore, the present number of this kind of financial intermediary might continue to exist for the next few years.

A completely different situation developed in the commercial banking market that will be in the centre of the following analyses because commercial banks offer universal services to private and corporate customers using branch networks:[35] At the end of the 1990s, the commercial banks were fiercely contesting for customers in almost all segments of the market. The period of institutional expansion came to an end in 1994, and has been followed by a concentration process since then. Although new banks still entered the market, there have been more and more mergers and liquidations since the middle of the 1990s. Responsible for this are asset valuation and mismanagement (see above), an increasingly tough competition for customers in a difficult economic environment, and the government's effort to promote mergers and take-overs rather than the market entry of new banks because of the privatisation process. The institutional developments in 1996 very well illustrate the highly dynamic environment of the commercial banking sector in the second half of the 1990s: Two mergers occurred

---

Securities Act served this transition during the 1990s. The types of credit institutions and their activities were modified along with these legal changes. The newly defined category 'specialised credit institutions' includes the Hungarian Development Bank (MFB), the Eximbank, the Land Credit and Mortgage Bank and the housing savings banks. All the other former specialised financial institutions had to be transformed into banks (NATIONAL BANK OF HUNGARY 1998, 73).

[34] The legal bases are Act CXIII of 1996 on building societies and Act XXX of 1997 on mortgage credit institutions and mortgage notes (see HUNGARIAN BANKING AND CAPITAL MARKET SUPERVISION 1997, 15).

[35] Almost all commercial banks have a branch network (see section 3). In 1998, only two specialised banks maintained three branches altogether, and two building societies had just started to establish a network of 17 consulting offices (KEREKES 1998). Many specialised banks offer their services through the branch networks of their parent company, such as OTP Building Society, or contract partners and thus do not need own branches.

between state-owned banks (Agrobank and Mezőbank, IBUSZ Bank and K&H).[36] A major part of state-owned Dunabank's infrastructure was acquired by the Dutch-owned ING Bank. The state-owned Iparbankház decided to withdraw from the market, MKBs privatisation was completed, and the subsidiaries of German-based Deutsche Bank and Dutch-based Rabobank started operation in 1996.

Altogether, the following processes triggered a major shift in the ownership structure of Hungarian banks:

- the liquidations of mainly small state-owned banks;
- the trend of state-owned banks merging into joint-venture or foreign-owned banks;
- the privatisation of state-owned banks;[37]
- the founding of new banks with FDI.

Figure 2:   Ownership structure among commercial banks (as % of total share capital)

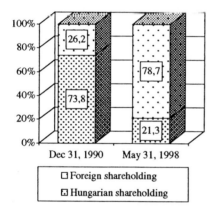

Dec 31, 1990   May 31, 1998

☐ Foreign shareholding
☒ Hungarian shareholding

Source: KEREKES 1998, 123-415; author's calculation.

In 1990, foreign ownership accounted for 26% of the commercial banks' aggregated share capital. By 1998 the respective number had risen to 79% (figure 2). At the end of 1990 15 of 26 banks were fully Hungarian-owned and only two banks were fully foreign-owned. Eight years later the relation was exactly the other way around: two fully Hungarian-owned and 15 fully foreign-owned banks with 30 banks in operation. In 25 of these 30 banks foreign inve-

[36]   These mergers, which were mainly based on the autonomous decision of the government as owner (SZELÉNYI and URSPRUNG 1998, 10), prepared the banks for privatisation in 1997 (Mezőbank was acquired by Erste Bank; a consortium formed by Irish life and Belgian Kredietbank became strategic investors in K&H).

[37]   The privatisation of state-owned *commercial* banks began in 1990 with ÁÉB. At the end of 1997, this privatisation process was almost finished (NATIONAL BANK OF HUNGARY 1999b, 95). The situation of Postabank is special because of the run on the bank's funds in February 1997 which might have been triggered off by rumours about its financial situation (see, for example, Bank&Tözsde VI (8), February 20, 1998). In order to save the important retail bank the state injected substantial financial aid. In 1998, 63% of Postabank's share capital was directly owned by the state and the state-owned MFB (see footnote 33; total Hungarian shareholding was 92.3%; KEREKES 1998, 374).

stors hold the majority of shares, and at the end of 1999 there was only *one* fully Hungarian-owned bank left (the privately owned Konzumbank). Since another bank with more than 90% Hungarian shareholding is probably going to cease operation in 2000,[38] Konzumbank, OTP and Postabank will be the last three commercial banks with Hungarian majority shareholding. In regard to the whole banking sector foreign ownership reached 60.4% of the aggregated share capital at the end of 1998.[39] This is an exceptionally high share by international standards, too.[40]

It is not easy to evaluate the dominance of foreign shareholding in Hungarian banks at the end of the 1990s.[41] On the one hand, given the world-wide prevalent capitalist economic order at the dawn of the 21st century, one could argue that the strong influence of foreign investors in the Hungarian banking system speeded up the development of a well functioning market economy. From this point of view the great advantage of FDI in the Hungarian banking sector results from the availability of sufficient capital. Corporations even benefit from promising conditions because joint-venture and foreign-owned banks undercut each other while fiercely contesting for customers, and these banks have the funds and the know-how to offer a wide range of high-quality services. Following this line of argument, the improved banking services have positive effects on corporations which in turn contribute to safeguarding jobs or even provide new jobs. Some would also argue that "the proliferation of foreign entities in Hungary has been a factor encouraging, even keeping under pressure, Hungarian-owned banks to improve the quality of their services,"[42] but this is not a very convincing argument, since there are not many Hungarian-owned banks left. On the other hand, one could argue that the banks' profits are not necessarily reinvested in Hungary because the majority of shareholders are foreigners. It is also doubtful whether the high-quality services are actually available to everyone, for example to Hungarian small and medium-sized enterprises (SME) or private individuals with medium or low income, and not only to joint-venture or foreign-owned companies and the high income strata of the population (see section 3).

However one would judge this aspect, fact is that during the first decade of the transformation process Hungarian-owned banks were almost completely ousted by joint-venture and foreign-owned banks because of their superiority regarding

- capitalisation;
- operational costs;
- no burden inherited from socialist times (non-performing assets);

---

[38]   Information received from the NBH in December 1999.
[39]   NATIONAL BANK OF HUNGARY 1999b, 95.
[40]   ERDEI 1998, 118.
[41]   For a summary of the pros and the cons regarding the market entry of foreign-owned banks in Hungary, see BUSCH and WEISIGK 1997.
[42]   SZELÉNYI and URSPRUNG 1998, 7.

- technological standards;
- know-how and expertise;
- strategies (see section 3).

The process of institutional concentration among the commercial banks will probably continue in the first decade of the 21st century. The low profitability of Hungarian banks may provide a further motive for concentration,[43] and institutional concentration among banks is an international trend that is driven by the desire to take advantage of the economies of scale and can be observed in Western Europe as well.[44] Experts consider about 20 commercial banks as the optimum for the Hungarian banking market.[45]

Among the savings co-operatives a concentration process can be observed since 1989. The number of savings co-operatives decreased from 262 in 1987 to 226 at the end of 1999 (figure 1). The number of mergers and liquidations increased in the second half of the 1990s because of the need to meet new capital requirements and problems of efficiency. This trend could not be compensated for by the founding of credit co-operatives from 1994 onwards. The savings co-operatives are very important for providing banking services in small villages (see section 3), but during the 1990s they were still characterised by low capitalisation, out-of-date technological standards, and a limited range of services.[46] While institutional concentration among co-operative credit institutions will probably continue for some years due to undercapitalisation, their overall situation seems to improve a lot since a new National Federation of Savings Co-operatives was established in 1998 in order to deal with common concerns such as the representation of interests, training of member co-operatives, strengthening of the financial market position, nation-wide marketing activities and an enhancement of supervisory activity and external auditing of savings co-operatives.[47]

In summary, one can more or less agree with LENGYEL (1994, 381-82) when he states that for the first years of political-economic transformation, the early decentralisation of the banking system going hand in hand with a gradual disinte-

---

[43]   HUNGARIAN BANKING AND CAPITAL MARKET SUPERVISION 1998, 19.

[44]   See EUROPÄISCHE KOMMISSION 1999. Among the most prominent recent examples of institutional concentration in Western European banking are the mergers of Hypo- and Vereinsbank in Germany or Bank Austria and Creditanstalt in Austria.

[45]   Statements of Managing Directors at OTP Bank and Hypobank (interviewed in 1996), at Budapest Bank and the HBCMS (interviewed in 1999) (all interviews in Budapest).

[46]   In the first two decades of their operation, savings co-operatives essentially operated as non-profit organisations. Thus, they lacked sufficient resources to accumulate assets and substantial reserves. The formation of profit-related capital and reserves only began in the 1980s and developed quite slowly (SZÖKE 1998, 587).

[47]   SZÖKE 1998, 587. The first step towards the integration of savings co-operatives was the foundation of the Hungarian Bank of Savings Co-operatives, the first of three central institutions, in 1989. By the end of 1997, the number of integrated savings co-operatives was 227.

gration of the centrally planned economy contributed to a "relatively smooth transformation of the political system in Hungary." Nevertheless, looking at the past events from the perspective of the year 2000, the legal and institutional developments in the Hungarian banking sector have been quite turbulent during the *whole* decade after the political transformation, especially after 1994. While the institutional re-organisation will probably continue for some time, it seems to have required thirteen years after the banking reform for the major legal and institutional innovations to be almost completed: a legal framework orientated to EU standards, the establishment and monitoring of safety regulations for the banks' operation, the privatisation of state-owned banks, institutional decentralisation (in certain segments followed by concentration) and institutional diversification. Accordingly, SZELÉNYI and URSPRUNG (1998) differentiate 'three stages of progress' towards the political aim of a banking system typical for market economies:

- growth period (1987-91);
- period of consolidation (1992-94);
- stabilisation (1995-97).

Considering the different political, legal and institutional developments in detail, the first decade of the re-emerging Hungarian banking market can be characterised as shown in figure 3. The following section adds further to these developments by analysing the banks' business and branch network strategies in regard to the nature of regional transformation and its significance for the banking sector.

Figure 3:   Important developments regarding Hungarian banking, 1980-1999

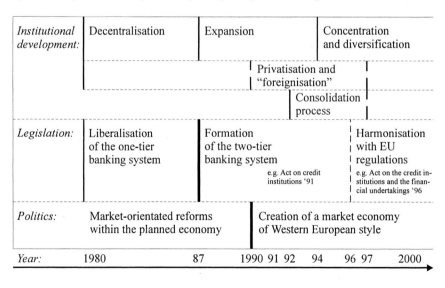

Source: Author's design.

3   Business activities and branch network strategies of Hungarian banks during the 1990s[48]

The development of bank branch networks in Hungary during the 1990s shows that more and more commercial banks established their own networks: While eight banks maintained at least one branch in 1987 (47%), there were 24 banks with at least one branch in 1999 (83%). The number of banks with large networks (at least 20 branches), rose from three to twelve banks in the same period of time (18% to 41%) (figure 1).

This development confirms that branch networks are very important elements in the business strategies of commercial banks. They enable the banks to get in contact with customers in places distant from the headquarter and to serve them all day long without too much effort for the clients. Acquiring clients in order to build up a customer base has, of course, top priority for almost all newly established commercial banks in the Hungarian banking market. The way they proceed depends in the first place on their main target groups. The size and business prospects of potential clients at different locations influence the basic decision whether it is profitable to establish a lot of costly branches or if it is sufficient to acquire and serve them in other ways (see section 3.1).

Thus, the first question to emerge is, which business and network strategies the individual banks followed, and how these strategies changed over time. From a geographical point of view, however, considering only the number and function of operative units is not enough to understand the development of the Hungarian banking sector during the 1990s. Based on the empirically and theoretically well founded fact that 'space makes a difference',[49] it can be assumed that the specific setting of a place does not only influence the decision where to open a branch, but its operation and profitability as well. Therefore, the spatial organisation of the branch network would have effects on the operation of a bank as a whole. The interrelation of business strategies, networking (regarding the number, functions and locations of branches) and the banks' profitability raises two further ques-

---

[48]   As indicated before, the focus of the following analysis will be on those banks who operated as commercial banks before 1996. In 1996 and 1997, new legislation laid down that former specialised financial institutions would have to be transformed into 'banks' during the next years (NATIONAL BANK OF HUNGARY 1998, 73). The concentration on the commercial banks is reasonable for the 1990s because they have been offering universal services to private and corporate customers, most of them using a branch network. While differing to a greater extent from specialised financial institutions, their kind of operation is basically comparable. Among the specialised financial institutions only Corvinbank, which merged with Konzumbank in 1997, had maintained a branch network for some years (six branches in 1996). In 1999, the building societies, Porsche Bank (two branches) and Rákóczi Bank (one branch) had at least one operative unit in addition to their headquarters (see footnote 35 as well).

[49]   See, for example, MASSEY 1984.

tions: W*here* have the *newly* established banks opened their branches, and how have the banks that *inherited* networks from socialist times reacted in time and space after the market came politically and legally into being?

Analysing the banking sector in the emerging Hungarian market economy from a spatial perspective does not only offer insight into the development of the banking sector as a driving force for the kind of market economy Hungary has been striving for, but for regional development as well:

- The presence and absence of bank branches express *current* regional dispari-ties in economic development because market-experienced banks strive to open their branches only in places that seem to be profitable, which usually is the case in economically prosperous regions.
- Bank branches can also be regarded as indicators of *future* development op-portunities: The local presence of a bank branch facilitates the provision of loans, capital investments and important information about business opportu-nities for residents and SME. In this way, bank branches foster consumption and investments in a region and thus potentially contribute to further eco-nomic development.[50] Since the absence of bank branches has opposite ef-fects, "geographical variations in access to the financial system deepen and accentuate prevailing levels of uneven development."[51]
- Since banks belong to the basic institutions of a market economy and are characterised by the above dialectic of *expressing* and *contributing to* re-gional development, the density and type of branch networks partly reflect the current stage of the economic transformation process.

Although the mere *distribution* of bank branches indicates whether people and SME at certain places could have access to a broader supply of banking services in general, the *conditions* under which the services are offered and their *quality* have also to be examined in order to differentiate who could actually benefit from certain kinds of bank branches.[52] Based on these considerations the actors among Hungarian commercial banks during the 1990s will now be characterised in more detail by studying their business and network strategies and analysing the corre-sponding regional organisation of their networks, including alternative financial infrastructure.

3.1   Competitive advantages, business activities and the size of networks

One of the basic processes regarding institutional development during the 1990s was the spreading out of foreign ownership in the Hungarian banking sector ('for-

---

[50]   See KLAGGE 1995; LEYSHON and THRIFT 1995.
[51]   DYMSKI and VEITCH 1992 quoted in LEYSHON and THRIFT 1995, 315.
[52]   See JÖNS 1996.

eignisation'). Accordingly, the ownership structure can be identified as an important aspect for characterising the institutional actors in this period of time.
Recalling the competitive advantages of banks founded with FDI, it becomes obvious that the banks with FDI were in many respects superior from the beginning
of their operation. In regard to business strategies, those banks that started operating in socialist times had to deal with customers they acquired at that time or
inherited from former institutions. These customers were often large unprofitable
corporations and clients from crises-prone sectors. In contrast, banks with FDI
could concentrate on highly profitable fields of activities and respective customers
(such as sound joint-venture and large companies, private banking). They were
also able to plan the number, size, function and location of their branches strategically, while most of the Hungarian-owned banks had to deal with branch networks
that were built up on the basis of completely different criteria than those important
in a highly competitive environment, for example, with regard to the technical
background of the branches or the branch locations.

Therefore, banks founded *without* FDI (I) and those founded *with* FDI (II) have
to be differentiated when analysing the Hungarian banking market in the 1990s.
Additionally, the banks founded without FDI can be differentiated in the former
large state-owned banks (I-1) and medium and small banks founded without FDI
(I-2). The banks in these groups differ substantially in regard to capitalisation and
customer base and thus their scope of activities. Since a lot of the Hungarian-
owned banks were partly or completely acquired by foreign investors in the
course of privatisation, a final differentiation will be made between those banks
who stayed Hungarian-owned (I-1a and I-2a) and those which attracted foreign
investors (I-1b and I-2b). The latter were able to benefit from the competitive advantages related to foreign direct investments and market expertise (before 1996
all foreign investors had come from established market economies).[53]

This classification of Hungarian banks, which was basically adopted from JÖNS
and KLAGGE (1997) is most useful for systematising the different institutional
actors in the 1990s. Accordingly, the numerical development of the different types
of banks very well illustrates the ousting of Hungarian-owned banks by banks
founded with FDI that dominated at the end of 1999 (figure 4). The acquisition of
Hungarian-owned banks by foreign investors, which started in 1989 among small
and medium-size banks and in 1994 among the large banks, was forced through
the privatisation process and lead to the second (I-2b) and third (I-1b) largest
groups of banks in 1999. The fact that just one of 29 commercial banks had no
FDI in 1999 indicates that the ownership structure was a suitable criterion for the
classification of banks in the first decade after the political transformation. In
future, other aspects will have to serve for differentiation. Those criteria, as the

---

[53]   In 1996 a Russian bank bought 100% of ÁÉB.

analysis will show, are the size of commercial banks and their market share in different segments of the market.

Figure 4:   Commercial banks by ownership structure, 1987-1999

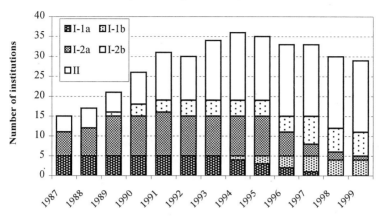

I: banks founded *without* FDI; I-1: former large state-owned banks; I-2: medium-sized and small banks; I-1a and I-2a: Hungarian-owned banks; I-1b and I-2b: banks acquired by foreign investors; II: banks founded *with* FDI

Source: See figure 1; author's compilation.

Figure 5:   Bank branches of commercial banks by ownership structure, 1987-1999

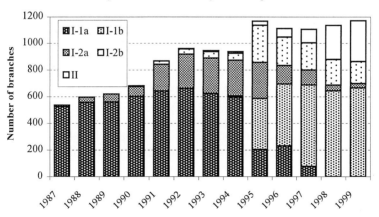

Source: See figure 1; author's compilation; for legend see figure 4.

3.1.1   Bank branch network developments

The development of the total number of Hungarian bank branches and the proportions of the different types of banks (figure 5) show that almost all operative units belonged to the large state-owned banks in 1987. While their number of branches fluctuated in the 1990s, the branch networks of small and medium-sized Hungarian-owned banks gained greater importance between 1991 and 1995, but decreased afterwards - parallel to the mergers, acquisitions and liquidations - to almost zero in 1999. The first banks founded with FDI started to open branches in 1990. The pace of the growth of their networks accelerated substantially at the end of the 1990s. For example, the number of branches operated by these banks grew about more than 2.5 times in 1998, and increased by another 20% in 1999. These developments and the fluctuating number of branches in total - increase until 1992, decrease in 1993-94, expansion in 1995 followed by a slight reduction in 1996-97 and another increase in 1998-99 - are the results of different developments that can only be identified by referring to the business and network strategies of the individual banks. It can only be stated by now that the high concentration of branches among the four former state-owned large banks - maintaining 57% of all branches in 1999 - seems to be a competitive advantage in the highly competitive environment at the end of the 1990s. This is because large networks guarantee such a high proximity to customers that can only be reached slowly by the banks which have to build up new networks. Whether the relatively slow pace of network development among the banks founded with FDI (type II) mirrors a disadvantage or is rather intended by the subsidiaries of market-experienced actors will become clear in the following paragraphs.

Considering the branch network development of individual banks between 1987 and 1999 (figure 6), the following types of banks can be identified by their networks:[54]

- Banks with networks since 1987. These banks were all state-owned before the middle of the 1990s and were then either privatised or taken over by foreign-owned banks (I-1 and I-2). All the respective networks continued to exist until 1999, but were modified in between.
- Banks that established networks either before the political transformation or afterwards, but which ceased to exist until 2000 (I-2a). All but one were medium or small state-owned banks characterised by undercapitalisation and accumulated bad debt. They either merged with other state-owned institutions as a preparation for privatisation, were partly taken over by other banks or liquidated. The exception is Takarékbank, the Bank of the Savings Co-operatives, which was privatised and continues to exist, but closed down its branches until 2000.

---

[54]   The statements about events in 2000 rely on personal information given by a banking expert (NBH) in December 1999.

- Banks that founded the first branches either in 1988 (one bank) or after 1990 (the rest) and have been expanding their networks in the second half of the 1990s (EKB merged with Citibank in 1999). They are all banks with FDI, although some banks were founded without FDI and partly or completely acquired by foreign investors later on (I-2b and II).[55]

At the end of 1987, OTP had the largest network by far due to its long monopoly of retail banking (429 branches). OTP was followed by the three banks created from the lending department of the NBH: K&H (51 branches), MHB (23 branches), BB (18 branches). All the other banks maintained fewer than ten branches. After the retail market segment was opened up for all commercial banks in 1989, OTP reduced its network by 13% until 1995 while almost all the other banks (I-1a and II-2a) extended their networks substantially until 1992. This first period of network expansion can be described as 'market euphoria' in which the banks strove for regional presence in order to get in contact with potential clients.[56] OTP, starting from a incomparably high basis, used the time for a reduction of the costly network, resembling the trend that could be observed in several Western European countries at that time.[57]

Between 1992 and 1995 the network expansion either stopped (Mezöbank, Konzumbank, MKB, Dunabank) or slowed down in almost all banks, some even reducing their number of branches again (K&H, MHB, Iparbankház). Responsible for this trend was a combination of different aspects: the portfolio crises (consolidation programs between 1992 and 1994) and related cautious business policies on the part of the banks; extremely high costs for establishing and even maintaining branches due, for example, to a high number of employees inherited from socialist times or the renewal of obsolete technical background; the influence of 'small-network' strategies announced by the banks with FDI at that time. Originally, the banks founded with FDI had planned to concentrate on highly profitable business activities such as the financing of multinationals, large joint-venture and Hungarian-owned companies and private banking and thus to serve a rather élite clientele from few branches in the major urban centres.

Nevertheless, a second period of network expansion began in the second half of the 1990s, based on the already large and medium-sized networks growing once more. In addition, the newly established networks of banks with FDI showed an accelerating growth. This development was favoured by the fact that at that time, the consolidation programs were finished, the need for provisions continued to diminish,[58] the number of commercial banks had reached its maximum and thus the competition had grown immensely. More and more banks with FDI set up

---

[55]   For the exception of Postabank, see footnote 37.
[56]   See JÖNS and KLAGGE 1997.
[57]   See, for example, LEYSHON and THRIFT 1995; EUROPÄISCHE KOMMISSION 1999.
[58]   SZELÉNYI and URSPRUNG 1998, 32.

Figure 6:   Branch network development of individual banks, 1987-1999

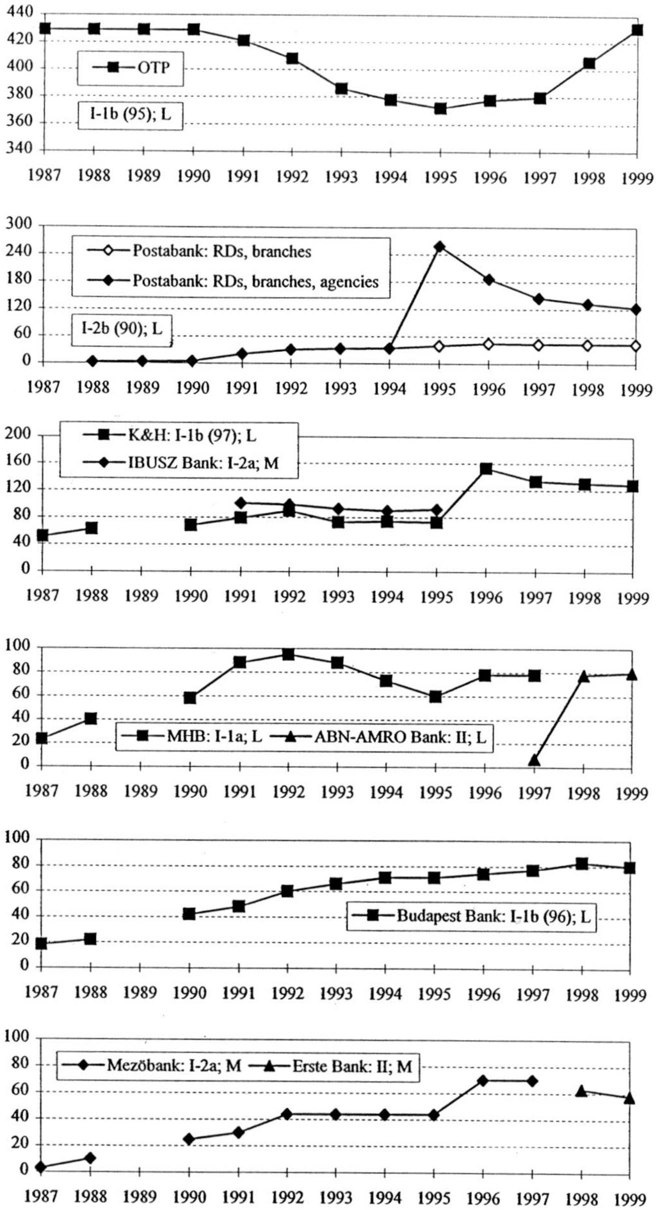

L = Large bank; M = Medium-sized bank; S = Small bank. For an explanation of the different types of banks (such as I-1b), see figure 4. Years in brackets indicate the date of acquirement by foreign investors.

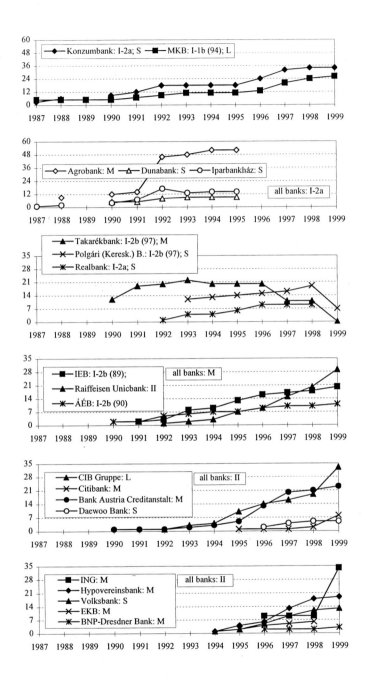

Source: See figure 1; author's compilation.

branches and became strong competitors for the Hungarian-owned banks outside Budapest, especially because of sophisticated products and high standards regarding service and technology. The battle for customers in the corporate segment strengthened so much that the banks started to look for alternative business opportunities in still more concentrated segments. Additionally, it had become clear among banks with FDI that liquidity for corporate financing is limited, and deposit taking one possibility to raise funds. Therefore, several banks reformulated their strategy and aimed at gradually increasing their presence in certain segments of the retail market as well as in lending to creditworthy small and medium-sized businesses by opening more new branches than originally planned.[59] Thus, an expanded branch network became an important competitive advantage in the race for spatially less concentrated customers.

Remarkably, the network expansion of the former state-owned large banks followed in each case the take over by foreign investors. This supports the observation of other authors that banks and companies with FDI have been very flexible and active in opening up new markets during the transformation process in Central Europe.[60] Thinking of the first period of network extension until 1992, which was backed by the state-owned banks, this flexibility and activity seems to depend to a high extent on sufficient funds and not so much on market experience per se. Postabank started the second period of network extension by opening 219 agencies in 1995. These agencies are not fully-fledged branches, but offer basic retail services in addition to the branches. In 1996, the other banks followed suit: OTP, which was just privatised, intended to secure and widen its large branch network that has been identified as a major competitive advantage in the second half of the 1990s. The efforts to maintain its market leader position as the largest retail bank in Hungary had to be enhanced because of other commercial banks' new ambitions in the retail market. Furthermore, the savings co-operatives announced the goal to double their market share within the next five years in 1998.[61] K&H extended its network through the merger with the retail-orientated IBUSZ Bank, while Mezőbank, Konzumbank and MKB are further examples of banks that joined the second period of branch network extension by opening new branches.

Among the banks with FDI the expansion of networks substantially accelerated in 1998-99 when CIB and ING started their offensives to become major players in the retail market. Six additional banks with FDI named the retail segment and

---

[59]   See 'strategy and future plans' of the individual commercial banks in KEREKES 1998, for example, pages 183 and 284. According to interviewees and JONES (1999, 42-43), the banks' strategies of "seeking out new retail deposits and cultivating the small business lending markets in the hunt for new businesses" resulted mainly from narrowing profit margin after blue-ship companies used the possibility to play off several client banks.

[60]   See, for example, BUSCH and WEISIGK 1997, 907; HAMILTON 1999, 140.

[61]   See the strategy and future plans of Takarékbank in 1998 (KEREKES 1998, 152).

network development top priorities for the coming years when asked in 1998.[62] For example, Raiffeisen Unicbank stated that it aims at a network of 50 branches in 2001, while ABN AMRO Bank, which acquired MHB in 1998, aims to develop a network of 120-150 branches until 2001.[63] Although most banks founded with FDI (62% of all commercial banks) changed their overall business and network strategies at the end of the 1990s, their proportion of branches reached only 26% at the end of 1999. After branches were identified as a major competitive advantage, the growth of branch networks accelerated, but the process of network development still seems to be a rather time consuming process. At this stage of the analysis it can just be assumed that money is not the only restrictive factor for network development among the foreign-owned banks. Later it will be shown that that the local circumstances existing at potential branch locations play an important role for the development of the bank branch networks (see section 3.2).

3.1.2   The development of business activities and market shares

The branch network developments of the individual commercial banks are strongly interrelated with the number of banks, their business activities and market shares. The corporate sector was contended at first because of much better business prospects than in the retail sector. The retail market attracted considerable attention as an opportunity to widen the banks' market share only when the number of commercial banks reached the maximum in the middle of the 1990s and more and more banks with FDI started to build up their own branch network in order to serve corporate clients. According to these developments the concentration of credit portfolio in the highly contended corporate sector was not high anymore at the end of the 1990s, while the retail market shares have just begun to become decentralised. The lending business to households has even just began to develop then (figure 7).[64] In 1997, OTP, the savings co-operatives and Postabank together still held 67.5% of the market shares in the household liabilities market, although their share was further declining compared to previous years.[65] Concerning the underdeveloped household lending sector, the concentration was extremely high in certain segments. For example, 95% of all housing credit was disbursed by OTP and savings co-operatives, while 97% of consumer

---

[62]   See KEREKES 1998.

[63]   See KEREKES 1998, 128, 400.

[64]   Retail services have been mainly concentrated on deposit taking, current account management and bank card services in the second half of the 1990s, while loan facilities to households - like loan facilities for small businesses in the corporate market - were still underdeveloped at the end of the 1990.

[65]   The HUNGARIAN BANKING AND CAPITAL MARKET SUPERVISION (1998, 25) points out that in 1997 nearly half of all credit institutions increased the share of household liabilities in their liability structures, but that the share of household liabilities was still below 10% among the liabilities of seventeen banks.

loans were made by four credit institutions and the savings co-operatives.[66] Altogether, primarily large and small banks have been active in the household market so far. In case of the medium-sized banks their proportion of the aggregated balance sheet is only reflected in the corporate market where they have been underscoring each other with offering the lowest interest rates for years. Although still 44% of all corporate loans were extended by four large banks in 1997,[67] "it was corporate financing where the most credit institutions participated in activities exceeding their market shares based on asset size."[68] Concentration on corporate deposits was much less than in the household market, too.[69]

As another result of the intensive competition, more and more medium-sized businesses got loans from the banks in the second half of the 1990s while the banks' business policy towards small companies is still rather reserved because of the risk involved.[70] The increasing engagement in lending to medium-sized and small companies at the end of the 1990s was not only encouraged by competition, but by the government as well. In order to support the underdeveloped SME sector in Hungary, the latter "is offering to pay up 70% of the interest on bank loans to small firms,"[71] but, nevertheless, a considerable credit risk still lies with the lending bank. Thus, financing small businesses will probably be one of the last segments of commercial banking that tight competition for market shares will go into. In order to win small businesses as customers, a large branch network will be a competitive advantage again.

Altogether, the concentration in the banking sector no longer counts as high in a European context since 1997.[72] To a considerable extent this can be put down to the expansion of almost all medium-sized banks.[73] In contrast, the group of small banks consists of much more heterogeneous institutions. Some have been producing extraordinary growth while others have hardly grown or even declined at the end of the 1990s.[74] A general problem of the Hungarian banking sector is still the banks' capitalisation that is rather low in international comparison. This becomes evident especially regarding the capital requirement of certain infrastructural investment projects in Hungary.[75] The low capitalisation might as well pro-

---

[66]   HUNGARIAN BANKING AND CAPITAL MARKET SUPERVISION 1998, 25.
[67]   HUNGARIAN BANKING AND CAPITAL MARKET SUPERVISION 1998, 28.
[68]   HUNGARIAN BANKING AND CAPITAL MARKET SUPERVISION 1998, 30.
[69]   See HUNGARIAN BANKING AND CAPITAL MARKET SUPERVISION 1998.
[70]   See NATIONAL BANK OF HUNGARY 1999b.
[71]   JONES 1999, 42.
[72]   NATIONAL BANK OF HUNGARY 1998, 74.
[73]   See NATIONAL BANK OF HUNGARY 1999b, 95.
[74]   NATIONAL BANK OF HUNGARY 1998, 74.
[75]   According to the HUNGARIAN BANKING AND CAPITAL MARKET SUPERVISION (1998, 59) the maximum amount of syndicate loan all Hungarian banks seen as a consortium could extend to a client without violating the relevant legal restrictions would be HUF 115 billion (= combined large risk assumption limit). Thus, their financing is impossible without co-operation between Hungarian and international banks or banking groups

vide another reason for ongoing institutional concentration among commercial banks in the future.

Figure 7:   Volume and market shares of credit institutions in the household and corporate sector market at the end of 1997 [76]

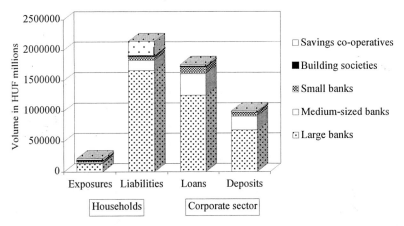

Source: HUNGARIAN BANKING AND CAPITAL MARKET SUPERVISION 1998, 23-30; author's compilation.

3.1.3   Future business and branch network strategies

At the end of the 1990s, the following banks can be differentiated by their business and branch network strategies:

- large banks who try to become a major player in all segments of the corporate and retail markets (including the mass market), and already maintain or strive for a large network with more than 80 branches;
- those banks who pursue broad activities in the corporate market, and, besides, aim at a reasonable stake in the more attractive segments of the retail market with establishing 50 to 80 branches;
- other banks who concentrated on foreign-owned, joint-venture and sound Hungarian-owned enterprises so far (often serving as second or third client bank), but have been trying to open up towards SME and private individuals with medium or high incomes recently. They aim at 25-30 branches, which is more or less equivalent to the number of counties in Hungary. Only a few banks continue to have their focus exclusively on corporate banking.

---

in consortia offering syndicate loans.

[76]   In this statistics the banks are classified according to its shares in the adjusted size of the sector (including specialised credit institutions, but not building societies): large banks: above 3%; medium-sized banks: between 1–3%; small banks: below 1%. The National Bank of Hungary defines slightly different categories.

In addition to the continuous development of the banks' branch networks for client acquisition both in corporate and retail market, other common business policy objectives named by the banks in 1998 as their 'future and business plans' were the following: [77]

- widening of the bank's retail product range (retail loan facilities etc.),
- developing investment banking facilities,
- enhancing the profitability and efficiency of the operative units,
- installing an integrated information technology infrastructure,
- training employees.

Many banks also have been providing or are developing direct banking services such as home and phone banking facilities; however, the recent trend in network development underlines that at the current stage of the transformation process local branches have a greater importance for *acquiring* business clients and private individuals than IT facilities. A major characteristic of the transition period is that banks have to make people *trust* in market-orientated banking activities, and this is best and mainly achieved through personal interaction. How fragile the environment still was at the end of the 1990s demonstrated the liquidity crises at Postabank in 1997 that resulted from a run on the banks funds. Nation-wide, people queued in Postabank branches in order to take their money out of the bank after rumours about financial problems had come up.[78] Direct banking facilities are, of course, a major competitive advantage *after* clients have been acquired and are more confident with the business opportunities offered by banks in a market economy (see section 3.2). Those commercial banks that do not have a branch network as yet or just maintain fewer than five branches are mostly small banks focusing on corporate business in certain sectors such as energy, transport and telecommunication, food, wholesale and retail business, or on financing bi- or multilateral activities between Hungary and certain regions such as ASEAN countries or countries in Western Europe. Banks operating on this basis are, for example, ÁÉB, Rabobank, and Hanwha Bank.

3.1.4   Summing up alternating waves of expansion and reduction

Regarding the quantitative development of bank branch networks in Hungary since 1987, the following main phases can be distinguished:

- a *'market euphoria'* until 1992 as the first period of network extension; this was dominated by the activity of Hungarian-owned banks; between 1987 and 1992 the total number of branches increased by 80%;
- a period of *network reduction* in 1993-94 due to the effects of the portfolio crises and high operational costs of branch networks; a stagnation in network

---

[77]   See KEREKES 1998.
[78]   See footnote 37.

development or a reduction of branches occurred at almost all Hungarian-owned banks, while banks with FDI started to build up their networks; the total number of branches was slightly reduced in 1993 and 1994;

- a period of *'reduction versus expansion'* between 1995 and 1997 caused by four overlapping processes: the continuing cautious network policies and reduction of branches among Hungarian-owned banks in 1995, mergers and liquidations among commercial banks with branches as well as the following reorganisation of enlarged networks through mergers, the beginning of a second period of network extension, encouraged by Postabank in 1995, and, finally, the steady opening of new branches by banks with FDI;

- a second period of *network expansion* becoming manifest in 1998 and even accelerating in 1999; this time the spreading out of branches was stimulated by growing competition for customers even beyond the most attractive clientele from the corporate sector, including more risky corporate lending activities, such as financing medium-sized and small companies, as well as different segments of the retail market; after a further reduction of branches among Hungarian-owned banks in 1998-99, there was only one Hungarian-owned bank left at the end of 1999 (which was privately owned).

One of the most eye-catching phenomenon in the above analysed branch network development is the fact that the Hungarian-owned banks were not able to use their advantage of large networks, which was even extended during the period of 'market euphoria': facing a growing competition of banks with FDI they had not enough time to build up a deficit in funds as well as operational skills and get rid of the obsolete technical background.[79] Only privatisation, which meant an increasing influence of foreign investors and improved financial bases, made the further extension of branch networks and thus the restoring of the former competitive advantage in those banks possible – mainly because of sufficient capital for bearing the high costs of network development.

Among the savings co-operatives, a similar trend to the network development of commercial banks can be observed: despite continuing institutional concentration since 1987 the number of co-operative branches decreased until 1995 and has been expanded again since then (1990: 1 771 branches; 1995: 1 658 branches; 1997: 1 757 branches).[80] This observation reinforces the conclusion that *alternating waves of expansion and reduction* are a major characteristic of individual and aggregated bank branch network developments during a period of economic transformation. They occur because banks are continuously operating in an area of conflict between customer acquisition and reduction of the operational costs. The

---

[79] This is true for the privatisation of banks as well: in Hungary itself, there was not enough time to build up capital in order to take over more than one state-owned bank with privately owned Hungarian capital. The pressure of well-capitalised foreign investors invading the market was too high.

[80] Author's calculation according to KEREKES 1991, 1996, 1998.

shifts in network strategies, that are accelerated by changes in the competitive environment, are based on the fact that the profitability of a branch is not completely predictable. Instead, it has to be *experienced* through business *practice,* and this identifies the bank branch network development as an *process of trial and error.* This does not only lead to the assumption that the changes in the size of individual branch networks will probably stay as dynamic as they were in the last ten years for at least another decade. It also raises the question whether such waves of expansion and reduction can be observed in established market economies as well? Although this question cannot be answered in this paper, some indications suggest that waves of expansion and reduction of bank branch networks are indeed not only a concomitant with the shift from a centrally planned to a market economy but a phenomenon that can be observed in established market economies as well, though to a smaller extent and over a longer period of time.[81] Consequently, a recent study on banking in Europe came to the conclusion that neither the development of IT facilities nor the growing number of bank mergers have yet changed the important role of traditional bank branches for (retail) banking.[82] This aspect has a lot to do with the great importance of face-to-face contacts for client acquisition, and the unique and dynamic character of places that is based on the activity of different actors.

---

[81]   Such indications are, for example, publications that investigate the impacts of the shift from periods of network expansion to massive network reduction in the UK and the US during the 1980s and 1990s (see LEYSHON and THRIFT 1995, 1997) or statistics about bank branch network development in different Western European countries (EU-15) (See EUROPÄISCHE KOMMISSION 1999): Between 1994 and 1997 the total number of commercial bank branches increased in Hungary by 8.6%. In the EU-15 states the development was quite heterogeneous during the same period of time: in some countries, the total number of banking units continuously increased between 1994 and 1997 (e.g. Portugal: 24.7%; Italy: 9.2%). In France and Spain a turn in the network development occurred after 1995 when an expansion ended and a reduction began, while the opposite development took place in the Netherlands (assumed that not a formal change in statistics occurred: see below). Finally, a reduction of banking units proceeded in most countries (e.g. Denmark: -3.0%; UK: -11.8%; Finland: -24.1%). In Germany, the number of all banking units was reduced by -6.9% between 1995 and 1997 (Author's calculation according to EUROPÄISCHE KOMMISSION 1999). The headline 'Less banks, more banking units' operating in Germany during the second half of the 1990s, published by the FAZ on January 18, 2000, is misleading, although it refers to the just quoted banking statistics. This is because these statistics include the branches of the German Postbank since 1995. (as part of the post offices they were existing before as well), but the dates of calculation were 1994 and 1997. Both categories of bank branches, i.e. Postbank branches and other banks' branches (1994-97: -3.7%), were reduced in the second half of the 1990s (Author's calculation according to DEUTSCHE BUNDESBANK 1999).

[82]   EUROPÄISCHE KOMMISSION 1999, 21.

3.2   Geographies of Hungarian bank branch networks and alternative financial infrastructure in the 1990s

In 1996, a branch network strategist, working in the headquarter of a middle-sized foreign-owned bank in Budapest, predicted that the strong competition for customers on the Hungarian banking market would be settled in 1999.[83] Three years later, the situation was actually very different. Competition had grown even more in 1998 and 1999 when several banks opened up toward less contested segments of the market such as financing medium-sized companies or doing business with private individuals having medium to high income. Those banks who already dealt with such customers turned to sound customers in fields regarded as even more risky, including lending to households or small companies. This shift in lines of business was almost always accompanied by a change in the banks' branch network strategy. While a rather cautious network policy prevailed among banks founded with FDI in the first half of the 1990s, and even a reduction of branches took place among originally Hungarian-owned banks after their 'market euphoria' had ended in 1992, an accelerated expansion of branch networks occurred in almost all banks after 1995.

In which way, if at all, has the location of the newly established or inherited branches mattered in this process? The argument is that analysing the spatial organisation of branches is necessary in order to achieve full understanding of banking in Hungary and the operation of different credit institutions. Thinking of the above described dialectic inherent in bank branches - i.e. *expressing* and *encouraging* regional development – it can also be expected to find some clues for reaching a better understanding of the regional transformation process.

3.2.1   Inherited networks

A common characteristic regarding the spatial organisation of the Hungarian banking system before and after the political transformation in 1990 has been an extremely high centralisation of headquarter functions in Budapest.[84] In 1990, all headquarters of the banks were found in Budapest, while in 1999 only one bank, a state-owned regional development bank in Miskolc, had its headquarter outside Budapest.[85] Three main reasons have been responsible for the centrally organised financial system in Hungary:[86]

- the historical dominance of Budapest within Hungary,

---

[83]   Interview conducted by the author in spring 1996.
[84]   For a detailed analysis of the regional distribution of jobs in Hungarian banking in 1980 and 1990, see JÖNS (1996).
[85]   This specialised credit institution, the Rákóczi Bank, had started operation in 1992.
[86]   See JÖNS 1996, 75; JÖNS and KLAGGE 1997, 36.

- the centrally managed socialist economy that made a concentration of decision-making necessary,
- the role of Budapest as the centre of innovation in the transformation process.

The only decentralised elements have always been the savings co-operatives. In addition to the savings co-operatives OTP maintained a nation wide branch network before 1990. The successor banks of the National Bank's banking department maintained branches outside Budapest as well, but, at the end of 1987, they maintained together less than a quarter of the branches operated by OTP at that time (97 versus 429). At this early stage of the re-established two-tier banking system, the smaller financial institutions, that were founded during the 1980s, had just started to build up their branch networks.

Between 1987 and 1990 the commercial banks' branch networks were enlarged by a considerable extent (+28.4%). According to socialist principles of public supply, almost all regions and sizes of settlements benefited to a certain extent from the establishing of new bank branches in this period of time.[87] Nevertheless, the highest number of branches and lowest population-per-branch ratios[88] continued to be concentrated in the major regional centres of the country that are more or less the large county capitals (figure 8). In the framework of the socialist regional development policy these major Hungarian towns had been decentralised development foci. They were situated along two 'rings of centrality' around Budapest.[89] The distance between Budapest and the urban centres of the first 'ring' was about 100 km, while those centres of the second 'ring' were located in about 200 km distance to Budapest. The country capitals in the eastern part of Hungary such as Miskolc, Nyíregyháza, Debrecen and Békéscsaba had been rather small urban centres in the less developed, agricultural part of pre-war Hungary, but in socialist times they became industrial centres through strong financial support by the government. At the end of the 1990s, the regional centres in the eastern part of Hungary equalled the other regional centres in many respects, not only in regard to the supply with banking infrastructure.

Hungary, like other socialist countries, had very low banking supply ratios. In 1990, slight west-east disparities existed that were preserved from pre-socialist times (figure 8). Pre-war disparities in banking infrastructure continued to exist in socialist times - though to a smaller extent - because the bank branch networks were mainly *reduced* when the one-tier banking system was created in 1947-50;

---

[87] See JÖNS 1996.

[88] The population-per-branch ratio is the inverse number of the branch-per-capita ratio. The first will be used in the figures of this article since the number of branches is quite low in comparison with the number of inhabitants, and thus this indicator is more vivid and better to handle. For a change, the term branch-per-capita ratio is used in the text as well. Both indicators facilitate the comparison to what extent a region or community is supplied with banking infrastructure.

[89] For the origin of these 'rings of centrality', see JÖNS 1996, chapter 3.

they were not completely reorganised.[90] The lowest branch-per-capita ratios in the north-east of Hungary were also influenced by a higher population density in the northern counties and a lower urban density in eastern Hungary.[91] Other disparities in the supply with banking infrastructure existed between different sizes of settlements.[92] The smaller the settlements' size the lower was the branch-per-capita ratio. The small towns and villages had the most disadvantageous starting position regarding the shift towards the market economy. For people living in small settlements neighbouring bigger towns, the lack of a banking infrastructure was less problematic than for people living in less urbanised regions such as north-east Hungary or other rural areas throughout Hungary.[93]

Figure 8:   The regional distribution of commercial bank branches in 1990

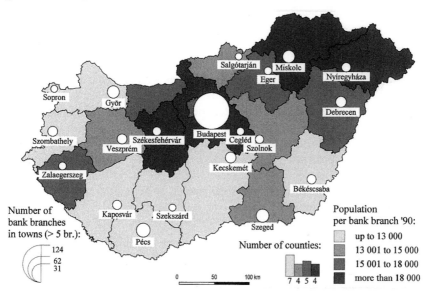

Source: See figure 1; population data from census 1990; author's compilation.

---

[90]   See ILLÉS 1994.

[91]   See JÖNS 1996.

[92]   The size of settlements is a very convenient analytical tool for examining the spatial organisation of bank branch networks in Hungary. This is because central planning was orientated to the settlements' population. In this hierarchical distribution system, the smaller villages were intentionally neglected (see, for example, MEUSBURGER 1995, 143).

[93]   Compared with Austria in 1980, the ratio of employees in banking per capita was three times lower in Hungary than in the adjoining market economy. While the ratio of job-per-capita in banking reached 1:3 between Budapest and Vienna, this proportion reached 1:6 in the villages with less than 500 inhabitants (JÖNS 1996, 60).

However, one cannot 'blame' socialist policy for creating regional disparities in banking because banking had almost no significance for socialist society. It was only when the turn to market principles occurred that a different value was put on bank branches and thus regional disparities in the supply of banking infrastructure started to make a difference for several actors in society. The access to money and related matters suddenly became a social necessity, for money governs a capitalist market economy. Thus, local accessibility to credit institutions, the essential provider of finance for people and SME, represented a competitive advantage. This is what HAMILTON (1995, 1999) hypothesised as a 'commodification' process characteristic for the transition to a market economy. In his words, this process "radically alters the perception, value, functions and influence of space."[94] According to HAMILTON (1999, 140), this re-evaluation of space affected natural and human capital resources as well as the location of a place.

3.2.2   Developments in the urban hierarchy

Until 1992 the *former large state-owned banks* established their new branches in all categories of settlements but the small villages with less than 5 001 inhabitants (figure 9). Subsequently, they reduced their number of branches so that the level of 1990 was reached again in all categories of settlements at the end of 1994. Mainly the upper levels of the hierarchical urban system benefited from the second period of network extension among the former large state-owned banks (Budapest: +30%, big towns: +16%, middle-sized towns: +11%).[95] This shows that criteria typical for banks in market economies began to influence the branch network strategies of the former large state-owned banks only in the second half of the 1990s, often not until they were privatised ('foreignised').

The developments among the *banks with FDI* (except Postabank which will be analysed separately) suggest *hierarchical top-down-diffusion* par excellence: Starting with establishing the first branches in Budapest in 1990, the banks with FDI appeared in the middle-sized towns in 1991, in the small towns only after the period of strong network expansion had started (1995), and, finally, in the smallest villages in 1997. This top-down diffusion illustrates the character of drawing-board planning regarding the branch network strategies of banks with FDI. Since they originally aimed at acquiring sound customers in the most attractive segments of the market - multinationals, large joint-venture and Hungarian-owned companies - they consequently went to the largest agglomerations of potential customers first. Only when the competition got harder at the end of the 1990s did the banks deviate from their original strategy and spread out into smaller municipalities with less and different business opportunities.

---

[94] HAMILTON 1999, 140.
[95] In the larger small towns (I) the expansion lasted only until 1996. It was followed by an even stronger reduction of branches (1994-98: -5%). In the smaller small towns (II) nothing changed (in sum), and the small villages gained only +9% until 1996.

How important the regional perspective is in this respect, can be very well illustrated by the fact that the first appearance of banks with FDI in small towns in 1995 is not very spectacular since these towns - Piliscsaba, Pilisvörösvár and Törökbálint - are all located within the agglomeration of Budapest. They are suburban municipalities that already belong to the county Pest (figure 11). In contrast, the branches founded by banks with FDI in small villages from 1997 onwards deserve special attention. They are the first expression of the turn towards *niche strategies* among foreign-owned banks that was discussed in the previous section in detail and can be labelled as an opening towards so far neglected segments of the market:[96] The first branch of interest is located in a village with only 3 234 inhabitants (1998) in the north-east of Hungary. The niche is a strategic and profitable location along a major transport route to the Ukraine and Russia where special border needs (duty, money exchange etc.) for truck drivers and other travellers offer very promising business opportunities. The other innovative branch location is less unusual since Fertőd (3 079 inhabitants in 1998) is located near the border to Austria. The most important aspect of its location is, however, the fact that Fertőd is the home of the famous baroque castle of Esterházy, a major cultural and tourist attraction with international reputation.

From this observation it can be concluded that *local knowledge* became more important for banks with FDI at the end of the 1990s when the major centres were already 'occupied' by almost all banks. At this time, competition had not diminished (see above) but strengthened, and drawing-board planning was not sufficient anymore. Creativity, local knowledge and a certain amount of risk became the most important aspects for creating a competitive advantage through strategic branch network locations.

Among the large retail banks, Postabank, the second largest retail bank behind OTP, took the first initiative to strengthen local presence and thus to close the gap towards the network size of OTP. By opening 219 agencies in addition to its 40 fully-fledged branches in 1995, Postabank increased its presence in the middle-sized and small towns considerably and thus created a competitive advantage before the contest for upper-level retail customers started. The fact that between one and three branches were closed in *each* of the six settlement categories before 1998 indicates that this extraordinary expansion of branch networks could not be sustained because the profitability of branch locations can only be experienced through business *practice*. Since the branch closures occurred in all sizes of settlements, the size itself seems to have little influence on the success or failure of a operative unit. The dynamic changes in the individuality of places seems to be responsible for this process (see below).

---

[96] Actually, only two of the six branches of banks with FDI located in small villages in 1998 have been newly founded. The other two had been taken over from Mezőbank (by Erste Bank) and from MHB (by ABN-AMRO).

Figure 9:   Branch network developments in the hierarchy of settlements, 1990-1998

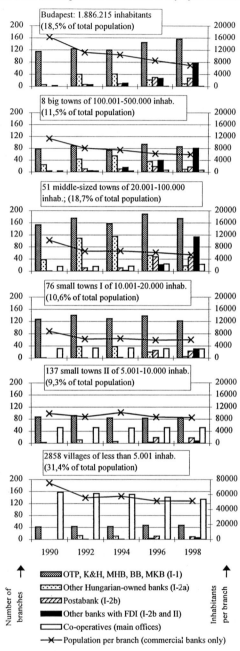

Source: Author's compilation; see figure 1; population data from census 1990 and KSH
1998.

In the small villages with 10 001-20 000 inhabitants the combination of a reduction of branches belonging to former Hungarian-owned banks (I), including the effects of institutional declining among medium-sized and small Hungarian-owned banks in the second half of the 1990s, and a relatively small number of new branches founded by banks with FDI (II) led to a rather insignificant increase of branches between 1994 and 1998 (+6.5). The benefit of overall network expansion of commercial banks was also comparably low in the small villages (+12.7%), while the other categories gained about a quarter of new branches in this period of time, and Budapest's branch networks even doubled (+49.7%).

The population-per-branch ratio shows that the *disparities between different sizes of settlements considerably decreased* since 1990. In more or less all categories, the development of the ratio followed the familiar pattern of improvement until 1992, stagnation or deterioration until 1994, and another improvement until 1998. In 1998, the most favourable ratio was found in the middle-sized towns (5 427 inhabitants per branch), the big towns (5 982) and the larger small towns (6 012). The ratio of Budapest (6 960) is incomparable because of the high population density and short distances between branches, but the ratio in the small towns with less than 10 000 inhabitants (8 540) already indicates a lack of supply of bank branches from the viewpoint of the population and SME. In the small villages the situation is much worse (51 501). The competitive disadvantage for actors in small villages and rural areas starts with the amount of time one needs to get to a branch and includes the access to information about profitable banking transaction or the clients' freedom of choice between competing banks. Although the minor role of commercial bank branches in the smaller villages is to a certain extent compensated by the activity of the savings co-operatives, the number of inhabitants per bank branch *and* savings co-operative was still three times higher in the small villages than in the middle-sized and small towns (ca. 5 000-5 800) in 1998.[97] Thus, disparities between rural and urban areas are among the most important disparities that can be observed in the geographies of banking in Hungary at the end of the 1990s. They hold the danger of 'financial exclusion'[98] in the small villages of rural areas (see below).

---

[97]   The savings co-operatives are the most important provider of financial services in villages with less than 5 001 inhabitants. Accordingly, the supply ratio for the small villages drops to 16 291 inhabitants per banking unit when taking the co-operatives into consideration. This ratio would even drop more when the branches of co-operatives were included, but the ratio would still exceed those of the other settlement sizes. Unfortunately, the main source of data (KEREKES 1990 etc.) does not provide the possibility to assign the savings co-operatives' branches to the settlement where it is located, but at least it shows how many branches a certain co-operative has.

[98]   See LEYSHON and THRIFT 1995, 1997.

3.2.3   Regional transformation and its mediators

The periods of branch network development outlined above are also connected with characteristic patterns regarding the choice of branch locations. Among the banks founded with FDI a diffusion can be traced from Budapest (1990) to the major regional centres (1992, 1994, 1996) (figure 10). According to the branch network development all regional centres were equally attractive as branch locations for banks founded with FDI until 1996. As a result of socialist business policy, the urban centres along the two 'rings of centrality' were agglomerations of industrial companies. Although many of them were in a bad shape, there were still enough business opportunities for foreign-owned banks in *all* parts of the country. Accordingly, FDI in industry was concentrated in these regional centres, too.[99]

The two 'rings of centrality' are *legacies* from socialist times and represent the first factor that strongly influenced the banks' network strategies. Striving for presence in each county, about 15 to 20 locations, more or less corresponding to the number of county capitals, were initially attractive for banks with FDI. Drawing-board planning seemed to be enough for serving the rather élite clientele targeted in the first period of network construction until 1996. Thus, the main question for the banks at that time was not 'Where to go?', but 'Where to open first?'. Only when the competition had grown enormously did niche strategies become more and more important (see figures 10 and 11). The banks started to open branches

- in small towns located in the agglomeration of Budapest where plants and other subsidiaries of multinationals were to be found (1996, 1998: for example Schöller, Quelle and Lekkerland in Törökbálint or Tchibo and Family Frost in Budörs);
- in other smaller towns of Transdanubia, particularly in the north-west and the western border region (1998: for example in Fertőd and Keszthely because of tourism and heritage or in Pápa and Ajka because of industry);
- in selected municipalities in the rest of Hungary, 'even' including the eastern part (1999: for example in Barcs and Tuszér because of transport routes).

These niche strategies followed each other (not necessarily in one bank). When the one 'hazardous business' was not enough to get a competitive advantage, the next, more risky one was dared or when the best locations of the first strategy were already occupied, other banks looked for new profitable locations. The second and third strategies only developed after the banks decided to widen their clientele in the corporate sector and to get more involved in retail banking. To a certain extent this was necessary because in 1996 the banks founded with FDI were *all* found in the same locations along the rings of centrality (figure 11). Thus, changes in business strategy *were not responsible* for shifts in branch net-

---

[99]   See, for example, ASCHAUER 1995, 71.

work strategies, but the regional organisation of the branch networks was a major *cause* of the shift in lines of business. This in turn produced changes of the network strategies. Therefore, the operation of the banking system and individual banks cannot be understood without taking account of the regional dimension.

The fact that drawing-board planning was not sufficient for successfully organising a branch network at the end of the 1990s leads to another conclusion related with the great significance of space for the operation of the Hungarian banking system and regional transformation. This refers to the bundle of criteria used by branch network strategists in order to decide 'Where to open a branch first?'. These criteria, named by branch network strategists of foreign-owned banks in 1996 and 1999,[100] show how important the local context is for the opening of a bank branch:

- regional economic structure; indicators are, for example: number of export-orientated companies, especially from the banks' owners home countries; number of joint-ventures and all companies; real income or other social-economic indicators regarding the population;
- human capital resources;
- accessibility from Budapest in terms of the time to get there and a save communication line;[101]
- co-operation of local authorities;
- availability of an appropriate office building;
- competitive situation in regard to the activities of other banks.

With the emergence of niche strategies, these criteria were expanded and partly replaced by more sophisticated local particularities such as the location along major transport routes or important cultural and tourist centres. In any case, the conditions found at a certain place deserve special attention. For example, the availability of human capital resources is most important for finding a potential branch manager and thus for the opening of a new branch. The higher qualified a branch manager, the more independent from the headquarter (and regionally 'embedded') his decisions can be. Against this background, the fact that highly qualified people tend to leave small towns and underdeveloped regions in order to succeed in more prosperous regions has a considerable influence on the opening and operating of bank branches in these regions. That the role of local authorities should not be underestimated as well, can be illustrated by the statement of a

---

[100] Interviews conducted by the author in different headquarters in Budapest.

[101] In general, the opening of a bank branch is a rather time consuming process. On average one year before the actual opening takes place a bank manager from Budapest has to go to the respective place one to two times a week in order to do the necessary arrangements. Therefore, places which are not easy to reach have second priority.

branch network strategists: "When they are not friendly, we go to another place. Why to fight if there is an easier way?"[102]

This leads to the following conclusion: First, small and middle-sized banks founded with FDI could not come to Hungary and exclusively rely on standard-ised criteria for network development. Since the end of the 1990s, *they have to pay close attention to the regional and local contexts* in order get a better chance to succeed in the current stage of banking competition. This close attention is nec-essary in order to save costs in network development. Nevertheless, the profitabil-ity of a branch is never completely predictable because of the dynamic changes in the individuality of places. Secondly, it is the *activity and availability of local decision-makers and actors that can influence the opening of a bank branch con-siderably*. After the branch has opened, it is a local actor itself and therefore also contributes to the role of the respective place or administrative unit within the changing urban hierarchy and regional structure of Hungary.

Although the individuality of places and the relation between places are con-stantly changing,[103] these dynamics are especially high during political-economic transformation. The break up of the rigid hierarchy between different sizes of set-tlements characterising the regional structure of the centrally planned economy in Hungary, means a complete revision of the interrelation between places. In this process of shaping new identities of places, the *mediating role of local human and nonhuman actors (actants)*[104] can be regarded as another decisive factor besides the earlier identified influence of legacies from socialist times. In the case of bank branch network development these mediations include the commitment of the mayor, the availability of a convenient office building, the 'co-operation' of a communication line, the mobility of highly skilled people, industrial investments, the bankers' decisions and more.

Considering the branch network development of banks with FDI in 1998 and 1999 again (figures 10 and 11) the preference for the north-west of Hungary is most striking. This is best explained by the prevailing contact patterns and poten-tials for interaction that almost reversed for Hungary after the political transfor-mation towards the market economy.[105] In socialist times, the country was focused on economic collaboration with Russia and the other CCE countries. Funds were centrally gathered and transferred to the east, especially the north-east. As banking was not significant for the economy, these contact patterns did not manifest them-selves in banking, but in industrial sites (for example in Borsod-Abauj-Zemplén).

[102] Interview conducted in the headquarter of a foreign-owned bank in spring 1996.
[103] See, for example, MASSEY 1984.
[104] These terms are used according to actor-network theory (see, for example, LATOUR 1999). Based on a understanding of agency as the capacity to have effects, actor-net-work theory restores agency to things. Since the term actor in other contexts tends to be limited to humans, the term actants is used in order to refer to humans and nonhumans (objects, things) at the same time.
[105] See, for example, HAMILTON 1995, 74.

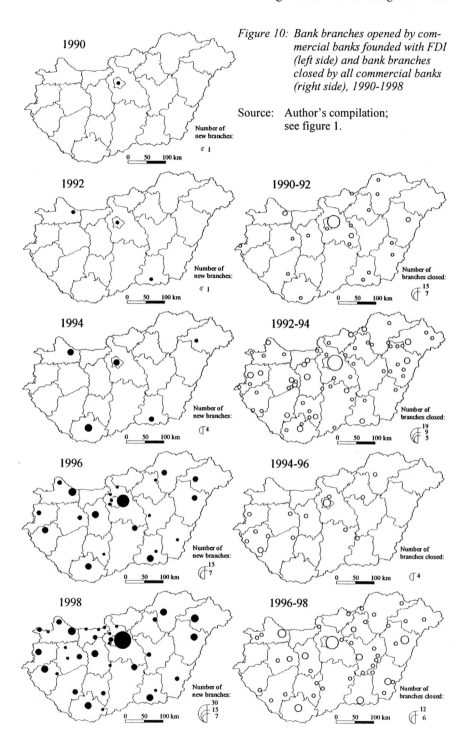

Figure 10: Bank branches opened by commercial banks founded with FDI (left side) and bank branches closed by all commercial banks (right side), 1990-1998

Source: Author's compilation; see figure 1.

With the turn towards the market economy and the wish to become a EU member state the new potential collaborators were found in Western Europe. Several foreign companies founded or took over plants in the western part of Hungary in order to use economic advantages in Hungary (salaries, taxes etc.) and short ways to the Western European market (for example in Győr-Sopron).[106] Thus, *changing contact patterns and potentials for interaction* make up the third aspect that strongly determines regional transformation of Hungarian banking.

Thinking of the foreign-owned banks' criteria for the opening of bank branches again, it becomes obvious that there are actually a few potential profitable branch locations outside the regional centres in the east, too, but these have been neglected by the banks so far (examples are Tokaj or Gyula). In general, interviewees stressed the comparably cost-saving procedure of opening branches in the eastern part of Hungary because of lower prices for real estate or lower salaries. Nevertheless, it seems that apart from economic disparities the *prevalent discourse about the 'depressed' and 'peripheral' east* makes the banks even more hesitant to open branches in locations other than the regional centres of eastern Hungary.

This is not the only way that *ideas* about different regions in Hungary influence the regional strategies of banks founded with FDI: JÖNS (1996) and JÖNS and KLAGGE (1997) pointed out that the branches of banks founded with FDI were found in all regional centres in 1996, but that they had different *functions*. Branches in the eastern county capitals concentrated on mobilising savings, while credit allocation was very restrictive in the eastern part of the country. Consequently, the banks transferred collected deposits via the headquarter in Budapest to branches in the Budapest agglomeration and northern Transdanubia in order to allocate credits to almost all kinds of companies. Assuming that two entrepreneurs, one located in Debrecen and the other in Győr, ask for credit in order to build up the same business that does not depend so much on a strategic location, the one in Győr would have a much greater chance to get this credit because the real differences in economic development are often passed on to the success of the respective business project, even if they would not matter much in these cases. As a result, the person in Debrecen would probably become mobile and move, for example, to Székesfehérvár in order to realise his project.

Thus, banks founded with FDI contributed to the deepening of uneven regional development in a twofold way: first, with opening those branches, which are founded outside the regional centres, mainly in small towns located in western Hungary, and secondly, with the division of labour just described - and generalised - between branches in different regions of Hungary. It is argued here that geographical imaginations reinforce real differences in economic development

---

[106] Examples are companies located in the agglomeration of Budapest such as those mentioned before or Opel in Szombathely or Philips in Veszprém.

Figure 11: Branch networks of selected foreign-owned banks, 1999

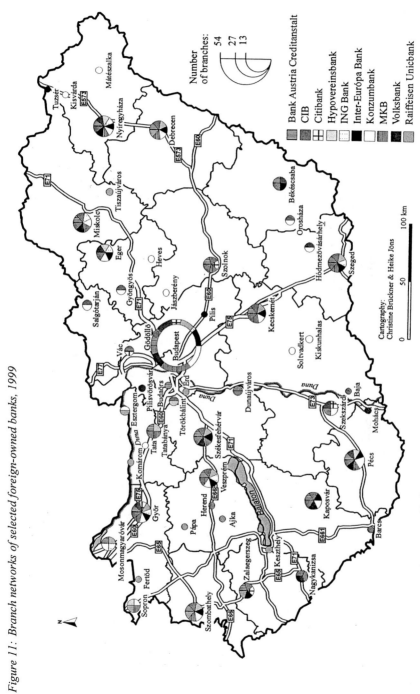

Source: See figure 1; author's compilation.

which in turn are responsible for both strategies. These results support John Agnew's argument that "people in the world use regional designations to make sense of the world and these draw on real differences between parts of the world but they cannot claim total fit to the world because they are based on ideas about regional differences that are not simply about those differences *per se* but also about ideas of how the world works."[107]

Finally, looking at the branch networks of the major retail banks with medium-sized and large networks that are not included in figure 11 (OTP, Postabank, K&H, ABN-AMRO Bank, Budapest Bank, Erste Bank), two more ways in which regional transformation has been mediated can be identified (figure 14): First, another contribution of *legacy* is expressed by the branch network of OTP that was inherited from socialist times. This is because there are still more branch locations occupied by commercial banks outside the regional centres in the eastern part of Hungary than would be expected from the behaviour of the banks founded with FDI. Nevertheless, OTP and other retail banks recently also started to encourage a deepening of the west-east disparities in access to banking by closing branches primarily in the eastern part of the country (figure 10). Other influences of legacy, even from pre-socialist times, are high population densities in the northern counties and lower urban densities in the eastern counties that add to the lower branch per population ratios in the eastern part of Hungary. Secondly, the influence of the banks' *agency* on regional transformation is expressed once more in the spatial *specialisation* of branches.[108] In the large bank branch networks in Hungary, the highest responsibility outside Budapest is centralised in four to eight regional directorates. With the decision to declare a branch location as regional directorate, the banks contribute to a further differentiation between the function of towns along the rings of centrality as regional centres. Other banks are mostly operating full-service branches and (specialised) sub-branches.[109] The hierarchy between branch locations within individual branch networks has been constantly changing during the 1990s in almost all banks because of ongoing changes in the individuality of places and their interrelations. In the long term, new sets of (still changing) relations that are different to socialist times will characterise the position and functional 'identity' of each place in the regional and urban hierarchy of the Hungarian market economy.

Thus, legacies, agency of several kinds of human and nonhuman actors, contact patterns and potentials for interaction as well as ideas about regional differences have been identified as determinants for uneven regional development in Hun-

---

[107] AGNEW 1999, 93. For the implications of this argumentation in regard to its critique of both realist and constructivist approaches, see AGNEW (1999).

[108] The spatial specialisation of bank branches is described by LEYSHON and THRIFT (1997, 217) as one of the major aspects of corporate restructuring in the UK during the 1990s.

[109] Postabank is an example for a four-tier system with regional directorates, full-service branches, agencies and post offices as external service units.

garian banking during the transformation process towards a market economy. Considering also the periods of network development identified before, the empirical investigations contributed to a better understanding of the opening and closing of branches (figure 10) as well as the distribution of commercial banking infrastructure in 1998 that can be regarded as the result of all kinds of network activity by commercial banks (figure 12). The most striking aspect of the spatial organisation of Hungarian bank branch networks in 1998 is the distinct west-east disparity regarding the population-per-branch ratio: The regional structure of Hungarian commercial banking was somehow polarised around the centre of Budapest (6 960) between Győr-Sopron, Vas, Zala and Somogy (ca. 6 700 to 7 300) along the western border region to Austria, Slovenia and Croatia on the one hand, and Borsod-Abáuj-Zemplén, Szabolcs-Szatmár and Hajdú-Bihar (ca. 10 600-11 700) in the north-east of Hungary, along the border to Slovakia, Ukraine and Romania, on the other hand. Nevertheless, the disparities between the different counties lowered between 1990 and 1998 because of the change in meaning of banking for society and the role of legacies that made all regions benefit from new branches. The change in meaning of banking for society resulted as much as the 'commodification' process described by HAMILTON (1999) and the earlier described mind-sets from the major change in *ideology* that became manifest in 1989/90. The investigations of this section have shown that all the different expressions of a comprehensive, ideology-based 're-evaluation of space' were very influential on the regional transformation of Hungarian banking.

Figure 12: Bank branches (1998) and population-per-branch ratios (1990-98)

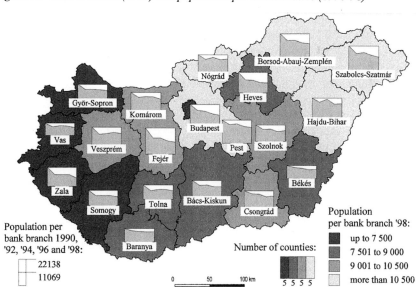

Source: See figures 1 and 9; author's compilation.

3.2.4   Alternative financial infrastructure and 'financial exclusion'

Comparing the total ratio of inhabitants per banking units in Hungary 1997 (3 227; including all types of credit institutions and branches of co-operatives) with the same ratio in market economies of Western Europe shows that *Hungary has reached the level of network density* in the EU countries after the first decade of transformation. In 1997, Hungary held a position below average within very heterogeneous ratios among the EU countries.[110] This fact underlines once more the importance of branches for the operation of Hungarian banks during the 1990s. Nevertheless, the geographies of banking showed an relatively high concentration of commercial bank branches on Budapest and the county capitals (figures 9 and 14). Only 13% of the 3 131 Hungarian municipalities (1998) had at least one bank branch or main office of a savings co-operative (figure 14). In some of the other 87% of municipalities branches of savings co-operatives or alternative financial infrastructure such as post offices and ATMs might have been available, but, however, figure 14 shows that there were immense regional disparities regarding the accessibility of banking units, the supply of high-quality services, and the opportunity to choose between different competing banks. The banking dilemma for people in rural areas can be illustrated by the fact that peasants have no chance to get credit from banks (land has only a low value), while the amount of money necessary for a new tractor or other investments often exceeds the credit limit of local savings co-operatives.

Although the ATM network was extended considerably between 1997-1999, it did not help to improve access to banking everywhere. The distribution of ATMs in 1999 showed once more distinct west-east disparities and an immense concentration in the regional centres, especially in the eastern counties (figure 13). In this respect, the card-accepting banking units were much more widespread due to the decentralised networks of the savings co-operatives that were operating about 60% of the Hungarian banking units in 1998. However, the availability of decentralised banking services does not guarantee the use of them.[111] First, people have to gain knowledge of it, secondly, they need to develop trust in banking transactions,[112] and thirdly, they have to get access to the respective services. While the first two aspects can only be achieved through face-to-face contacts, the latter requires a minimum amount of income or own funds. Thus, IT can only serve as a supple-

---

[110] The ratio of Hungary was similar to those in the UK (3 681), in Sweden (3 514) and Finland (3 101) (Author's calculation according to EUROPÄISCHE KOMMISSION 1999, and recent population statistics from the respective national statistics offices).

[111] For example, 25 banks and 144 savings co-operatives issued bankcards at the end of 1998, but the number of cards reached only 2.9 million (NATIONAL BANK OF HUNGARY 1999a, 14). This resembles less than one third of Hungary's population or the number of inhabitants in the nine largest Hungarian towns.

[112] Gaining trust in banking is one of the major tasks for people of the emerging market economy of Hungary (see page 23). For the general relation of virtual money and trust, see LEYSHON and THRIFT 1997, 30.

mentary service at this stage of the transformation process in Hungary. Neither for the customers nor for the banks' acquisition of clients can IT replace the traditional bank branch in the medium range, but, however, it will become a competitive advantage for banks in the long-term perspective.[113]

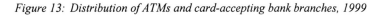

Figure 13: Distribution of ATMs and card-accepting bank branches, 1999

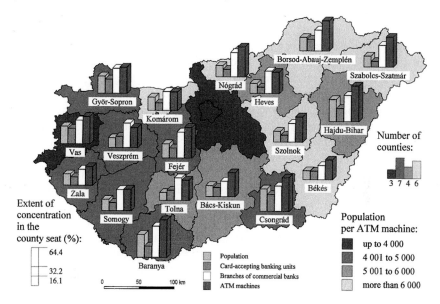

Source: NATIONAL BANK OF HUNGARY 1999c; see also figures 1 and 9; author's
          compilation.

Taking into consideration that retail services, in general, are still not as developed as in Western European countries, and that the market share of savings co-operatives was only 5% altogether (10-15% were reached in some products of the household business branch),[114] the danger of a two-tier banking community becomes quite obvious: one represented by predominately foreign-owned banks, offering sophisticated services for large and medium-sized companies as well as people of the higher social strata; the other consisting of Hungarian-owned co-operative credit institutions with rather low capitalisation, offering mainly basic

---

[113]   Since a demand for sophisticated banking services already exists among companies and
        people of the higher income strata, the banks introduced and substantially extended
        phone and home banking services, ATMs, small banking units and electronic branches
        in shopping centres at the end of the 1990s (NATIONAL BANK OF HUNGARY 1999a, 15).
[114]   SZÖKE 1998, 587.

banking services to the public:[115] "There are still no signs of banks, which intend to increase their shares in the retail market, wishing to open up towards lower-income strata. [...] In addition to the upper middle class, banks would prefer to see strata still belonging to the middle class having lower but regular incomes, among their clients."[116]

In Hungary, there is a danger of 'financial exclusion'[117] in many small towns, small villages and rural areas, especially in the eastern part of the country but also in parts of Transdanubia. Furthermore, there is a great inequality in access to banking within towns or communities between people with none or low income and those with at least medium income (especially, because local co-operative credit institutions are missing in bigger towns, and the banks are concentrating on people with medium and high income). According to LEYSHON and THRIFT (1995), financial exclusion could be countered by promoting "alternative financial infrastructure which can, in particular, supply the basic banking facilities and low-cost loans to low-income households."[118] For Hungary, this means in the first place a strengthening of co-operative credit institutions as the most important provider of finance in rural areas. That the sector of co-operative credit institutions is not only underdeveloped in regard to capitalisation can be illustrated by two more facts: first, there were much more co-operative credit institutions working in Hungary before World War II than today;[119] secondly, the Hungarian supply ratio regarding commercial banks (8 858 inhabitants per branch) was even better than the same ratio of Germany in 1997 (9 931), while the respective ratio regarding co-operative credit institutions was more than twice as high in Hungary (5 077 versus 2 130 in Germany).[120]

Some other strategies for avoiding further economic and social marginalisation of people in places threatened by 'financial exclusion' were already implemented by the government; on the one hand by founding a regional development bank in Miskolc, and on the other hand by introducing different project-related state funds through the Hungarian Development Bank and the Eximbank.[121] The breaking up of prejudiced ideas about the eastern part of Hungary 'as a whole', and trying to keep motivated people in the region and to support their activities could be helpful for lessening the deepening of uneven regional development, too.

---

[115] The metaphor of the two-tier economy was adopted from JONES (1999) who applied it to the whole Hungarian economy.
[116] NATIONAL BANK OF HUNGARY (1999a, 14).
[117] LEYSHON and THRIFT (1995, 341) defined this term as "processes that prevent poor and disadvantaged social groups from gaining access to the financial system."
[118] LEYSHON and THRIFT 1995, 336. Examples for this infrastructure are post offices, community development banks, credit unions or non-financial retail outlets.
[119] ILLÉS 1994, 168.
[120] Author's compilation according to DEUTSCHE BUNDESBANK (1999).
[121] See JÖNS 1996. The constructions of both state-owned banks are available in almost all commercial banks and savings co-operatives.

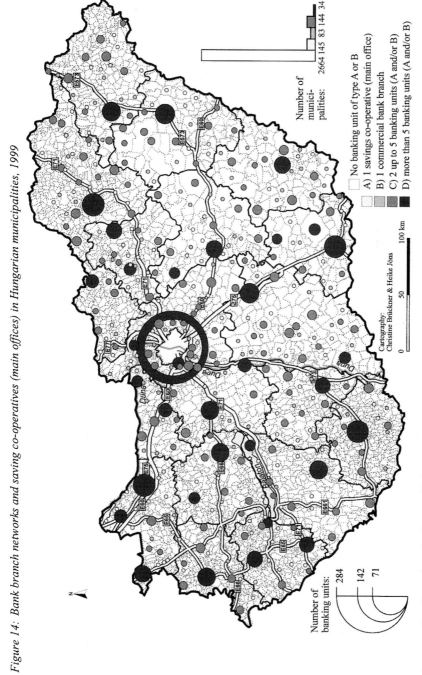

Figure 14: Bank branch networks and saving co-operatives (main offices) in Hungarian municipalities, 1999

Source: See figure 1; author's compilation.

For the next decade, two main developments for the spatial organisation of Hungarian banking (and thus the operation of the banking sector) can be expected:

- the founding of foreign bank branches as independent institutions with a preference for branch locations in the western border region,[122]
- a period of network stagnation and reduction after the tough competition is over; these prospects emphasise the important role of co-operative credit institutions and thus local initiatives for future banking in Hungary.

4   Conclusion and regional transformation theory

Examining the development of the Hungarian banking system between 1987 and 1999 reveals a bundle of complex interrelated dynamic changes regarding legislation, institutions and the banks' business and branch network strategies. Thus, the *first* result is a comprehensive periodisation of different developments related with Hungarian banking in the first decade of political-economic shift from a centrally planned towards a Western European style capitalist market economy (figure 15). Compared with preceding decades, Hungarian banking experienced extraordinary instabilities in time and space during this period. The main reason for this instability was the major ideological change that became manifest in 1989/90. This transformed the meaning of banking within Hungarian society. In some respects, the preliminary 'results' of these major changes seem very often to be comparable with the state of affairs at the 'starting point', although the qualities of the respective state of affairs are completely different (figure 15).

The *second* result is that one cannot understand Hungarian banking in the 1990s without examining the spatial organisation of branch networks as well as the significance of the individuality of place for the opening, operating and closing of banking units. The basic principle of this finding was already pointed out by MASSEY (1984) in regard to the spatial organisation of production: "It is not, then, just a question of mapping social relations (economic, sociological or whatever) *on to* space. The fact that these relations *occur over* space matters. It is not just that 'space is socially constructed' [...] but that social processes are conducted over space."[123] Furthermore, understanding local particularities and their interrelations within capitalist production is central for the new regional geography outlined by MASSEY (1984). LEYSHON and THRIFT (1997, 189) pointed in a similar direction when they stated in regard to banking that "each monetary network takes

---

[122] The founding of branches by foreign banks without having a subsidiary bank in Hungary was allowed in 1998, but it has not happened yet. It can be expected to happen when the legal status of these branches will have to be harmonised with EU standards (right now, they are basically handled like subsidiaries) (NATIONAL BANK OF HUNGARY 1999b, 93).

[123] MASSEY 1984, 56.

up space and space is constitutive of each network." The essential interdependencies between space, place and banking in Hungary during the 1990s, which came to light in this article, can be summed up as follows:

Figure 15: Transformation in time and space: the case of Hungarian banking

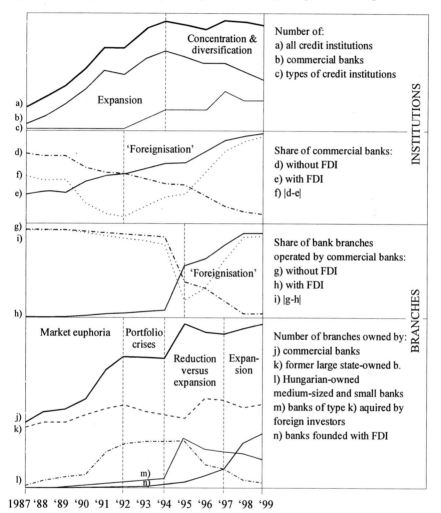

1987 '88 '89 '90 '91 '92 '93 '94 '95 '96 '97 '98 '99

Source: Author's design.

- The major shift in the business strategies of several commercial banks at the end of the 1990s, which meant an opening towards so far less attractive segments of the market, did not simply *lead* to different branch network strategies. They *resulted* from the interrelation of an increasing number of commercial banks, growing competition, *and* the spatial organisation of branch networks at that time. In 1996-97, the competing banks found themselves all seeking for the same customers in the same places. One way of looking for new business opportunities was to go to other places, and thus they had to suit the business activities to the spatially less concentrated customers.

- At the current stage of transformation, marked by high competition, most commercial banks have to branch out in other places than those that are most likely profitable branch locations in order to get a competitive advantage for acquiring new clients. Thus, creativity, local knowledge and a higher amount of risk replaced drawing-board planning as main branch network strategy for most Hungarian banks at the dawn of the 21st century.

- The bank branch network development is characterised by waves of network expansion and reduction. These dynamics depend on the banks' ongoing conflict between cost reduction and client acquisition, and the dynamic individuality of place. The latter is characterised by constantly changing relations of different actors (actants) and makes the profitability of a branch location not completely predictable. Banks have to *try out* through business *practice* whether or not a place is profitable enough to maintain a branch. Thus, bank branch network development can to a certain extent be regarded as a process of trial and error. These dynamics, which are probably characteristic for banking in market economies, seem to be exceptionally high during a period of political-economic transformation such as the one monitored in Hungary because interrelations between places are completely reorganised.

- Traditional bank branches are indispensable for client acquisition in current political-economic transformation in Hungary. People have to learn to cope with the principles of the market, including the opportunities, function, value and limits of banking transactions, and this can only be achieved by establishing branches as places of face-to-face contacts. Currently, IT serves as an additional intermediary, but it will gain growing importance in the future.

- In the first decade after the political transformation in Hungary, commercial banks founded with FDI contributed to a deepening of uneven regional development in a twofold way: first, with opening those branches, which are founded outside the regional centres, mainly in small towns located in western Hungary, and secondly, with a spatial specialisation of branches in different regions of Hungary. Thus, there is a danger of 'financial exclusion' in many small towns, small villages and rural areas of Hungary, especially in the eastern part of the country. Furthermore, there is a great inequality in access to banking within towns or communities between people with none or low income and those with at least medium income.

- The ongoing branch network expansion among commercial banks at the end of the 1990s is forced by tough competition for clients. Currently, banks are *dependent on people and enterprises* as potential customers. A period of network reduction can be expected when the run on clients will be decided, competition will have slightly decreased, people will have got acquainted with banking transactions and related business opportunities, and banks have tried out the value of different branch locations. Then, *people will be dependent on the access to banking infrastructure*, so that a reduction of costly branches can be 'dared' from the banks' views. In view of these prospects and current processes that can be characterised as 'financial exclusion', the strengthening of the underdeveloped co-operative banking sector should be an important task in Hungary for the near future.

In order to understand these processes theoretically, a combination of MASSEY's (1999, 28) understanding of space as 'a product of interrelations', as 'the sphere of the possibility of the existence of multiplicity' and as being 'always in a process of becoming', and the concept of actor-network theory as outlined by Bruno LATOUR (1999) and others can be very helpful.[124] Since the latter regards the world as a complex web of strongly connected human and nonhuman actors that are constantly involved in network building processes, regional transformation can be understood as the result of a collective of place-specific actor-networks that are continuously changing through events under the influence of either a few or complex bundles of mediators. From this perspective, sets of complex associations between the different mediators are responsible for the interlinking of places.

As mentioned above, the interrelations between mediators and thus between places were completely rearranged in Hungary during the 1990s because of the change in ideology and the new role of banking. Which mediators, or - in terms of actor-network theory - 'actants', were responsible for regional transformation of Hungarian banking during the 1990s and thus for the deepening of west-east disparities in banking infrastructure is revealed in this article (figure 16). The most important were such heterogeneous aspects as the agglomeration of potential customers, the degree of population densities, legacies of socialist and also pre-socialist times, contact patterns and potentials for interaction, appropriate legislation, the commitment of the mayor and other decision-makers of local authorities, the availability of a convenient office building, the co-operation of a save communication line, the accessibility of a branch location from Budapest, the availability and mobility of highly skilled people or the presence of competing banks. Also responsible in the complex web of mediators were geographical imaginations about Hungary in which existing differences in regional development were reinforced by potentials for economic interaction and an ideology-based differentiation and evaluation of recent and former contact patterns. These geographical imaginations resulted from a re-evaluation of space under the influence of market-orientated criteria and the political-economic orientation towards EU countries.[125]

---

[124] For the application of actor-network theory to geographies of finance in other contexts, see THRIFT 1996 and LEYSHON 1997.

[125] See HAMILTON 1999.

As one of their effects, they supported the banks' hesitation to open new branches in other locations than the regional centres of eastern Hungary, as well as the introduction of regionally specialised branches. At the same time, legacies regarding contacts and potentials for interaction with the adjoining eastern countries (Russia, Ukraine, Romania) have been underestimated. It can also be assumed that the melange of facts and fictions contributed to migration from the eastern regions to Budapest or 'the west' which in turn led to a lack of human capital resources in economically less attractive regions. In this way, important links within place-specific actor-networks were eliminated. Links that would be necessary for the banks' decision to open a bank branch there. Thus, the mediations of human actors, institutional actors, nonhuman actors as constituted, for example, by legacies or local infrastructure, contact patterns, potentials for interaction and geographical imaginations were mingled in the process of regional transformation of Hungarian banking during the 1990s (figure 16). For each bank branch it would be possible to follow a chain of associations between the specific actants or mediators that are responsible for its existence. One missing or disappearing link could already prevent the establishment of a new branch or force the closure of an existing branch.

However, granting ideas a capacity to have effects would mean that according to actor-network theory they are actants themselves, but the fact that actor-network theory only distinguishes between human and nonhuman actors already indicates that this would have broad implications. Without referring to actor-network theory, AGNEW (1999) suggested that "[r]egions both reflect differences in the world and ideas about differences."[126] While he discusses general philosophical consequences of this concept,[127] which was supported by the findings of this study, the implications of granting ideas the capacity to act for actor-network theory are to be exemplified in the following in order to apply the resulting concept to regional transformation theory: The main argument is that actor network theorists have failed to consistently disentangle all types of mediators that are responsible for network building processes since they left out ideas in the concept of humans and nonhumans. According to actor-network theory,[128] the involvement of nonhuman actors in human interaction and the exchange of properties between human and nonhuman actors enables the stabilisation of social relations and thus the formation of the collective we are living in.[129] This argument leads to the theory's 'symmetrical principle' between human actors on the one side and nonhuman actors on the other side with both kinds of (hybrid) actors having the poten-

---

[126]   AGNEW 1999, 93.
[127]   The main philosophical consequence of conceptualising regions this way is a critic of both realist and constructivist approaches (see AGNEW 1999).
[128]   The technical terms such as 'actants', 'human actors', 'nonhuman actors', 'collective', 'historicity' and 'hybrid' are used according to actor-network theory (LATOUR 1999).
[129]   See LATOUR 1999, 198.

tial to act (figure 17a).[130] However, in the final analysis this conceptual symmetry of actor-network theory is not consistent since what is referred to as humans in actor-network theory means much more than what counts as nonhumans: On the one hand, nonhumans (also named quasi-objects, inscriptions or immutable mobiles) seem always to belong, at least partly, to the material world.[131] On the other hand, humans show all the attributes of nonhumans as well, especially the (socio)materiality that is vividly expressed in the human body, but they are also able to deal with ideas that can be regarded as immaterial pendant to nonhumans. It can be argued that disentangling what is happening on the side of human beings would be necessary for achieving full understanding of network building processes (such as involved in regional transformation), and that the complex attributes and skills associated with humans such as body, brain, mind, spirit, language, knowledge or culture facilitate competencies that are always superior to those of nonhumans. As a result, much more is happening on the side of the humans, and their mediation appears to be indispensable for "socializing nonhumans to bear upon the human collective."[132] What follows from this is an unequally distributed power-balance between human and nonhuman actors, and thus an asymmetrical relationship (figure 17b).

Nevertheless, it seems to be possible to establish a real conceptual symmetry within actor-network theory taking into consideration the mediating role of what I called 'suprahuman actors' in figure 16: individual experiences, stereotypes, mind-sets, language and all kinds of other stored information, ideas, images and thoughts that strongly influence human agency and thus are actants themselves. This introduction of 'suprahuman actors' establishes a consistent symmetrical principle centred around the most hybrid and dynamic entities known so far, i.e. human beings (figure 17c). Understanding human actors as being entities in which suprahumans and nonhumans are tied together in a very dynamic way[133] and that actively tie together other actants, it becomes obvious that their competencies, because of the combinational power, are ultimately greater than those of singular suprahumans or nonhumans. Nevertheless it is important to stress that within the suggested concept all three types of actants still have the same responsibility for action, and can all be hybrids in regard to their historicity.[134]

---

[130] See, for example, LATOUR 1999, 182; MURDOCH 1997, 331.

[131] See, for example, LATOUR 1993, 79, 138.

[132] LATOUR 1999, 296.

[133] There are, for example, no inherited or 'new' human actors thinkable (figure 16); humans just have more links with either inherited or new non- and/or suprahumans.

[134] This argument is crucial because it strongly supports the most important achievements of actor-network theory. For actants' responsibility for action, see LATOUR 1999, 183. The notions of sociomaterial hybridity and the historicity of things are developed in LATOUR 1999, chapters 5 and 6, see especially 212-214.

Figure 16: Understanding regional transformation from the perspective of banking

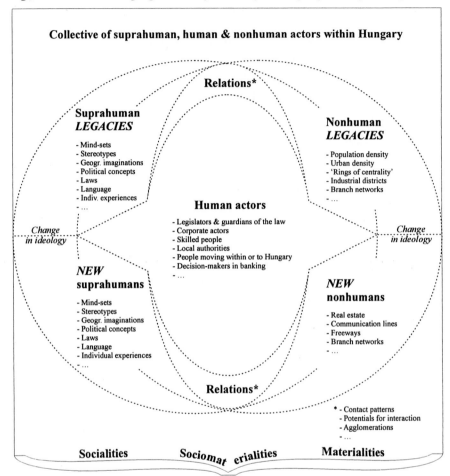

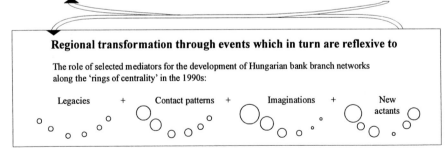

Source: Author's design.

*Figure 17:   Extending actor-network theory by introducing a third type of actant and a
               different understanding of humans*

a) The symmetry of actants within actor-network theory according to LATOUR (1999)

b) The asymmetry of actants within actor-network theory implicit in LATOUR (1999)

c) A new concept of actants within actor-network theory: the perspective of humans

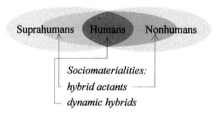

Source: Author's design.

Thus, applying actor-network theory to regional transformation in Hungarian
banking reveals two major inconsistencies in the theory. First, there is a misinter-
pretation of the role of humans (and certain nonhumans) due to an insufficient
concept of hybridity; and secondly, there is a suppression of the immaterial coun-
terpoint to (socio)material nonhumans. My main argument is therefore that inter-
actions cannot only be stabilised and extended through time and space by the use
of matter, but through the involvement of memory, too. The contribution of what I
called 'suprahuman actors' to interaction in general or regional transformation in
particular can be as important as that of nonhuman and human actors. Nevertheless, the involvement of human actors is indispensable for forming actor-networks
such as bank branch networks. Only in this sense, do humans differ from the other
actants.[135]

---

[135] For the full background and implications of this reformulation of actor-network theory,
       see JÖNS (forthcoming).

In 2000, the mediation of human, nonhuman and suprahuman actors resulted in a rather polarised regional structure of Hungary between the north-west and the north-east with Budapest as the dominant centre in the middle. Actually, the new sets of relations between the millions of unique places or 3 131 administrative units in Hungary are much more complex, of course. They are constantly shaped in the - endless but becoming less dynamic - 'construction' of a new urban hierarchy and regional structure of the emerging market economy of Hungary. This construction is the expression of the sum of actions performed by human, nonhuman and suprahuman actors. Influencing the mind-sets about the eastern part of Hungary by stressing positive aspects of this regions such as future potentials of interaction along the eastern border region could be one way of contributing to a better business environment and more bank branches outside the regional centres in this region.

Some of the phenomena identified in this article as decisive for uneven regional development in Hungarian banking during the transformation process towards a market economy, have been addressed in the context of transformation by other authors as well.[136] Nevertheless, the relation of these factors are hardly discussed, and the active participation of mind-sets seems to be underestimated. Prevalent ideas about the eastern part of Hungary seem to reinforce existing differences in 'place-specific factor endowments'[137] considerably. The value of the proposed concept of human, nonhuman and suprahuman actors presented in figures 16 and 17 can be illustrated by the fact that the "key factors in uneven regional development of post-socialist Europe," summed up by DINGSDALE (1999, 148), can be 'reduced' or 'abstracted' to the mediation of these three types of actants [main responsibilities are indicated in brackets]:

- western capital and ideas imported through direct foreign investment; [humans, suprahumans and relations, especially mind-sets and potentials for interaction]
- corporate location strategy; [humans, nonhuman legacies and suprahumans]
- sectoral and enterprise restructuring; [humans, nonhuman legacies and suprahumans]
- the slowness of governments to formulate regulatory policy and regional policy; [human actors]
- the local impact of indigenous private and public entrepreneurship; [human actors]

---

[136] See, for example, HAMILTON 1999. The role of legacy from pre-socialist and socialist times is among the best explored phenomena in regional transformation theory (see, for example, FAßMANN 1999, 16; HAMILTON 1999, 140).

[137] This term is used by HAMILTON (1999, 140).

- deterrence of environmental legacy; [nonhuman legacies and suprahumans, i.e. action takes place somewhere else][138]
- tourism and transport; [nonhuman legacies, new nonhumans, human actors]
- economically motivated cross border co-operation. [suprahuman legacies, new suprahumans, human actors]

Thus, the empirical findings of this article seem to fit into existing reflections on regional transformation theory, if somewhat slightly modifying and reorganising them. Furthermore, these investigations seem to have revealed and dissolved an asymmetry implicit in actor-network theory. The introduction of 'suprahuman actors' established a consistent concept of actants centred around human beings as the most hybrid and dynamic entities known so far. The author suspects that the actants identified as decisive for understanding regional transformation of banking could be valuable for understanding other aspects of regional transformation as well. The best way to check this assumption would be further research into regional transformation of banking in other post-socialist countries or of other social dimensions in Hungary that is based on the proposed concept of actants.

References

AGNEW, J. (1999): Regions on the Mind does not Equal Regions of the Mind. Progress in Human Geography 23 (1), 91-96

ASCHAUER, W. (1995): Bedeutung und regionale Verteilung von Joint-Ventures in Ungarn. In: MEUSBURGER, P., KLINGER, A. (eds): Vom Plan zum Markt: Eine Untersuchung am Beispiel Ungarns. Heidelberg

BALASSA, Á. (1996): Restructuring and Recent Situation of the Hungarian Banking Sector. Budapest (= NBH Workshop Studies 4)

BOGNÁR, K., FORGÁCS, A. (1994): National Bank of Hungary: 70 Years of Existence (1924-1994). Budapest

BOTOS, K. (1998): The Status and Prospects of the Hungarian Banking System. In: VORST, K. S., WEHMEYER, W. (eds): Financial Market Restructuring in Selected Central European Countries. Aldershot, 83-101

BUCH, C. M. (1997): Opening Up for Foreign Banks: How Central and Eastern Europe Can Benefit. Economics of Transition 5 (2), 339-366

BUSCH, U., WEISIGK, D. (1997): Das Engagement ausländischer Geschäftsbanken in Ungarn. Bank-Archiv (11), 901-910

---

[138] Deterrent industrial sites were found in all parts of Hungary at the end of the 1980s. It can be suspected that the suprahuman 'results' of the re-evaluation of space such as the discourse about the 'depressed' East, and the new contact patterns made a difference in the quality of deterrence in different parts of the country.

CSÉFALVAY, Z. (1994): The Regional Differentiation of the Hungarian Economy in Transition. GeoJournal 32 (4), 351-361

DEUTSCHE BUNDESBANK (ed.) (1999): Bankenstatistik Dezember 1999. Wiesbaden (= Statistisches Beiheft zum Monatsbericht 1)

DINGSDALE, A. (1999): New Geographies of Post-Socialist Europe. The Geographical Journal 165 (2), 145-153

ERDEI, T. (1998): Circumstances and Prospects. In: KEREKES, G. (ed.): Hungarian Financial and Stock Exchange Almanac 1997-1998. Budapest, 117-119

EUROPÄISCHE KOMMISSION (ed.) (1999): Banken in Europa: Daten 1994-1997. Luxemburg (Panorama der Europäischen Union)

FAßMANN, H. (1999): Regionale Transformationsforschung: Konzeptionen und empirische Befunde. In: Pütz, R. (ed.): Ostmitteleuropa im Umbruch: Wirtschafts- und sozialgeographische Aspekte der Transformation. Mainz (= Mainzer Kontaktstudium Geographie 5), 11-20

HAMILTON, F. E. I. (1995): Re-evaluating Space: Locational Change and Adjustment in Eastern and Central Europe. Geographische Zeitschrift 82 (2), 67-86

HAMILTON, F. E. I. (1999): Transformation and Space in Central and Eastern Europe. The Geographical Journal 165 (2), 135-144

HUNGARIAN BANKING AND CAPITAL MARKET SUPERVISION (ed.) (1998): Annual Report 1997. Budapest

ILLÉS, I. (1994): The Banking System and Regional Development in Hungary. In: HAJDÚ, Z., HORVÁTH, G. (eds): European Challenges and Hungarian Responses in Regional Policy. Pécs, 167-171

JONES, C. (1999): Foreign Affairs. The Banker 149 (881), 41-43

JÖNS, H. (1996): Bankwesen und Regionalstruktur beim Übergang von der Plan- zur Marktwirtschaft in Ungarn: Eine wirtschaftsgeographische Analyse der Transformation des ungarischen Bankwesens zwischen 1980 und 1996. Heidelberg (Diplomarbeit, unpublished)

JÖNS, H. (forthcoming): Räumliche Mobilität und grenzüberschreitende Kooperation in den Wissenschaften: Deutschlandaufenthalte US-amerikanischer Humboldt-Forschungspreisträger aus einer erweiterten Akteursnetzwerkperspektive. Heidelberg

JÖNS, H., KLAGGE, B. (1997): Bankwesen und Regionalstruktur in Ungarn: Eine Analyse der Filialnetzstrukturen und -strategien aus regionalwirtschaftlicher Perspektive. Wien (= ISR-Forschungsberichte 16)

KEREKES, G. (ed.) (1990) [1991, 1992, 1994, 1995]: Magyar Pénzügyi és Töszdei Almanach 1990 [1991, 1992-1993, 1993-1994, 1994-1995]. Budapest

KEREKES, G. (ed.) (1996) [1998]: Hungarian Financial and Stock Exchange Almanac 1995-1996 [1997-1998]. Budapest

KLAGGE, B. (1995): Strukturwandel im Bankwesen und regionalwirtschaftliche Implikationen: Konzeptionelle Ansätze und empirische Befunde. Erdkunde 49, 285-304

KLAGGE, B. (1997a): Ausländische Direktinvestitionstätigkeit im ungarischen und tschechischen Bankensektor: Empirische Analysen und Erklärung der Differenzen. Bank-Archiv 45 (8), 597-605

KLAGGE, B. (1997b): Internationalisierung des Bankwesens in Osteuropa: Die ausländischen Direktinvestitionstätigkeit im ungarischen und tschechischen Bankensektor im Spannungsfeld zwischen nationalen Bedingungen und der internationalen Niederlassungspolitik multinationaler Banken. Münster (= Wirtschaftsgeographie 12)

KSH (ed.) (1998): A Magyar Köztársaság Helységnévkönyve: 1998, január 1. Budapest

LATOUR, B. (1993): We Have Never Been Modern. Cambridge

LATOUR, B. (1999): Pandora's Hope: Essays on the Reality of Science Studies. Cambridge, London

LENGYEL, I. (1994): The Hungarian Banking System in Transition. GeoJournal 32 (4), 381-391

LENGYEL, I. (1995): From Plan to Market: The Case of the Hungarian Banking System. In: TYKKYLÄINEN, M. (ed.): Local and Regional Development During the 1990s Transition in Eastern Europe. Aldershot, Brookfield, Hong Kong, Singapore, Sydney, 109-118

LEYSHON, A. (1997): Geographies of Money and Finance II. Progress in Human Geography 21 (3), 381-392

LEYSHON, A., THRIFT, N. (1995): Geographies of Financial Exclusion: Financial Abandonment in Britain and the United States. Transactions of the Institute of British Geographers NS 20, 312-341

LEYSHON, A., THRIFT, N. (1997): Money/Space: Geographies of Monetary Transformation. London, New York

MASSEY, D. (1984): Spatial Divisions of Labor: Social Structures and the Geography of Production. New York

MASSEY, D. (1999): Power-Geometries and the Politics of Space-Time: Hettner-Lecture 1998. Heidelberg

MEUSBURGER, P. (1995): Spatial Disparities of Labour Markets in Centrally Planned and Free Market Economies: A Comparison Between Hungary and Austria in the Early 1980's. In: FLÜCHTER, W. (ed.): Japan and Central Europe Restructuring: Geographical Aspects of Socio-Economic, Urban, and Regional Development. Wiesbaden, 67-82

MURDOCH, J. (1997): Towards a Geography of Heterogeneous Associations. Progress in Human Geography 21 (3), 321-337

NATIONAL BANK OF HUNGARY (ed.) (1991) [1994, 1998, 1999b]: Annual Report 1990 [1993, 1997, 1998]. Budapest

NATIONAL BANK OF HUNGARY (ed.) (1999a): The Hungarian Banking Sector: Developments in 1998. Budapest

NATIONAL BANK OF HUNGARY (ed.) (1999c): The Payment Card Business in Hungary: Based on Figures for the First Half of 1999. Budapest

STATE BANKING SUPERVISION (ed.) (1991a): New Banking Act in Hungary I. December 1, 1991. Budapest

STATE BANKING SUPERVISION (ed.) (1991b): New Banking Act in Hungary II. December 1, 1991. Budapest

STATE BANKING SUPERVISION (ed.) (1997): New Legislation on Financial and Capital Market in Hungary 1997. Budapest

SZELÉNYI, E., URSPRUNG, J. (1998): The Hungarian Two-Tier Banking System: The First Eleven Years (1987-1997). Budapest

SZÖKE, A. (1998): The Rural Bank: Past, Present and Future of Savings Co-operatives. In: KEREKES, G. (ed.): Hungarian Financial and Stock Exchange Almanac 1997-1998. Volume I. Budapest, 584-587

THRIFT, N. (1996): Spatial Formations. London

WARDENGA, U. (ed.) (2000): Transformationsforschung: Stand und Perspektiven: Beiträge einer wissenschaftlichen Tagung. Europa Regional 8 (3/4). Leipzig

WASS VON CZEGE, A. (1987): Wirtschaftsentwicklung und -system in Ungarn. Südosteuropa. Zeitschrift für Gegenwartsforschung 36 (7/8), 371-430

International Competitiveness: Impacts of Foreign Direct Investment in Hungary and Other Central and East European Countries

Gábor Hunya

1 Introduction[1]

A comparison of the per capita GDP levels in 1999 and 1990 reveals that three of the first-group EU accession countries - Poland, Hungary and Slovenia - surpassed their 1990 GDP levels, thus overcoming the shock of transformation. Other countries are still below the 1990 level. The Czech Republic, for example, experienced a setback in economic growth in 1997-1999, following higher growth rates in the mid-1990s. Estonia suffered from a strong transformational recession in the early 1990s due to the separation from the Soviet economic system and, more lately, from the Russian crisis in 1998-1999.

Hungary embarked on an impressive economic growth path in 1997 with GDP growth rates of 4.5-5% in the first three years and about 5.5% in the year 2000. Economic growth is sustainable as it has not lead to a deterioration of the current account or to an increase of indebtedness. The most specific features of the Hungarian growth path is that it relies on a massive increase of output, labour productivity and exports in the manufacturing sector. The competitive position of the country has substantially improved on the EU market and the composition of exports shifted to skill intensive products.

More than in any other country in Central and Eastern Europe the most important positions in the Hungarian economy are held by foreign subsidiaries. Output growth and structural change in the manufacturing sector is predominantly carried by the subsidiaries of multinational companies established in Hungary during the 1990s. The aim of this paper is to present evidence for the relative uniqueness of the Hungarian growth path in comparison with other transition countries. Hungary, Poland and Slovenia, the three countries with stable economic growth for the last few years, and good prospects, show very different levels of attractiveness to foreign direct investment (FDI). In Hungary, growth has been primarily due to the success of export-oriented foreign investment projects. In Slovenia, growth was related to a high degree of integration into European networks, mainly not

[1] Research underlying this paper was partly undertaken with the support from the European Union's Phare ACE Programme 1997 (P97-8112-R). The content of this publication is the sole responsibility of the author and it in no way represents the view of the EU Commission or its services.

through foreign direct investment. Polish growth was predominantly led by domestic demand, generating increasing imports but less exports, the trade gap being financed by both foreign direct investment and loans.

The theoretical framework for the analysis is provided by the concept of international competitiveness. International competitiveness of countries is understood as the performance of industries and firms on world markets. The aim of this paper is to find out how the inflow of foreign direct investment and the performance of foreign affiliates influence the international competitiveness of industries in Central European EU-accession countries. The analysis focuses on five transition countries: Estonia, the Czech Republic, Hungary, Poland and Slovenia (CEEC-5). These are the most advanced among the transition countries in terms of per capita GDP, penetration by foreign direct investment and economic transformation. They have association agreements with the EU which basically means free trade for non-food manufactured goods and the possibility to join the EU. In 1998, these countries started accession negotiations ahead of other candidate countries. Although the term 'first-wave accession candidates' is no longer used, since negotiations have started also with the other associated countries, the gap between the two groups persists.

Competitiveness between countries is discussed by looking at the ability to attract foreign direct investment. This paper shows why countries differ in terms of attractiveness towards foreign investment and with regard to the intensity of FDI penetration and structural change in manufacturing. It analyses the superior productivity of the foreign sector and the international competitiveness by looking at market shares in the European Union. The contribution of foreign direct investment to earnings is shown by the different profit rates in the foreign and the domestic sector.

2   Competitiveness of industries and countries – the role of foreign direct investment (FDI)

Competitiveness of countries (nations) as defined by TRABOLD (1995, 182) includes the ability to sell, the ability to attract and the ability to adjust - all of these factors leading to the ability to earn. These components can be measured by specific economic indicators and brought into relationship with foreign direct investment and the performance of foreign affiliates in a country.

- The ability to sell in terms of international competitiveness means the ability to export. The market shares on the main export markets and their development can be taken as the basic indicators of international competitiveness.
- The ability to attract refers to attracting activities and investments from abroad. Attractiveness for foreign investment is the summary effect of location factors in the country. Although other forms of international capital flows

can also be important, a basic indicator of attractiveness is the size of annual inflows of foreign direct investment (FDI) and FDI stocks. The share of the foreign sector shows the degree of FDI penetration, and the importance of the foreign sector in the economy. In this paper, this aspect will be measured by various indicators, such as assets, employment, sales, exports and investments.

- The ability to adjust can be measured by the speed of structural change. Through structural change the country changes its product and export specialisation in order to increase its capacity to earn. Structural upgrading means a shift to higher value-added, higher technology products which generally allow for higher earnings.

- The ability to earn is shown by the per capita level and increase of GDP. GDP growth, when compared with that of other countries, expresses whether a country is catching up or falling behind. The ability to earn is less specified at the industry or company level. Value added does not function as a real success indicator. It is rather the profit rate of the industry or of the company that can be used as a success indicator. The long term perspective of both country-level GDP and industry- or firm-level profits can be increased by innovation, adaptation and learning. These skills can also be imported, most rapidly through foreign direct investment.

The link between firm-level and country-level competitiveness was established by PORTER (1990). He argues that industries and companies can be competitive if the national environment and government policy supports companies' profit-earning and innovative efforts. Firm-level competitiveness depends on production factor costs, demand conditions, firm strategy and firm networking (clusters). The environment in which the firm operates is shaped by government policies, chances, opportunities and the international business environment. Internationalisation of markets opens up new opportunities for firms and leads to alliances, among them foreign direct investment. It demands that governments set policy targets and use policy tools in an internationally competitive environment partially regulated by multinational agreements.

Foreign direct investment can be understood as a competitiveness factor in two senses - as an indicator and as a factor of competitiveness. The approach of TRABOLD (1995) is limited to the indicator function: the level of foreign direct investment in a country expresses its competitiveness as a business location. In the approach of PORTER (1990) and DUNNING (1993), international production itself appears as a primary factor of international competitiveness.

Direct investments increase a country's competitiveness in several ways. The impact appears primarily at the company level and can be identified also at the industry level. Foreign investors bring knowledge, technology, investment and access to new markets and thus upgrade the competitive advantage of companies and industries. Foreign multinationals integrate host country firms into international networks where companies join efforts to support their competitive positions. Foreign direct investment is thus increasing the ability to sell. The speciali-sation of foreign affiliates can be different from the domestic ones and therefore shift the production structure. Through technology inflow and market access, for-eign direct investment increases the ability to adjust to market developments and technological change. Foreign direct investment, however, has to be supported by endogenous knowledge resources, which are not ubiquitously available (see chapter by Peter Meusburger).

A country has either absolute cost advantage and/or relative factor cost endowment advantage, which can be brought to effective use (in internationally competitive terms) through foreign direct investment. Foreign direct investment can increase the allocative efficiency in a country by improving the distribution of production and investment among industries. It can be of a comparative-advan-tage-augmenting type pointing out that cost-advantage-seeking foreign direct investment goes into those manufacturing industries for which the target country has superior factor endowments and thus upgrades the host country's comparative advantage (OZAWA 1992; MEYER 1995). At the microeconomic level, the indus-trial efficiency impact of foreign direct investment can be proven. The targeted firm gets access to the technological, organisational and managerial skills con-centrated in multinational enterprises. Future economic growth will be influenced by the pace and scope of technology transfer of foreign investors and by spill-over effects of foreign direct investment to the domestic firms of the target country. Both depend to a large extent on the capabilities of the host country. Countries with little FDI penetration may fall back in economic development if domestic firms are too weak. Spill-over is the spread of knowledge from superior foreign companies to domestic companies. The speed and intensity of spill-over can be increased by networking and other forms of learning.

3   Competitiveness in terms of attracting FDI

The CEEC-5 have been net direct capital importers like other medium-developed countries. They have been the most important targets of foreign direct investment in Central and Eastern Europe. They use the inflow of investment means, technol-ogy and skills as a vehicle of economic modernisation. The volume of foreign direct investment in a transition country is an expression of a country's advance made in terms of transformation. Foreign firms reinforce economic behaviour patterns in conformity with international standards, most notably those of the EU. Multinational enterprises have integrated economies of Central and East European countries into the EU at the microeconomic level to various degrees. The process of ownership-based integration is most advanced in Hungary, followed by Estonia and the Czech Republic, while Poland is catching up. Slovenian companies are less integrated in terms of capital ownership but have close links through company networks. The competitive position of accession countries will be influenced by further FDI flows during the accession negotiations.

The inflow of foreign direct investment to the CEEC-5 was USD 11.8 billion in 1998, a substantial 3 billion increase over the previous years. It increased further to USD 14.8 billion in 1999 due to higher inflows in the Czech Republic (tables 1 and 2). The per capita or per gross fixed capital formation amounts of foreign direct investment in most of these countries are similar to large FDI receiver emerging markets in Latin America and South-East Asia. The second group of accession countries has much lower FDI inflows per capita, but due to low domestic investments, foreign direct investment can be significant compared to the level of gross fixed capital formation. The latter indicator does not show the contribution of foreign direct investment to investments as most of the foreign direct investment is take-over and not new capital formation. FDI stocks above 30% of GDP in Hungary, Estonia and the Czech Republic are significantly high in comparison to international levels. In these countries, the foreign sector is a decisive factor in forming economic development. Poland and Slovenia have about half that amount which makes them similar to the second group of accession countries.

Table 1: Foreign direct investment flows in CEECs, 1992-1999, as recorded in the balance of payments, USD million

	1992	1993	1994	1995	1996	1997	1998	1999	1999 Inflow per gfcf* %	1999 Inflow per cap** USD
Czech Republic	1 004	654	869	2 562	1 428	1 300	2 720	5 108	36.2	497
Estonia	82	162	215	202	151	267	581	305	.	212
Hungary	1 471	2 339	1 147	4 453	2 275	2 173	2 036	1 944	16.6	194
Poland	.	1 715	1 875	3 659	4 498	4 908	6 365	7 270	18.2	188
Slovenia	111	113	128	176	186	321	165	181	3.9	92
Total (5)	.	4 983	4 234	11 052	8 538	8 969	11 867	14 808		

*   Infl per gfcf = inflow in 1999 as a per centage of gross fixed capital formation;
** Infl per cap = inflow in 1999 per number of population in USD.
Remarks: Estonia: equity capital cash + reinvested earnings + loans; Czech Republic: equity capital cash + in kind + reinvested earnings from 1998; Hungary: equity capital cash + loans from 1996; Poland : equity capital cash + in kind + reinvested earnings + loans - on a transaction basis; Slovenia: equity capital cash + in kind from 1997.

Source: National banks of respective countries.

Table 2:   Foreign direct investment stock in CEECs, 1992-1999 year-end balance of payments, USD million

	1992	1993	1994	1995	1996	1997	1998	1999	1999 Stock/ GDP %	1999 Stock per capita
Czech Republic	2 889	3 423	4 547	7 350	8 572	9 234	14 375	16 246	30.4	1 580
Estonia	.	419	696	955	1 026	1 148	1 822	2 467	35.4	1 710
Hungary	3 435	5 585	7 095	11 926	14 958	16 086	18 517	19 276	39.4	1 919
Poland	1 370	2 307	3 789	7 843	11 463	14 587	22 479	26 075	17.0	674
Slovenia	.	954	1 326	1 759	2 069	2 297	2 907	2 684	13.7	1 352
Total (5)	.	12 688	17 453	29 833	38 088	43 352	60 154	66 748		

Remarks: Estonia: equity capital + reinvested earnings + loans; Czech Republic: equity capital cash + in kind + reinvested earnings from 1997 + loans from 1997; excluding privatisation revenues; Hungary: equity capital cash + loans from 1996; Poland: equity capital cash + in kind + reinvested earnings + loans - on a transaction basis; Slovenia: equity capital + reinvested earnings + loans.

Source: National banks of respective countries.

Manufacturing is the most important target of foreign investors, with the exception of Estonia where it comes third. Only in Poland and Slovenia could manufacturing attract more than 50% of the FDI stock because the privatisation in the tertiary sector is slow. In the Czech Republic manufacturing attracted 46% of the foreign direct investment and also trade and banking were significant investment targets. Hungary stands out with high foreign direct investment in the electricity and gas services as well as real estate and business services investments. The more even spread of FDI targets is undoubtedly due to the general advance in privatisation. But there is no explanation why real estate and other business services have such a high share in Hungary and such a low one in Poland. In the case of Estonia, the low share of foreign direct investment in manufacturing reflects both the weakness of this sector in the Estonian economy and the strength of the country as a regional transport and financial centre. The following analysis in this paper focuses on the manufacturing sector, which is by no means representative for the processes in foreign direct investment as a whole. The prominent position of foreign direct investment in manufacturing can be seen in its role as a means of technology transfer and producer of export goods.

The common primary investor and trading partner of the CEEC-5 is the European Union. Companies from Germany, neighbouring EU countries, together with USA-based multinational enterprises (MNEs), are the most important investors, with significant investment differences among the countries. Germany leads investment in the Czech Republic, Hungary and Poland, the USA is the second highest investor in Hungary and Poland. Austria is a prominent trading partner and investor for Slovenia, Hungary and the Czech Republic. Estonia has intensive regional links to Sweden and Finland. The proximity of Central and East European Countries to the European Union has stimulated market-seeking investments by EU-based multinationals and more recently efficiency-seeking greenfield investments. The ongoing corporate-level integration of Central and East European companies into European corporate networks provides a stimulus for EU enlargement.

In the *World Competitiveness Yearbook* (INTERNATIONAL INSTITUTE FOR MANAGEMENT DEVELOPMENT 2000) competitiveness is measured by economic indicators, technology indicators and executive surveys. Internet technology, mobile phone availability, investment in research and in education as well as a liberal economic environment are seen as indicators of growing competitiveness. It is maintained that information technology allows for countries with a geographic disadvantage to participate in the global economy and compete more successfully than before. In the year 2000, the USA ranked first, Finland came third and the Netherlands fourth, Switzerland fifth and Ireland seventh after being eleventh in 1999. The CEEC-5 are well down the list but have the most competitive countries among their primary foreign investors. This is a distinctive feature of these more attractive FDI target countries if compared to second-tier Romania and Bulgaria, where companies of lower technological levels have high shares in FDI inflows.

4   Government policies related to FDI

In DUNNING's theory (1993), foreign direct investment flows are 'shaped' by three sets of factors. First, the ownership advantages, second, the locational advantages, and third, the internationalisation advantages. Locational advantages represent those advantages that make production in the given place more profitable or advantageous, from the point of view of the investor, than exporting the product from a foreign production unit to the given market, or locating new production capacities to a third country. The economic policy of a given FDI-recipient country can influence the relative locational advantages. Once foreign firms are present in a country, they have a distinctive impact on the host country's economy in the field of sourcing, competition, ownership relations and economic policy.

Characteristics of locations appear in the form of general and FDI-specific conditions. General conditions involve the overall stability and development pattern of the economy, the skills of the labour force as well as the general regulatory framework such as the tax system. General conditions were partly outlined in the two previous sections, a more detailed analysis goes beyond the scope of this paper. Investment and FDI policies and investment incentives will be outlined in section 4.1. There is a further incentive specific to Central and East European Countries - privatisation - that is unique to transformation economies in its scope and substance (section 4.2).

4.1   General investment incentives

Three of the countries in question are already OECD members and all of them are preparing for EU accession. These international treaties restrict discriminatory policies and demand equal rights for domestic and foreign firms. However, Central and East European Countries differ widely in terms of their governments' attitude towards foreign investors, the general level of corporate income tax, the system of tax and customs allowances as well as in terms of direct investment promotion. Labour market policy and regional policy offer further investment incentives. Economic policy in several Central and East European Countries has recently shifted from stabilisation to growth promotion including FDI incentives. A major stimulus for the introduction of lower taxes and of investment incentives is the international competition for foreign direct investment. As their mobility increases, international investors, more than before, compare locations looking at the cost of entry and the cost of production. Hungary has the most complex incentive scheme, ranging from tax and customs allowances to R&D- and infra-structure-related subsidies (table 3). Corporate tax has been low in Hungary, while it has been lowered in the Czech Republic and Poland, and completely abolished in Estonia in 2000. Countries that have for a long time suffered from low levels of foreign direct investment, such as Slovenia, have introduced attractive incentive schemes.

Even if the tax and incentive system is the same for domestic and foreign investors, there can be a difference between the capacity of firms to make use of the incentives. As will be shown later in this paper, the differences between foreign and domestic firms are huge in terms of size, efficiency, access to financing and so on. Small and medium-size domestic firms cannot meet the minimum investment and employment requirements to become eligible for tax breaks or to receive direct investment incentives. It is mainly large foreign investors who benefit. The result can be illustrated by the indicators for the Hungarian manu-facturing industry: foreign affiliates produce 86% of the pre-tax profit but pay only 59% of the corporate tax. This is partly the result of the policy preference provided to large investors, and partly the result of tax holidays provided to for-eign investors before 1996. The gap between domestic and foreign companies can widen due to unequal access to incentives.

A special incentive in Hungary is the system of customs-free zones. The uniqueness of the Hungarian system is, that companies can become a customs-free zone with no geographic limitation. The most important conditions are: an area over 2000 m², at least 80% of production going to exports, and technical facilities for customs control and statistical observation according to the rules of customs procedures.

Table 3: Review of measures for the support of the inflow of foreign direct investment in four CEECs, as of early 2000

	Hungary	Czech Republic	Poland	Slovenia
Taxes	- 18% corporate tax - 20% dividend tax	- 31% corporate tax	- 32% corporate tax	- 25% corporate tax - 1.5% withheld tax
Incentives	- Corporate tax relief for up to 10 years for investment of at least USD 40 million and more than 500 employees. - Corporate tax relief for up to 5-10 years	- Corporate tax relief for up to 10 years - Criteria - investment of USD 10 million, at least 50% goes to production sector, 40% of the investment goes to new machinery	- Tax deduction up to 30% of investment amount from the tax base: conditions e.g. revenue from export is over 50%, buying patents, ISO 9000, pharmaceutical industry	- Job creation support scheme - Possible negotiation about financial support of the government
Special incentives	- For regions with more than 15% unemployment - Corporate tax relief for up to 5 years for investment in production - Direct subsidies to innovation centres, industrial parks, local business development	- Location in a customs-free zone - Job-creation grants (up to USD 3000 per each new job) - Training grants (up to 50% of the costs) - Provision of low-cost building land and/or infrastructure	- Full tax allowances in selected regions for investment projects of at least ECU 0.4 million	- 10% corporate tax in free zones (also some other benefits - e.g. another reduction of the tax base by investment, for job creation or training)
Customs regime, Free zones	- Customs-free zone status for export-oriented companies	- Duty-free imports of new machinery related to projects exceeding CZK 10 million - Customs clearance - drawback system	- Duty-free import of machinery - Duty-free import of new investment goods - Duty-free special zones	- Duty-free import of the new machinery under OECD list 84 and 85 - Customs-free trade zones

Source: Author's design.

According to legal regulations, customs-free zones are allowed to be founded by foreign and domestic investors but the Hungarian government's primary aim was to attract foreign direct investment. From the second half of the 1990s, the number of customs-free zones began to increase, hosting mainly the suppliers of multinational companies in green-field investments. In 1997, 89 industrial customs-free zones had a licence, in 1998, this figure increased to 109 zones. In 1999, the number of customs-free zones was 116 of which 101 were manufacturing zones. The customs-free zones were operated by 70 companies, some of these companies such as Philips, Videoton, Flextronics Group and others, operated more than one customs-free zone.

The activity of Hungarian customs-free zones is strongly concentrated. About 70% of them are engaged in the production of components and finished telecommunication goods, electronics, informatics, car industry and engineering. These are also the most dynamically growing activities of free trade zone companies. 15% produced textiles, clothing and shoes. The remaining 15% are shared by the plastic, timber and leather industry, chemistry, the food industry, glass and pharmaceutical industry. 58% of the customs-free zones operate in the Western part of the country, 17% in Budapest and its surroundings, and only 25% East of the Danube. In 1999, 64% of the exports by companies with customs-free zone status were located in just two counties, Győr-Moson-Sopron and Fejér. Spatial concentration is increasing and this contributes to the advantage of the north-west and Central regions over other parts of the country. Customs-free zones played a leading role in the expansion of foreign trade over the last few years. Their share in the Hungarian exports increased from 18 to 43% between 1996 and 1999, and in import from 14 to 30%.

The Hungarian customs-free zone regime is perhaps the only incentive scheme which has led to well documented location advantages. Despite the wide range of incentive schemes in the various countries, the efficiency of these policy tools have not been properly investigated. However, the effects of incentives cannot be separated from other locational factors.

Table 4: The share of industrial free trade zones in the Hungarian foreign trade (%)

	1996	1997	1998	1999
Export	18,1	26,2	36,0	43,0
Import	13,9	19,8	25,2	30,6

Source: Ministry of Economic Affairs.

4.2   The role of privatisation policy in generating FDI[2]

The different speed and ways of privatisation contributed a great deal to the differences of economic performance among the Central and East European countries (table 5). Overall, it seems economically desirable to privatise rather fast, by selling the state-owned firms to investors who would restructure, capitalise and run them efficiently. Conducting such sales is just one and often not a popular or feasible policy option. This is because privatisation is also a political process fraught with conflicts between various policy objectives and vested interests which result in time-consuming searches for legal and political compromises. However, privatisation can be analysed in purely economic terms, considering its impact on overall capital formation, budget revenues, balance of payments, as well as on microeconomic performance (restructuring, efficiency improvement and upgrading management practices etc.). Economic and political constraints in the first years of transformation, such as shortage of domestic capital and vested interest of workers and managers, curtailed the possibility of privatisation by sale. The free distribution of property was considered an easy, fast and just way of privatisation.

Table 5: Share of the private sector* in value added, CEEC-5, 1990 and 1998 (%)

	Czech Republic		Estonia		Hungary		Poland		Slovenia	
	1990	1998	1990	1998	1990	1998	1990	1998	1990	1998
Total (GDP)	12.3	77.3	10	70	25	85	30.9	69.9	15	50
Industry	.	83.3	.	.	.	87.7	18.3	69.1	.	.

*   Private sector means majority private ownership.

Source: National statistics and EBRD.

There is a marked difference between fast privatisers - the Czech Republic, Estonia and Hungary - and slow privatisers - Poland and Slovenia. In the Czech Republic, a period of intensive privatisation and foreign direct investment in 1991-1995 was followed by two less intensive years. Foreign direct investment and privatisation picked up again in 1998 and 1999. While privatisation is nearing its end now, foreign acquisitions in the private sector become more important. In Hungary, foreign direct investment and privatisation went hand in hand until 1997, but in the last two years foreign direct investment has been almost exclusively unrelated to privatisation. The example of Hungary indicates that the inflow of foreign direct investment can continue after privatisation is over. In Poland and Slovenia, privatisation was slow until 1996 and so was foreign direct investment.

---

[2]   This chapter summarizes the findings of HUNYA 2000a.

After 1996, inflows of foreign direct investment accelerated and the share of FDI revenues in privatisation also grew significantly (table 6).

Table 6: Privatisation and foreign direct investment

	1990-1996		1997-1999	
	Forex rev. in total privatisation rev., %	Forex priv. revenue in FDI, %	Forex rev. in total privatisation rev., %	Forex priv. revenue in FDI, %
Czech Republic	15	80	80	50
Estonia	60	33	60	70
Hungary	63	47	40	20
Poland	low	20	medium	40
Slovenia	low	low	low	low

Remarks: Estonia first period: 1993-1996. – Foreign exchange (forex) revenue in total; privatisation revenue could not be calculated for Poland in the first period as the value of non-cash privatisation could not be measured. Based on the relative role of various modes of privatisation, a very rough estimation could be made: 'low' means less than one quarter, 'medium' means between one quarter and one half, and 'high' means above one half. – In Slovenia the way of privatisation does not allow for a calculation of foreign shares.

Source: Own calculation and estimation based on data from: ZEMPLÍNEROVÁ and MARTIN 2000 (for the Czech Republic); Estonian National Bank (for Estonia); ÁPVRT - Hungarian Privatisation and State Holding Company (for Hungary); DURKA 1999 (for Poland).

The economic aspects of privatisation became increasingly important in the second half of the 1990s, when the drawbacks of slow privatisation and of voucher schemes became quite clear. Privatisation by sales was discovered as an important source of budget revenues, of foreign currency inflows and as an essential ingredient of corporate restructuring. Current account deficits became a significant problem for Hungary in 1993, for the Czech Republic and Romania in 1996, and for Poland in 1998. The earlier the deficit problems surfaced, the sooner the country opted for revenue-generating modes of privatisation and FDI-friendly policies. Generally, sale to foreign strategic investors has proved to be the most efficient way of privatisation. This lesson had to be learned by the Czech Republic, by Slovenia and a number of second-tier accession countries that have tested other methods of privatisation or have delayed privatisation until recently.

Privatisation contracts can be seen as policy tools to attract foreign direct investment. They not only envisage to maximise state revenues but may also ensure that the new owner is a respected international investor interested in the long-term development of the acquired firm.

4.3   Companies' position after foreign take-over

Companies turned into subsidiaries of multinational corporations (MNCs) may prosper provided they are assigned a proper position in the international corporate network and given access to new technology and capital. Their success depends on three important conditions:

- The subsidiary's initial position in the network of the multinational corporation. This is determined by the privatisation contract, the intention of the investor and the human resources (level of skills) available in the subsidiary. The scope of decision-making in the subsidiary, brand name and product specialisation are determined at this initial point.
- Efforts of the subsidiary to upgrade its position in the multinational corporation and to acquire new technologies and skills. The subsidiary must improve its competitive position on a restricted but very competitive market within the multinational corporation.
- The long-term attractiveness of the business location. The target country must maintain economic stability and growth, as well as adhere to investor-friendly economic policies in order to keep investors when labour costs increase.

Government policies can play a role in promoting research and development, attracting headquarter functions, and supporting education and learning. Such policies can affect the type of activities assigned to the affiliates: either technology-based or assembly-based. The latter predominate in Central and East European Countries, especially among greenfield investments. Affiliates originating in privatisation acquisitions may be different as they often retain some local suppliers and market shares. But they may stay at a lower technology level than new greenfield investments. In addition, locally integrated affiliates are less footless than globally integrated ones and can have a more secure future. The difference between the two types of firms may diminish with time. Both of them have to become more technology-based to compensate for diminishing labour cost advantages.

5   Characteristics of FDI penetration

The extent of FDI penetration is shown by the share of foreign investment enterprises (FIEs) in nominal capital, assets, value added, employment, sales, export sales, investment outlays and profits derived from the income statements and/or tax declarations of companies. Some of these indicators - nominal or own capital, sales or output, employment and investment outlays - are available for all countries (table 7).[3] The role of foreign investment enterprises has increased for all five countries in regard to almost all indicators over the period 1996-1998. As capital indicators are not standardised, the most widespread common indicators - sales and employment - are discussed in more detail below. A comparison of the development of FDI penetration over time can be made for 1994-1998, keeping in mind the distortions caused by shifts from the domestic to the foreign sector.

Table 7:   Share of foreign investment enterprises (FIEs) in main indicators of manufacturing companies, 1996 and 1998 (%)

	Equity capital		Employment		Investments		Sales		Export sales	
	1996	1998	1996	1998	1996	1998	1996	1998	1996	1998
Czech Republic	21.5[1]	27.9	13.1	19.6	33.5	41.6	22.6	31.5	15.9	47.0
Czech Republic, adjusted							18	25		
Estonia	43.5[1]	40.1[1]	16.8	20.8	41.8	32.9	26.6	28.2	32.5	35.2
Hungary	67.4[2]	72.7[2]	36.1	44.9	82.5	78.7	61.4	70.0	77.5	85.9
Poland	29.3	43.2	12.0	26.0	30.6	51.0	17.4	40.6	26.3	52.4
Poland, adjusted							14	32		
Slovenia	15.6	21.6	10.1	13.1	20.3	24.3	19.6	24.4	25.8	32.9

1) Czech Republic 1996 and Estonia: Own capital. – 2) Hungary: Nominal capital in cash. Adjusted: Czech Republic and Poland adjusted for size limit by increasing the indicators for DEs by 20%.

Source: HUNYA 2000b.

The highest share of foreign investment enterprises by all indicators was reached by Hungary in each year since 1993. 70% of manufacturing sales come from foreign investment enterprises, which employed 45% of the manufacturing labour force in 1998. The second place is occupied by Poland with 41% of sales and 26% of employment. The Czech Republic comes next, with 32% and 20% respectively. The difference between Hungary on the one hand and the Czech Republic and Poland on the other was three times in 1994 and narrowed to two times in 1998.

---

[3]   Companies with some foreign share in their nominal or equity capital, foreign investment enterprises (FIEs), were selected from national databases containing data on the income statements of companies. The remaining companies are classified as domestic enterprises (DEs). Estonia is a special case where only majority FIEs were included in the newly generated database. In the case of Hungary in 1997-1998, and Slovenia, the coverage was limited to companies with at least 10% foreign ownership, which corresponds to the internationally accepted definition of FDI. For the Czech Republic and Poland, companies with even lower foreign shares had to be included. In Hungary and Slovenia, only very small ventures may fall out. Data for the Czech Republic cover only companies with 100 or more employees. Data for Estonia cover companies with more than 20 employees for 1996-1998, but for 1995 the limit is 50 employees. In Poland, companies with more than 50 employees were included.

The most dynamic increase has been recorded in the Czech Republic. In Slovenia and Estonia FDI penetration was lower and increased more slowly than in the other countries.

Table 8: Sales, share of foreign investment enterprises in manufacturing (%)

	1993	1994	1995	1996	1997	1998	1998/1994
Czech Republic	11.5	12.5	16.8	22.6	27.2	32.1	325
Estonia	.	.	20.1	26.6	27.1	28.2	140
Hungary	41.3	55.4	56.1	61.4	66.1	70.0	126
Poland	14.5	17.4	23.6	31.9	36.0	40.6	233
Slovenia	.	16.9	17.6	19.6	21.1	24.4	144

Source: HUNYA 2000b.

FDI penetration in the Czech Republic almost doubled between 1994 and 1996 by most indicators and expanded dynamically also in the following two years. The foreign sector showed a rapid expansion not only in terms of capital and sales but also in terms of employment. 50 000 new manufacturing jobs were created in, or shifted to the foreign sector, while the domestic sector lost 85 000 jobs in 1994-1996. By 1998 the share of foreign subsidiaries in total manufacturing employment reached one fifth. The sales shares of foreign investment enterprises increased in the period of overall recovery in Czech manufacturing following the first transformational recession. In the period 1994-1996, sales of foreign investment enterprises increased by 130%, sales of domestic enterprises by 14% (in current USD terms). Although ownership shifts cannot be identified, it seems that the foreign sector was an important driving force of the recovery in the mid-1990s. The upswing of car sales due to the success of the car manufacturer Škoda after being acquired by Volkswagen has been the most important example. In the period 1996-1998, the Czech economy underwent a second transformational recession. The causes were linked to the overvaluation of the exchange rate and slow progress of restructuring. The reactions of the domestic and the foreign sectors to the stabilisation measures were completely different. In this period the production of domestic companies increased only by 6.5% in nominal terms, which means a decline in real terms. Foreign companies' sales however increased by 73% in the same period. The foreign sector maintained its dynamism, and relied more on foreign markets replacing domestic enterprises on the Czech market. The competitiveness problem due to the overvalued exchange rate affected domestic companies more than foreign investment enterprises which had more opportunities to increase prices. The expansion of the foreign sector by number, size and sales of companies was very dynamic in the 1996-1998 period, replacing and outperforming domestic companies. But shifts from the domestic to the foreign sector were not very frequent.

FDI penetration indicators for Estonia reached, by 1996, the third highest level among the countries under discussion. This was mainly the result of the fast opening and privatisation after the introduction of the currency board in 1993. But the performance increase of foreign investment enterprises after 1996 was slow. The country remained behind Poland and was overtaken by the Czech Republic. The foreign sector in Estonia did not grow much by adding new companies but by the expansion of existing foreign investment enterprises. The small country, having little experience with modern industries, has not yet gained importance for export processing.

FDI penetration in Hungary's manufacturing took place much earlier. Already in 1994 the foreign investment enterprises share in nominal capital reached 60%, and it has increased only slightly since then. The same applies to the employment share of foreign investment enterprises, which has stagnated at 37% since 1994, partly due to the changes of computing the number of employees. The investment share of foreign investment enterprises came close to 80% in 1994 and increased only slightly in the subsequent years. It seems that FDI penetration in Hungarian manufacturing has already reached a level where no further dynamic increase can be expected. There is nevertheless still very intensive foreign direct investment activity in the form of capital increase in existing foreign investment enterprises, and the number of important greenfield projects is growing. Sales, and especially export sales, illustrate by which the share of foreign investment enterprises increased. They show the fastest growth between 1994 and 1998. This indicates that the intensive investment activity of the first half of the 1990s established competitive production capacities that can increase sales both in Hungary and abroad more rapidly than those of Hungarian-owned companies which had been lagging behind in terms of restructuring.

FDI penetration in Poland reached the second highest level among the five countries in 1998 by all indicators. Employment, sales and export shares of FIEs doubled between 1996 and 1998. Among the five countries, Poland experienced the most rapid expansion of the foreign sector. An upswing in privatisation stimulated foreign take-overs. Greenfield investments were attracted by the rapidly growing domestic market. While economic growth on the whole was strong, its main driving force changed from newly established domestic small and medium enterprises to foreign affiliates.

Slovenia had the lowest FDI penetration by all indicators in 1998. The gap in comparison to the other four countries grew between 1996 and 1998. Still, the shares of FIEs have increased constantly since 1994. The Slovenian economy has maintained a strong international competitive position mainly by successful domestic-owned companies.

6 Competitiveness in terms of structural change and its relationship to FDI

Data on FDI penetration is available for 23 industries, and for a number of indicators. The most important common indicator available for all CEEC-5 is *revenues from sales* (table 9). This is preferred to the less widely available *equity capital* to express FDI penetration by industries. Comparing the industry distribution of FDI penetration by equity and sales, the trends indicated by the two sets of data are the same. The difference between industries in terms of FDI penetration tends to grow over time.

In the Czech Republic, only the tobacco industry and the production of motor vehicles have absolute foreign control with over 80% of sales produced by foreign investment enterprises. The tobacco monopoly was sold to a foreign investor who is now almost the only producer. In the motor industry this is the result of the Škoda–Volkswagen deal that was followed by a number of take-overs and greenfield investments of supplier firms. The radio and TV set producing industries represent the third Czech industry with majority foreign control over sales. There are a few other industries with intensive foreign investment activity in which domestic enterprises are still in the majority: electrical machinery, non-metallic minerals, rubber & plastic, and some more foreign investment enterprises with a very low share in several industries such as 'other transport equipment', coke & petroleum, basic metals, and leather (below 10% of sales). While most of the foreign capital is concentrated in a few successful industries and companies, the major part of the Czech economy is still plagued by slow restructuring in domestic-owned companies that emerged from voucher privatisation.

In Hungary, industries fall into three categories in terms of FIEs' shares in sales. Low foreign shares are below 50%; medium shares range from 50% to 70%; high shares are above the average of 70% (1998). The lowest foreign share in an industry is 33% (furniture, manufacturing not elsewhere classified - n.e.c.) - a share that would be above-average in other countries. Industries where the majority of the production comes from domestic-owned companies are: clothing, wood, publishing and printing, basic metals, fabricated metals, medical instruments, other transport equipment, furniture and manufacturing n.e.c., and recycling. Light industries and metal industries are declining industries with poor market prospects both in Hungary and abroad, therefore they are avoided by investors. These are also less knowledge-intensive industries in which the presence of foreign direct investment is usually low. The two more sophisticated industries, instruments and other vehicles, have certain problems connected to the less successful privatisation of main companies. Industries such as food, textiles, leather, rubber and plastic, machinery and equipment n.e.c. industries show a medium FDI penetration, so foreign and domestic companies have almost equal shares. These branches mainly represent less knowledge-intensive and declining industries. The industries with high foreign shares in sales (above 70%) are tobacco, paper, coke & petroleum, chemicals, other non-metallic minerals, office machinery, electrical machinery, radio and TV sets, and motor vehicles. In these industries domestic firms' production is negligible and thus comparisons between the foreign and the domestic sectors do not make sense. Among these branches we find some low-technology branches with stable domestic markets, and high-technology, knowledge-based industries set up by foreign investors.

Table 9:   Industries with significant above-average shares of foreign investment enterprises in sales, 1994, 1996, 1998 (%)

Czech Republic	1994	1996	1998	
	.	.	94.6	Tobacco
	60.0	66.9	82.1	Motor vehicles
	37.2	43.8	45.2	Rubber and plastic
	23.7	45.6	44.5	Non-metallic minerals
	13.2	32.0	48.1	Electrical machinery
	(4.8)	35.9	57.8	Radio and TV sets
	3.3	26.5	38.3	Manufacturing n.e.c.*
	12.5	22.6	31.5	Manufacturing total
Hungary	1994	1996	1998	
	99.6	99.2	100	Coke and petroleum
	99.5	98.7	95.7	Tobacco
	78.4	82.7	79.9	Electrical machinery
	72.0	84.8	96.9	Motor vehicles
	70.0	71.8	48.6	Other transport equipment
	61.0	79.0	82.8	Radio and TV sets
	(53.7)	78.7	83.6	Chemicals
	55.4	61.4	70.0	Manufacturing total
Poland	1994	1996	1998	
	86.9	94.1	96.7	Paper, paper products
	8.4	90.7	95.3	Tobacco
	49.9	82.5	89.9	Motor vehicles
	52.4	66.7	81.8	Radio, TV sets
	46.0	55.6	60.4	Manufacturing n.e.c.*
	26.7	54.6	56.7	Rubber and plastic
	17.4	31.9	40.6	Manufacturing total
Slovenia	1994	1996	1998	
	100.0	100.0	100.0	Tobacco
	64.5	82.3	83.1	Transport equipment
	42.9	35.4	48.1	Paper
	.	40.4	42.6	Radio, TV sets
	.	21.3	26.1	Machinery n.e.c.*
	16.9	19.6	24.4	Manufacturing total

* n.e.c. = not elsewhere classified

Source: HUNYA 2000b.

The distribution of manufacturing sales between industries in Hungary shows a concentration in the food industry, accounting for 19.1% of total sales and 15.2% of foreign investment enterprises' sales, as well as the motor vehicle industry, with 13.4% of total sales and 18.5% of foreign investment enterprises sales. While the share of the food industry in both total and foreign-sector sales declined, the motor industry is the main winner of structural change. Coke and petroleum as well as chemicals have had high but declining significance in the industrial structure since 1996. Industries that gained shares in FIE's sales were, apart from motor vehicles, office machinery, radio and TV set manufacturing, and - to a lesser extent - clothing and basic metals production. Out of 22 industries, only these five gained, the others lost relative significance in the sales structure of foreign investment enterprises. This is also true for the total manufacturing sector as domestic enterprises have had a low share and lower dynamics than foreign investment enterprises (an exception is the category 'Other transport equipment'). A radical shift of the Hungarian industrial structure took place in favour of the more knowledge-intensive industries. This shift was driven by foreign direct investment. In the early privatisation and domestic-market-driven period the penetration of foreign capital occurred in all industries. In the later stage, export-oriented greenfield investments dominated and foreign direct investment was concentrated in a few industries.

In Poland, more than 40% of the manufacturing sales in 1998 originated from foreign investment enterprises. There are large differences of FIE shares by industries. Industries form groups with significant discontinuity in their FIE shares. High penetration (over 70%) occurred in the paper industry, tobacco industry, motor vehicles, radio and TV set production. These are industries where domestic companies are almost non-existent. The main difference compared to Hungary lies in the absence of foreign direct investment in coke & petroleum – a result of different privatisation policies – and in the more prominent presence of the wood industry – a natural assets-based advantage of Poland. FIE sales ratios between 50% and 60% can be found in Polish manufacturing n.e.c., rubber and plastic, publishing and printing, and electrical machinery. These are the industries where the foreign and the domestic sectors are in real competition and the higher productivity and profitability of foreign investment enterprises matter the most. The superiority of foreign investment enterprises is proven by the data from all these industries. The next group consists of industries with FIE sales ratios between 45% and 30%, which includes among others the food industry. Here the position of foreign investment enterprises differs even stronger from the domestic enterprises. There is a typical cherry picking situation as the rate of profit in domestic food companies is close to nil but rather high among foreign investment enterprises. The last group of industries has a foreign share of less than 20%.

Except for Estonia, the manufacturing of motor vehicles has over 80% FDI penetration in the CEEC-5. The car industry was attracted to this region both by unsatisfied domestic demand and by favourable conditions for low-cost production. Also tobacco manufacturing is usually foreign-owned as only big international companies can cope with the brand names and promotion costs of this industry. In the Czech Republic and in Hungary, electrical machinery production also has a high rate of FDI penetration. In the other three countries the paper industry has become both a foreign-controlled branch and a major export industry. High FDI penetration in the chemical industry is specific to Hungary due to the pharmaceutical industry which is one of the most internationalised world-wide.

The size of FDI penetration in the CEEC-5 depends on industry-specific features, and on the characteristics of the privatisation policy. Foreign direct investment in Central and East European countries follows world-wide patterns in regard to the corporate integration of industries. Technology-intensive electrical machinery and car production as well as activities with relatively stable domestic markets, e.g. in the beverages and tobacco industries, are the main targets of foreign capital. Privatisation by sales attracted foreign direct investment to all industries in Hungary, but only to a few in other countries. A foreign participation remained relatively low in branches with great structural difficulties and oversized capacities, such as the steel industry.

7   Productivity growth in the foreign and the domestic sectors

Labour productivity in FIEs is on average as much as two times higher than in domestic enterprises. In this respect, there was no significant difference among the CEEC-5 in the mid-1990s. But countries' situations differed in terms of productivity dynamics in the 1994-1998 period (table 10). The gap between foreign investment enterprises and domestic enterprises increased rapidly in Hungary until 1996; then it stabilised. In 1998, foreign investment enterprises were 2.9 times more productive than domestic enterprises. In Poland, the productivity gap increased from 1.5 to 1.9 in the 1994-1998 period. A stable 1.9 times gap was characteristic of the Czech Republic between 1995-1998. A decrease of the productivity difference to below 2 took place in Slovenia. The productivity gap is now very similar in the Czech Republic, Poland and Slovenia.

The extremely high productivity gap in Hungary shows, on the one hand, the gain foreign ownership means to the economy, and on the other hand, it demonstrates an unhealthy duality between the booming foreign sector and the stagnating domestic sector. The productivity gap, however, did not grow much after 1996 when the second transformational recession came to an end. As will be shown in the analysis of industrial branches, most of the gap results from the different sectoral distribution of domestic enterprises and foreign investment enterprises. In many industries the domestic sector is so small that it makes little sense to compare it with the overwhelming foreign sector.

Table 10:   *Sales per employee, foreign investment enterprises in % of domestic enterprises in manufacturing, 1993-1998*

	1993	1994	1995	1996	1997	1998	1998/1994
Czech Republic	209.1	186.3	190.5	193.7	188.8	189.0	101
Estonia	.	.	240.7	188.1	160.1	150.2	62
Hungary	151.4	209.0	259.9	281.8	278.9	286.7	137
Poland	158.7	154.5	156.9	185.1	184.5	194.4	126
Slovenia	.	240.9	228.0	217.8	198	197	82

Source: HUNYA 2000b.

The convergence of labour productivity between domestic enterprises and foreign investment enterprises in Slovenia and especially in Estonia may indicate some spill-over effects coming from foreign firms. In Estonia, this process is very fast and can be related both to the very liberal conditions in the economy and the absence of highly productive advanced industries both in the foreign and the domestic sectors. In the Estonian case, this process is not related to the lead of foreign investment enterprises in terms of labour productivity, which is specific to the CEEC-5, but rather to extremely large differences in labour productivity. In OECD countries, the productivity advantage of foreign investment enterprises compared to the average productivity of the manufacturing industry is only 30%. The smaller and more specialised the foreign investment enterprises sector, the larger its lead over the average productivity in the country. Higher productivity of subsidiaries is due to lower labour input and to narrow specialisation, as well as to the absence of top management and research functions. In addition, in transition economies, foreign investment enterprises usually represent a special quality in technology, management and marketing, that is more developed than in domestic, especially state-owned enterprises. The productivity advantage exists both in technical terms and in terms of higher output value due to higher sales prices. Higher prices can be achieved by better marketing, the use of western brand names, and so on. If the foreign investment sector is very different from the domestic one, the two segments of the economy may find it difficult to co-operate. Thus, the foreign sector may function as an enclave. In such a case, direct spill-over effects do not exist, while indirect spill-over takes place through the income and knowledge of individual employees. The learning process going on in domestic-owned companies may with time lead to narrower FIE/DE gaps.

Endowment with capital is higher in the sector dominated by foreign enterprises than in the domestic-owned enterprises. This may confirm the expectation that foreign investors use more recent, capital-intensive and labour-saving technology. It also reflects the concentration of foreign direct investment in manufacturing branches with high capital intensity. The lead of foreign-owned enterprises in terms of capital intensity is especially pronounced in Hungary where capital-intensive industries (e.g. steel industry, oil refineries) were more accessible to foreign investors than in the other countries. The duality of performance in the manufacturing sector appears in two respects:

- the dichotomy of modern, foreign-dominated industries on the one hand and traditional branches with both domestic and foreign companies on the other. In Hungary, the nine foreign-dominated industries represented 50% of manufacturing sales in 1998;
- in the branches of industries with both foreign and domestic companies, a comparison of indicators shows that the foreign sector is more efficient and more export-oriented than the domestic sector.

This duality between foreign- and domestic-dominated industries appeared in all countries and has been growing over time. The difference of performance between the foreign- and the domestic-owned companies in the same industry is largest in Hungary and smallest in Slovenia.

8   Competitiveness on EU markets

Enterprises dominated by foreign investment have high and growing shares in export sales. The outstanding export performance relative to sales indicates that foreign investment enterprises are more export-oriented than domestic firms (tables 11 and 12). In Hungary, foreign investment enterprises account for 86% of manufacturing exports. The difference of export intensity (exports/sales) between the domestic and the foreign sectors has been growing. Export intensity in the domestic sector was 22% in both 1994 and 1998, but it increased from 37% to 56% in the case of foreign investment enterprises. The shift of foreign investment enterprises to exports has accelerated in recent years when more export-oriented, assembly-type greenfield investments started production. The domestic sector's export volume remained the same, USD 2.4 billion, in both 1993 and 1998, while exports from the foreign sector increased from USD 5.8 million to USD 14.6 million.

The export share of foreign investment enterprises in the Czech Republic (47% in 1998) was just about half the Hungarian rate. The increase is nevertheless impressive, considering that the share of FI-enterprises in export sales was only 16% in 1994. Also the export intensity lead of foreign investment enterprises over domestic enterprises increased very rapidly. In 1994, foreign investment enterprises were only 1.3 times more export-oriented than domestic enterprises, but in 1998, it was already 1.9 times more. Estonia is a different case: although foreign investment enterprises are more export-oriented than domestic enterprises, the gap is only 1.4 times and has not grown with time. In Poland, more than half of the export sales were provided by foreign investment enterprises in 1998 as a result of the rapid increase over the previous four years. The export intensity lead of FI-enterprises over domestic enterprises by a factor of 1.6 did not change much over time. Polish domestic enterprises and foreign investment enterprises are both more domestic-market-oriented than in other countries. This has to do with the size of the country and the rapid increase of domestic demand in the mid-1990s. Slovenia is a strongly export-oriented country where both domestic enterprises and foreign investment enterprises have a high proportion of export sales in total sales. The gap between the two increased from the 1.3 fold to the 1.5 fold over a four-year period. Still, foreign investment enterprises provide only one third of the export sales. In the two smallest and most export-oriented countries, Estonia and Slovenia, foreign investment enterprises have the smallest role in selling abroad.

Table 11:   Export sales, share of foreign investment enterprises in manufacturing exports, 1993-1998 (%)

	1993	1994	1995	1996	1997	1998	1998/1994 %
Czech Republic	14.9	15.9	.	.	41.9	47.0	296
Estonia	.	.	25.4	32.5	32.1	35.2	139[1]
Hungary	52.2	65.5	68.3	73.9	83.3	85.9	131
Poland	36.1	26.3	33.9	40.5	45.1	52.4	199
Slovenia	.	21.1	23.2	25.8	28.0	32.9	156

[1]1998/1995, %.

Source: HUNYA 2000b.

Table 12:   Exports per sales, foreign investment enterprises in % of domestic enterprises in manufacturing, 1993-1998

	1993	1994	1995	1996	1997	1998	1998/1994 %
Czech Republic	134.0	132.3	.	.	.	187.5	142
Estonia	.	.	135.1	132.7	127.5	137.9	102[1]
Hungary	155.3	152.9	168.6	177.8	255.8	259.9	167
Poland	333.0	168.3	166.5	146.0	146.8	161.8	96
Slovenia	.	131.7	141.7	142.5	145.6	152.1	115

[1]1998/1995, %.

Source: HUNYA 2000b.

The competitiveness on EU markets can be measured by the share of each country in the EU's imports, and the development of EU imports between 1995 and 1998 (table 13). Successful exporters in Central and Eastern Europe increased their export volumes (EU-15 imports) and market shares dynamically - in the first place Hungary, which has the highest FDI penetration. It is followed by Estonia, a small country with small export volumes. The medium range is formed by the Czech Republic. Low export dynamism and stagnating market shares characterise Poland and Slovenia. Exports to the EU increased due to reorientation and to overall export dynamics. Reorientation of trade took place mainly in the early 1990s; after 1995 it was significant only in Estonia.

Table 13:   Market shares of Central and East European Countries in EU-15 extra-EU imports, 1995-1998

	Czech Rep.	Estonia	Hungary	Poland	Slovenia
Market share 1995 (%)	1.85	0.17	1.54	2.49	0.93
Market share 1998 (%)	2.35	0.25	2.33	2.55	0.88
Market share change (% point)	0.50	0.08	0.79	0.06	-0.05
Market share change (%)	127	147	151	102	95
Export volume change (%)	64.4	93.8	96.3	32.9	22.7
Share of FIEs in export sales, 1998 (%)	47.0	35.2	85.9	52.3	32.9
FIE: export sales/sales, 1998 (%)	57.2	50.9	56.0	27.8	72.3

Source: Eurostat Comext database and HUNYA 2000b.

The relationship between market share development and FDI penetration is most obvious in the case of Hungary and Slovenia. The rapid market gains of Hungary were the result of the restructuring and market-conquering activity of foreign investment enterprises. Slovenia recorded low foreign direct investment, a low share of foreign investment enterprises in export sales, and a loss of EU-market shares. Estonian exports increased as fast as in Hungary, Czech exports developed at medium high speed, while Polish export shares stagnated. FDI penetration in the latter three countries is very similar to each other, if we correct for the discrepancies in data coverage, thus the very different market share dynamics cannot be explained by the presence of foreign investors. Poland has the strongest FDI penetration among them and has the worst export performance. A possible explanation for this is that foreign investment in Poland is more domestic-market-oriented, as indicated by export sales as low as 28% per sales compared to over 50% in the other countries. Estonian exports also depend mainly on the performance of domestic-owned companies as the FDI penetration of the manufacturing sector is low.

Market share developments at the industrial branch level show which industries have gained or lost competitiveness between 1995-1998 (table 14). In the case of the Czech Republic, half of the 22 industries gained shares and half of them lost shares. The major winners were the industries that produce motor vehicles, electrical machinery n.e.c., fabricated metals and paper, and the printing & publishing industries. The main losers were the light industries (categories 15-20) as well as the industry with the highest market share, non-metallic minerals. The shift of exports is towards high value-added products. Both industries with the highest gains are dominated by foreign capital, while industries that lost market-shares show generally lower FDI penetration.

Table 14:   Imports of the EU-15 from selected Central and East European countries by industry: market share gain and market share loss in the top 3 industries

Czech Republic NACE industry	Gain % points 1994-1998	Market share 1998 (%)	FIE share in exports 1998 (%)
34. Motor vehicles	4.08	6.84	88.2
31. Machinery n.e.c.*	2.22	5.17	60.2
28 Fabricated metals	2.13	9.43	36.9
	Loss % points		
19. Leather	-0.63	0.35	10.8
26. Non-metallic minerals	-0.30	9.99	43.5
18. Wearing apparel	-0.15	1.39	32.4
Other high market share industries			
20. Wood	-0.15	4.76	52.2
25. Rubber and plastic	1.47	4.49	60.1
22. Publishing, printing	2.08	4.43	29.0
Hungary NACE industry	Gain % points 1994-1998	Market share 1998, %	FIE share in exports, 1998, %
34. Motor vehicles	3.79	8.83	99.1
31. Electrical machinery	2.57	5.67	92.7
30. Office machinery	2.37	2.67	99.9
	Loss % points		
16. Tobacco	-0.88	0.0	100.0
22. Printing, publishing	-0.35	0.66	31.1
27. Basic metals	-0.23	1.37	59.4
Other high market share industries			
28. Fabricated metals	0.40	3.77	62.5
26. Non-metallic minerals	0.44	3.03	64.9
Poland NACE industry	Gain % points 1994-1998	Market share 1998, %	FIE share in exports, 1998, %
31. Electrical machinery	1.54	3.55	74.7
32. Radio, TV	0.84	1.43	96.43
21. Pulp, paper	0.69	2.92	94.7
	Loss % points		
26 Non-metallic minerals	-1.34	6.94	44.4
27 Basic metals	-0.81	3.59	14.6
18 Wearing apparel	-0.63	6.27	46.0
Other high market share industries			
20. Wood	0.19	8.54	59.8
28. Fabricated metals	0.47	8.30	42.3
34. Motor vehicles	0.41	4.13	95.7

* n.e.c. = not elsewhere classified

Source: Eurostat Comext database and HUNYA 2000b.

Estonia can be characterised by generally increasing market shares (three exceptions are industries with low exports rates). Market shares were particularly gained in the main industries of specialisation such as wood, clothing, textiles, fabricated metals and radio & TV-set production. However, a generally backward export structure and a lack of export-oriented manufacturing branches puts limits on future export growth. In the case of small countries such as Estonia or Slovenia, economies of scale cannot be developed enough to put up export-oriented subsidiaries.

In Hungary, motor vehicles, electrical machinery and office machinery were the major industries gaining market shares, all are totally foreign controlled. The losers were basic metals and publishing which represent industries with predominantly domestic ownership.

Poland is a country with almost stagnating market shares, although more industries gained market shares (13) than in the Czech Republic (11). But both gains and losses of market shares are of a small magnitude showing that structural change has been slow. Gaining industries such as electrical machinery and radio & TV set production are among the market share winners in other countries too. Together with the wood industry these are almost completely foreign controlled. Motor vehicles have a relatively small share and little gains of market shares, showing that the big foreign direct investment coming into this branch is mainly attracted by the large and expanding domestic market. Significant losers, such as non-metallic minerals, metals and clothing production, point towards growing disparities between the prosperous production and exports of foreign investment enterprises, and shrinking production and exports of domestic firms. Poland seems to have a problem with international competitiveness in most industries. Although it shows the second highest FDI penetration rate measured by sales after Hungary, this has not contributed much to the export performance.

In the case of Slovenia, the loss in market shares affects a wide range of industries, among them traditionally strong ones with high market shares such as paper, clothing and non-metallic minerals. Market-share winners such as metal products, electrical machinery and printing & publishing are industries with low FDI penetration. Those with the highest FDI penetration, motor vehicles, paper and radio & TV set production, have by and large stagnating market shares in the EU-15.

The analysis of the data reveals that Hungary has had a clear competitiveness gain due to foreign direct investment penetration. Estonia also had a competitiveness gain but less linked to foreign direct investment. The competitiveness gain of the Czech Republic is less than of the former two countries, but it mainly results from foreign direct investment. Poland has strong FDI penetration with little effect on overall competitiveness, while Slovenia has the most severe international competitiveness problem since it is losing market shares in the EU. This can be a result of the relatively low FDI penetration and low inflow of foreign direct investment. The modern branches, even under foreign control, do not develop fast enough to generate structural change and win new markets.

9   The ability to earn and its impact on growth at the industry level: profitability and investment propensity of foreign and domestic firms

The rate of profit (profits per sales) is higher in foreign investment enterprises than in domestic enterprises, thus a high share of profits in Central and East European Countries is produced by foreign investment enterprises (tables 15 and 16). 92% of the profit in Czech manufacturing in both 1996 and 1998 was earned by foreign investment enterprises. This indicates once again the generally difficult financial position of domestic enterprises. The low rate of profit in the domestic sector will further curtail investment and delay restructuring. The highest profit rates in the Czech foreign investment sector were achieved in the tobacco, rubber and plastic, and non-metallic minerals industries. The major lossmaker in both the domestic and the foreign sectors in 1998 was the production of 'other transport equipment'. The difficult start but final success of the Škoda-Volkswagen company is reflected in the profit development of the motor vehicle industry: in 1993 and 1994, huge losses were booked. In 1997 and 1998, the car industry became profitable and produced one quarter of the manufacturing industry's profits.

The profit share of foreign investment enterprises peaked in Hungary in 1996 with 90% and has declined slowly since then. In the last two years, increasing profits in the domestic sector, despite declining sales shares, point to positive results of restructuring. In the mid-1990s, there were several industries with negative aggregate profit. In 1998, however, they disappeared in the domestic sector but not in the foreign sector. The major producer of profits in 1998 was the motor vehicle industry, followed by chemicals and office machinery. High-technology industries with the highest amounts of foreign direct investment were the main profit generators with profit rates of 10% or above. Profits per sales were generally low at the time of recession in the mid-1990s, while in the past two years profit rates have reached high levels in Hungary. But foreign investment enterprises in the industries 'basic metals', 'other transport equipment' and 'recycling' have made losses on average. The risk of failure persists mainly in the case of privatised companies. The bad situation of the 'other transport equipment' industry is not unique. It received foreign direct investment at the early stage of transformation in the Czech Republic, Hungary and Poland, but became problems in later years. One possible cause of this is low investment in public railways.

In Poland, the profitability gap between foreign investment enterprises and domestic enterprises has grown rapidly. In 1993, the foreign sector made losses, but it became profitable a year later and increased its share in profits as well as its rate of profit. In 1996, foreign investment enterprises received 41% of the profits, in 1998, this share reached 66%. The rate of profit (profit per sales) equalised between the domestic and the foreign sectors at 5% in 1996. Since then a growing gap in favour of foreign investment enterprises has appeared. Some industries remained lossmakers even in the foreign sector: textiles, leather, metals, and other transport equipment. Profit rates diminished for both sectors in 1998, reflecting overall economic difficulties and the slowing down of economic growth. In the foreign sector the clothing production remained the only industry where the profit rate was 10%, in the domestic sector 'office machinery' had a similar rate. The motor vehicles industry had very low profits in the case of foreign investment enterprises and losses in the case of domestic enterprises. The mostly domestic-market-oriented Polish car industry is doing significantly worse than the export-oriented Hungarian and Czech car industries.

Table 15: Profits, share of foreign investment enterprises in manufacturing 1993-1998 (%)

	1993	1994	1995	1996	1997	1998
Czech Republic	4.6	0.2	26.9	92.5	70.3	92.1
Estonia	.	.	loss	loss	25.6	59.2
Hungary	.	.	63.3	89.7	89.7	88.8
Poland	loss	2.4	23.6	40.6	43.9	66.0
Slovenia	.	17.8	21.0	21.9	21.2	24.9

Source: HUNYA 2000b.

Table 16: Profits per sales in the foreign investment enterprises (FIE) and the domestic enterprises (DE) in manufacturing 1994-1998 (%)

	1994		1996		1998	
	FIE	DE	FIE	DE	FIE	DE
Czech Republic	0.1	13.0	5.0	0.1	6.4	0.2
Estonia	.	.	4.8	-3.4	0.9	0.3
Hungary	.	.	5.8	1.1	8.0	2.4
Poland	0.6	4.8	5.3	3.6	3.6	1.3
Slovenia	3.1	3.0	3.3	2.8	3.8	3.7

Source: HUNYA 2000b.

Companies in Estonia made losses in 1995 and 1996, domestic and foreign alike. In the past two years, both sectors recorded profits, but 59% of all profits were generated in the foreign sector. In 1998, profits per sales decreased due to difficulties following the Russian crisis. In Slovenia, profit rates showed a lead in the foreign sector for 1996 which had almost vanished by 1998. This is another sign of the rather balanced relationship between the two sectors in this country.

Foreign investment enterprises are more active than domestic firms in terms of investment activity (table 17). Investment per assets and investment per sales show a clear lead of foreign investment enterprises over domestic enterprises. This is a confirmation of the importance of foreign direct investment in economic growth and restructuring. Investment data suggest that foreign investors rapidly restructure the acquired manufacturing firms and make further investment to expand activities. As a result of stepped-up investment activities, the weight of foreign investment enterprises in Central and East European manufacturing will grow in the future even in the absence of new projects. Investment outlays per sales for foreign investment enterprises in Slovenia are not better than for domestic enterprises, which is another proof for the strength of the domestic sector in this country.

Table 17:   Investment outlays, share of foreign investment enterprises in manufacturing 1993-1998 (%)

	1993	1994	1995	1996	1997	1998	1998/1994
Czech Republic	25.3	26.9	27.4	33.5	31.9	41.6	155
Estonia	.	.	.	41.8	27.1	32.9	79 (/96)
Hungary	58.9	79.0	79.9	82.5	78.3	78.7	100
Poland	.	30.6	41.0	45.6	49.9	51.0	166
Slovenia	.	.	14.0	20.3	23.3	24.3	174 (/95)

Source: HUNYA 2000b.

In 1996-1998, foreign investment enterprises increased their share in investments in three countries, the Czech Republic, Poland and Slovenia. The small decrease in the case of Hungary is in line with the general recovery of the domestic sector after 1996. If 1994-1998 is taken into consideration, the 79% shares of foreign investment enterprises in manufacturing investments remained flat. In the case of Estonia, the recovery of domestic-sector investment was much more pronounced. In key industries, such as food, wood and clothing, a clear lead of domestic firms is visible. The lack of high-technology industries prohibits large differences between sectors.

10 Conclusion and policy implications

The positive link between FDI penetration and various components of international competitiveness can be demonstrated in the case of five first-tier EU accession countries. This is true both at the aggregate and the sectoral levels. It is obvious that the activity of a strong foreign sector in manufacturing increases international competitiveness. In 1994-1998, GDP growth, productivity growth, structural change, and profit rates were higher in countries with a stronger presence of foreign direct investment. Economic policy can support the long-term attractiveness of a country for foreign investors by strengthening its locational advantages. Exchange rate policy and wage policy have to support cost competitiveness, and fiscal and other investment-related regulations and incentives should be attractive compared to neighbouring countries.

The deeper the FDI penetration, the faster the speed of structural change: Hungary was first, followed by the Czech Republic and Poland in the period 1996-1998. This is relevant both for the change of output structure and the country's exports to the EU.

The size and industry distribution of FDI penetration depends on industry-specific features and on the characteristics of the privatisation policy. Foreign direct investment in Central and Eastern European countries follows the world-wide characteristics in the corporate integration of industries; technology-intensive 'electrical machinery' and 'car production' are the main targets. Foreign direct investment helped Central and Eastern Europe to shift their product structure to become more similar to the more developed EU countries. This may give further impetus to economic growth and narrow the development gap between the more advanced Central and East European Countries and the EU.

Foreign presence remained relatively small in branches with great structural difficulties and oversized capacities, such as the steel industry. Privatisation is not enough to set restructuring of these industries in motion. Sectoral policy and financial restructuring is necessary to make companies attractive for foreign takeovers.

Foreign capital has penetrated activities with relatively stable domestic markets, e.g. in the 'beverages' and 'tobacco industries'. Profit rate differences point to the abuse of monopoly positions especially in the tobacco industry. Competition policy is especially important in countries hosting large multinationals.

A disparity between foreign- and domestic-dominated industries appeared in all countries and is growing over time. Within the manufacturing sector, disparities can emerge in two respects,

- between modern, foreign-dominated industries on the one hand and traditional industries with both domestic and foreign companies on the other, while foreign direct investment concentrates increasingly in a few technologically more advanced industries;
- a foreign-domestic gap within the industries with both foreign and domestic companies.

There are some indications for a slow productivity and profitability catch-up in sectors with both foreign and domestic companies, but at an aggregate level, this gap grows due to the faster growth of totally foreign-owned industries.

The dichotomy of productivity and profit rates between the foreign- and the domestic-owned companies in one and the same industry is largest in Hungary and smallest in Slovenia. In Slovenia, the balanced relationship between the domestic and the foreign sector is coupled with a low average rate of FDI penetration and the relatively low presence of technology-intensive industries. This might indicate a slow rate of technological progress and few spill-overs.

Foreign subsidiaries can perform better but not behave independently of the general conditions determining corporate income. Profit rates in the economy usually differ between the foreign- and the domestic-owned companies, but they usually develop in the same direction as a response to overall economic conditions. The alarmingly low profit rate of domestic enterprises is a problem especially in the Czech Republic. It is also becoming a problem in Poland, while profit rates are generally low in Estonia. Relief from the corporate tax to attract investors may be of little value in countries with poor profit expectations. Incentives may rather increase foreign direct investment by targeting the costs of investment: regional and employment policy measures, customs allowances, industrial parks.

The gap between domestic and foreign companies can widen due to unequal access to investment incentives despite the national treatment principle. Large investors, for example, usually benefit more than small ones from economic policy measures, while small and medium-size domestic firms cannot meet the minimum investment and employment requirements to become eligible for tax breaks or to receive direct investment incentives. Thus, the implementation of a specific small and medium-size enterprise policy along with investment promotion incentives would mainly serve the domestic-owned firms.

References

ANTALÓCZY, K. (2000): FDI Policy and Incentives in Hungary at the End of the Nineties. Paper prepared in the framework of the PHARE-ACE research project P97-8112-R, mimeo

ÁPVRT, Hungarian Privatization and State Holding Company, http://www.apvrt.hu/

BALDWIN, R. et al. (1997): The Costs and Benefits of Eastern Enlargement: The Impact on the EU and Central Europe. Economic Policy (24), 127-176

CSÁKI, Gy., ÁKOS, M. (1998): The Ten Years of Hungarian Privatization (1988-1997). Privatization and State Holding Company, Budapest

DURKA, B. et al. (1999): Foreign Investments in Poland 1999. Foreign Trade Research Institute, Warsaw

DUNNING, J. (1993): Multinational Enterprises and the Global Economy. Addison Wesley Publications, Wokingham

ÉLTETÖ, A. (2000): Foreign Direct Investment in Hungary at the End of the Nineties. Paper prepared in the framework of the PHARE-ACE research project P97-8112-R, mimeo

FAGERBERG, J. (1996): Technology and Competitiveness. Oxford Review of Economic Policy (3), 39-51

HUNYA, G. (2000a): Upswing of Privatization-related FDI in CEECs in the Late 1990s. The Vienna Institute Monthly Report (5), 2-8

HUNYA, G. (2000b): FDI Penetration and Performance in CEECs 1994-1998: Database on Foreign Investment Enterprises in the Czech Republic, Estonia, Hungary, Poland and Slovenia. The Vienna Institute for International Economic Studies (WIIW): prepared in the framework of the PHARE-ACE research project P97-8112-R, mimeo

INTERNATIONAL INSTITUTE FOR MANAGEMENT DEVELOPMENT (2000): World Competitiveness Yearbook, Lausanne

KNELL, M. (2000): Foreign Direct Investment and Productivity Spillovers in the accession countries. Paper prepared in the framework of the PHARE-ACE research project P97-8112-R, mimeo

KRUGMAN, P. R. (1996): Making Sense of the Competitiveness Debate. Oxford Review of Economic Policy (2), 17-25

MEYER, K. E. (1995): Direct Foreign Investment, Structural Change and Development: Can the East Asian Experience be Replicated in East Central Europe?. London Business School, CIS-Middle Europe Centre, London (= Discussion Paper Series 16)

OZAWA, T. (1992): Foreign Direct Investment and Economic Development. Transnational Corporations 1, 27-54

PORTER, M. E. (1990): The Competitive Advantage of Nations. Macmillan, London

REILJAN, J., HINRIKUS, M., IVANOV, A. (2000): Key Issues in Defining and Analysing the Competitiveness of a Country. University of Tartu, Faculty of Economics and Business Administration, Tartu (= Working Paper 1)

ROJEC, M. (2000): Foreign Direct Investment in Slovenia: Major Characteristics, Trends and Developments in 1997-1999. Paper prepared in the framework of the PHARE-ACE research project P97-8112-R, mimeo

TRABOLD, H. (1995): Die internationale Wettbewerbsfähigkeit einer Volkswirtschaft. DIW, Berlin (=DIW-Vierteljahresheft 2), 169-183

UNCTAD (1997): World Investment Report 1997. UN, New York and Geneva

VARBLANE, U. (2000): Foreign Direct Investment in Estonia: Major Characteristics, Trends and Developments in 1993-1999. Paper prepared in the framework of the PHARE-ACE research project P97-8112-R, mimeo

WYZNIKIEWICZ, B. (2000): Contribution of FDI to Polish Economic Growth in the Nineties. Paper prepared in the framework of the PHARE-ACE research project P97-8112-R, mimeo

ZEMPLÍNEROVÁ, A. and J. MARTIN (2000): Impact of FDI on the International Competitiveness of Manufacturing: The Case of the Czech Republic. Paper prepared in the framework of the PHARE-ACE research project P97-8112-R, mimeo

The Spread of Entrepreneurship in Eastern Europe

Tibor Kuczi and György Lengyel

1 Introduction

Former communist countries in Eastern Europe differed remarkably in their attitude to private enterprises. In some countries the population did have a chance, however limited, to get acquainted with entrepreneurship many years before the political change because the private economic sector had not been fully abolished and the second economy had been tolerated by state authorities. In these countries entrepreneurial skills and experience had already developed before the official introduction of a market economy. In other countries, private production and private trade had been repressed or practically abolished for decades, and thus the role and weight of the second economy was negligible. This paper discusses whether these pre-existing differences have led to new disparities in the 1990s regarding the recruitment of new entrepreneurs. It also analyses the main factors that influence the development of entrepreneurship in former socialist countries.

At the time of the political change three markedly different theories were presented to define who would become entrepreneurs in the rising market economy (CSITE 1998; KOVÁCH 1998). The first claimed that people whose families (parents, grandparents) were entrepreneurs between the two World Wars would be more likely to become entrepreneurs than others since they had the chance to acquire values, knowledge and behavioural patterns necessary for entrepreneurship. This view is primarily associated with Iván SZELÉNYI (1998), who formulated the idea of 'interrupted embourgeoisement'. He argues that the bourgeois values developed between the World Wars were not annihilated during the decades of socialism but 'lay dormant' and 'reawakened' after the political change.

Others argue that members of the former nomenclature would predominantly become entrepreneurs since they had the power and networks to convert their political capital into economic capital. This view suggests that entering the private economic sector is not primarily conditioned by entrepreneurial knowledge, experience, skills or values but by political and economic power that helps to acquire indispensable information and to influence decisions related to privatisation.

The third argument states that the recruitment of entrepreneurs is decisively influenced by occupation, level of education, gender, age, and personal value systems. In addition, the extent and frequency of networks is regarded as an aspect of paramount importance. This refers mainly to relatives and friends in positions of power playing a significant role in raising the capital and labour necessary to become an entrepreneur. The fact that one has the confidence, trust and solidarity of a wider community, either of a smaller settlement or a professional group, may largely contribute to someone's entry into business. The findings of a comparative international research project entitled 'The spread of entrepreneurship in Eastern Europe' and additional data analysis related to entrepreneurship in Bulgaria, Hungary, Poland, Russia and Serbia revealed that in Russia and Bulgaria, the chances of becoming an entrepreneur are principally determined by the size of political capital, that is, developments in these countries can best be explained by the second theory (although educational achievement and professional skills are further significant explanatory factors). The situation in Hungary and Poland is better explained by the third theory, while Serbia differs due to the political conflicts that characterised Yugoslavia in the 1990s.

2   Different ways of recruiting entrepreneurs

Spatial differences in the recruitment of entrepreneurs are related to different historical legacies, different political priorities and, as a result, variations in the begin and velocity of the private sector's growth. In Hungary, the emergence of the private sector began in the early 1980s (RÓNA-TAS 1997; LAKI 1998). In 1982, a government order allowed the establishment of economic associations. Both skilled workers and university graduates were given a chance to enter the private sector, as well as to make a career in state-run enterprises, banks or public administration. In 1980, 3.4% of Hungary's working population was employed in the private sector. This shows that the development of the market economy and the emergence of small entrepreneurs started before the early nineties. The private sector, however, was politically controlled, even if the legal framework became gradually more and more liberalised. Some of the most influential business ventures took off at the beginning of the 1990s.

After the change of the political system, the opportunities to become an entrepreneur increased significantly. Two surveys of small- and medium-scale private enterprises carried out in 1988 and 1993 reveal that the social composition of entrepreneurs did not change as a result of the political transformations although their number increased remarkably from 11% in 1990 to 20% in 1992 (see KUCZI and VAJDA 1991; KUCZI, LENGYEL, NAGY and VAJDA 1991; CZAKÓ, KUCZI, LENGYEL and VAJDA 1994). However, a substantial change was observed in regard to the proportion of university graduates, which increased from 20% in 1988 to 27% in 1993. The stability of the social composition of the entrepreneurs in these five years is mainly to be explained by the fact that economic conditions have become harsher due to the recession that resulted in more difficult conditions for entering the business sphere (LEVELEKI 1994). Furthermore, diploma-holders and workers trained in certain trades began moving towards the private sector. Often they started their careers in the state sphere. They would probably have remained there, had their jobs not become insecure (KELEMEN 1999).

Neither of the two empirical surveys proved that the involvement of parents or grandparents in entrepreneurship had any influence on someone setting up a business on their own. The 1988 survey showed that 8% of the urban population had artisan or merchant fathers, with a mere extra 3% among entrepreneurs. By 1993, the influence of the parental entrepreneurial background was further diminished. The main factors motivating a person's decision to become an entrepreneur were therefore the level of educational attainment and the former employment career.

Since in Hungary the foundation of private enterprises had been allowed before the political change took place, those who set up on their own in the 1980s could choose between the alternatives of private entrepreneurship or a career in state-owned instituions. In Russia, the year 1988 marked the beginning of establishing private ventures on a mass scale. Therefore, the latter coincided with the collapse of former career opportunities. As a result of uncertainty, those employed in the army, the large state-owned enterprises and the political bureaucracy began to orientate themselves towards the private sector.

Three phases can be distinguished in the development of the private sector in Russia (RADAEV 1997). The first lasted from 1988 to 1990. The main characteristic of this stage was the transformation of political capital into economic assets (i.e. 'political embourgeoisement'). This period was favourable for launching enterprises because the state sector had withdrawn from several fields of the market (CONNOR 1991), and thus there were almost no competitors. The second phase lasted from 1990 to 1992. Its main feature was the liberalisation of economic policy and the legalisation of various entrepreneurial forms such as joint ventures, individual private enterprises, closed joint-stock companies, and 'small enterprises'. The more favourable legal conditions, however, coincided with a deterioration of economic conditions, when a 35% tax was introduced and the interest rate of loans rocketed to 50%. Those who entered the private sector during this phase included, first of all, ministerial officials and discharged army officers. The third phase began in 1992. This period saw the privatisation of large enterprises and the government's effort to control institutional changes. In this period, state-owned companies were transformed into joint-stock companies, with over 40% of state-run firms being privatised until 1994.

Analysing the three phases in the development of the private economy, one finds that in Russia social origin had almost no relevance in the recruitment of entrepreneurs, while membership to the nomenclature (top leaders of party organisations) had the greatest influence. Nearly all private ventures originated in former power positions converted to economic assets, with less than 3% of the working population being able to set up on their own and relying on their own resources. Three groups of the nomenclature can be distinguished:

1. Functionaries of the party and the communist youth organisation; however, this group is narrow and has almost no significance.

2.   State officials; this group is larger and more open than the party bureaucrats.
3.   Economic nomenclature; their role is decisive, especially in privatisation.

In Serbia (BOLCIC 1997), there were no entrepreneurs in the original sense of the word before 1990, only small retail traders. In 1990, only 62 000 persons were employed in the private sector, amounting to 2.4% of active earners. In 1952, for example, this figure was 3.3%. Until recently, entrepreneurship in Serbia only had an appeal to those with low levels of educational achievement. However, since the entrepreneurial act was passed in 1989, the private sector grew rapidly. Between 1990 and 1994 there was an eight-fold rise in the number of registered private companies. This fast growth produced a rapid change in the ratio of state and private firms. Private companies accounted for 77% of all enterprises in 1990 and 93% in 1993. It must be noted, however, that similarly as with the Hungarian situation, the size of the private firms was extremely small (95% of the private enterprises in Serbia employed only one to five people), and that the proportion of private firms is misleading in two regards. On the one hand, many registered companies do no business at all. On the other hand, there is a very high number of unregistered firms. According to 1994 estimates, 54% of the country's production derived from the hidden (informal) economy. Nevertheless, the state-run companies preserved their economic power: 98% of the capital was still concentrated in state hands at that time.

In spite of the fact, that the Russian, Serbian and Hungarian data are diverse and hence hardly comparable, they reveal quite clearly that the three countries have embarked on three markedly different paths towards the emergence of entrepreneurs. In Russia, the new strata of entrepreneurs were recruited from the political and economic nomenclature. Consequently, the rigid hierarchy based on the sharp differentiation of positions typical of socialist times seems to remain in place or to get reproduced under market conditions.

In Hungary, most entrepreneurs were recruited from the middle-class strata. During the 1980s, the opportunities to obtain more favourable social positions had increased, and many Hungarians acquired the skills and experiences necessary for a market economy, to establish networks, and to accumulate capital for launching a venture (LENGYEL 1999). This group includes not only managers with university degrees but also skilled workers with marketable trades. This is important to note because the rigid social hierarchy based chiefly on position and schooling slackened in Hungary during the 1980s. This enabled the skilled workers to become entrepreneurs and thus to compete with the university graduates in financial, and often cultural terms. Furthermore, university graduates founded types of enterprises which would have reduced their prestige in earlier times. For example, teachers and engineers opened shops or bought trucks to become forwarding agents. After the political change, the group of middle-class entrepreneurs was enlarged by those members of the nomenclature who could take advantage of the privatisation of state enterprises (RÓNA-TAS 1994). It is noteworthy that there is no significant difference in the magnitude of capital or economic influence between the two groups.

The social composition of entrepreneurs seems to differ radically in Russia and Serbia. In Russia the lower and middle-strata of entrepreneurs are missing completely. Those Russians entering business do this in order to improve their financial situation, not for the purpose of subsistence. Their level of education is conspicuously high: 82.3% of the Russian entrepreneurs had university diplomas, 12.6% had graduate degrees. The Serbian society was not sufficiently prepared to cope with the new conditions of the market economy after 1989, including legal frameworks for economic independence on mass scale. Before 1989, not enough social capital and financial capital had been accumulated. This might explain why the majority of capital was still in state hands (98%). Private ventures were very small, working with minimal capital. The recruitment basis of entrepreneurs was very wide, and many people of limited means founded businesses to escape unemployment. In Serbia, two entrepreneurial groups have to be sharply distinguished: one is the self-employed using entrepreneurship to escape unemployment, the other, a small elite, was recruited from the nomenclature. In Russia, the composition of entrepreneurs appears to be unipolar, predominated by the highly qualified economic nomenclature.

These variations result from differences in the historical development of the two countries. The situation in Serbia can be explained by the theory of interrupted embourgeoisement. In spite of the fact that there was no extensive private sphere in industry and services in the interwar period, there were many small farms which encouraged the emergence of basic entrepreneurial skills bequeathed from generation to generation. This may explain why those losing their jobs try to become entrepreneurs. In Russia, however, several generations passed without experiencing the private economy, so the theory of interrupted embourgeoisement does not apply.

The Hungarian situation resembles that of Serbia (and hence differs from Russia) in the sense that entrepreneurial skills also evolved in the peasant enterprises (and in the strata of artisans and retail traders) during the interwar years, and were passed down from generation to generation in socialist times. In Hungary, however, the effect of *interrupted* embourgeoisement was considerably weakened by the fact that due to the early reforms financial and social capital, as well as experience with the market economy had been acquired by those in favourable positions (with higher education, more marketable knowledge, better connections). As a result, the group of Hungarian entrepreneurs is more stratified and consists of a broad stratum of small and middle-scale entrepreneurs including both an elite of owners who had founded their private firms in the early 1980s or belonged to the economic nomenclature, and a large number of the self-employed.

The above analysis of the three countries is, however, highly hypothetical since the available data are sporadic, and not always comparable. Further research is needed to understand the similarities and differences between these countries. As the data for Poland and Bulgaria are even more sparse, these countries were excluded from the analysis. However, one might suggest that the Polish development is similar to the Hungarian, while the Bulgarian situation seems to be closer to the one in Russia.

3   The inclination to become an entrepreneur

3.1   Disparities in the inclination to become entrepreneur

Due to the quality and quantity of information gathered by the OMNIBUS survey in 1993, a reliable picture about the inclination to become an entrepreneur can be outlined.[1] The questionnaire was compiled in co-operation with all participating countries to assess entrepreneurial inclinations and their correlation with variables such as gender, age, education, occupation, employment stability, presence or absence of managerial experience, entrepreneurship of parents or grandparents, activity in the second economy, number of friends, readiness to move in hope of a better job.

In each of the countries nearly the same share of the population felt inclined to become an entrepreneur: 24.1% of the Bulgarian respondents, 28.3% of Hungarians, 32.0% of the Poles and 30.3% of Russians answered "yes" to the question "Would you like to be an entrepreneur?" The Serbian figure was only 16.9%, as a large proportion of those surveyed chose the option "it depends".

Table 1: *Entrepreneurial inclination by countries in 1993*

	Bulgaria	Hungary	Poland	Russia	Serbia
Yes	24.1	28.3	32.0	30.3	16.9
It depends, don't know	31.5	14.5	19.5	30.1	33.0
No	44.4	57.3	48.4	39.6	50.1
Sample size	1 341	1 196	1 140	1 197	767

Source: OMNIBUS survey, 1993.

The high share of Hungarians and the relatively low proportion of Russians and Bulgarians who were *not* aiming at being an entrepreneur in 1993 can be explained by the fact that Hungarians had got to know earlier the difficulties of becoming an entrepreneur and sustaining a business, and thus apparently had less

[1]   The survey includes Bulgaria with 1 463 respondents, Hungary with 1 500, Poland with 1 224, Russia with 1 315, and Serbia with 800 respondents.

illusions than Russians or Bulgarians. Right after the political turnover fewer Hungarians had rejected the idea of entrepreneurship than in 1993. When asked in a more concrete way "Do you plan to launch a venture within a year?", about the same proportion of the respondents answered "no" or "I don't know" (about 85%) and "yes" (about 10%) in each country.

3.2   Entrepreneurial inclinations by age

The distribution of entrepreneurial inclinations by age is quite similar in the five countries. The highest rate of positive answers to the question "Would you like to be an entrepreneur?" was in the young age group. As age increased, entrepreneurial ambitions decreased, though the decline is not steady in each case.

The greatest correspondence is in the youngest age category, with less than 3% divergence between the countries.[2] A thought-provoking similarity exists between Poland and Hungary on the one hand, and Bulgaria and Russia on the other. In the latter two countries entrepreneurial ambitions decrease steadily with the increase of age, the rate of those replying in the affirmative in the age bracket above 40 being below 30%. In the former two countries, the decrease is not steady, some older age groups preceding the youngest ones. The higher share of potential entrepreneurs among the middle-aged Hungarian and Polish people may also indicate that in these countries the possibility of economic independence is considered to be a real alternative by a broader strata of the population.

Table 2: Entrepreneurial inclinations by age in the five countries under study (%)

| | "Yes, I would like to be an entrepreneur" | | | | | |
| | a g e | | | | | |
	- 20	21-30	31-40	41-50	51-60	60 -
Bulgaria	48.5	41.3	38.1	29.1	14.8	5.5
Hungary	47.5	52.2	40.2	35.5	13.6	8.6
Poland	50.0	38.0	34.4	38.9	31.3	12.2
Russia	50.5	38.6	33.9	26.3	14.9	6.3
Serbia	28.1	28.8	22.8	16.9	9.2	1.4

Source: OMNIBUS survey, 1993.

In Poland, the ambitions of the older generations are even more intensive than in Hungary. One third of the age group between 51 and 60 answered in the affirmative. One possible explanation for this is that in Poland unemployment reinforces entrepreneurial aspirations, which encourages older people to consider founding a business as a way out.

---

[2]   The divergence of Serbian figures must partly be ascribed to the way how the survey was conducted.

3.3   Entrepreneurial inclination by gender

The distribution of entrepreneurial inclination by gender was nearly identical in four of the five countries. 30-35% of men and 20-25% of women (28% in Poland) answered "yes" to the question "Would you like to be an entrepreneur?". The Serbian data, again, show different results with a lower "yes" rate and higher "it depends" rate than elsewhere.

Several factors may account for the differences in entrepreneurial aspirations of women. The first factor is the varying social prestige of entrepreneurship. In Hungary, the rate of older people and women among entrepreneurs increased in times when the prestige of entrepreneurship was lower. From the second half of the 1980s onwards, however, the share of men among Hungarian entrepreneurs began to grow, with women mainly being represented in typically 'feminine' entrepreneurial jobs such as 'catering' or 'boutiques'. The rise in the social acknowledgement of entrepreneurship entailed a decrease in the proportion of women. Another influencing factor is the rate of unemployment. The higher the unemployment, the larger the tendency to become self-employed. This tendency might verify the fact that in Poland, where the jobless show an above average interest in entrepreneurship, there is little difference between the entrepreneurial inclinations of men and women.

Table 3: Entrepreneurial inclinations by gender in 1993 (%)

| | "Yes, I would like to be an entrepreneur" gender | |
	male	female
Bulgaria	34.4	20.0
Hungary	36.3	21.7
Poland	31.9	28.1
Russia	35.3	25.3
Serbia	17.8	14.7

Source: OMNIBUS survey, 1993.

3.4   Entrepreneurial inclination by level of educational achievement

After the political change in Eastern Europe the role of the missing bourgeoisie had to be played by the higher educated middle classes from which most entrepreneurs were recruited (RÓNA-TAS and LENGYEL 1997). In each country there is a strong correlation between the level of educational achievement and the inclination to become an entrepreneur. The level of educational achievement of entrepreneurs is high above the national average.[3] Secondary schools (grammar schools) or special vocational schools seem to be the decisive level of schooling to become a successful entrepreneur.

Table 4 reveals that the difference between those with primary schooling and those who finished secondary school is considerably larger than the difference between those with secondary education and university degrees. The divergence between the latter two is minimal in every country, with university graduates being slightly less driven to become entrepreneur than those with secondary education.

One of the major findings of the survey is that while there are significant differences between the countries in the social composition of the private economic sector, these differences are far smaller with regard to entrepreneurial inclinations. In the future, the recruitment basis of entrepreneurs may become more similar in all Central and Eastern European countries.

Table 4: Entrepreneurial inclination by level of educational achievement (%)

| | "Yes, I would like to be an entrepreneur" Education | | |
	Primary school or less	Secondary school	University
Bulgaria	14.4	37.5	39.7
Hungary	17.4	35.1	37.5
Poland	22.9	34.7	28.6
Russia	*	24.4	30.7
Serbia	12.1	28.7	14.9

* No data available

Source: OMNIBUS survey, 1993.

3.5   Entrepreneurial inclination by occupation

The occupational categories used in the OMNIBUS survey differ between the various countries. Therefore, it is better to concentrate on a few well-identifiable categories of identical contents. Old-age pensioners have little interest in entrepreneurship. Their positive answers to the question "Would you like to be an entrepreneur?" (2-7%, with the exception of Poland at 20.3%) lagged far behind the average (the average is 27%).

A more differentiated picture emerges regarding manual workers. In Hungary, the positive answers given by unskilled and semi-skilled workers lies 5-6% below

---

[3]   For example, in Hungary the educational level of entrepreneurs has further increased in the past five years, which is due to the entry of many young graduates choosing this professional career. To do justice, it must be noted that in the actual sense of the word they are not entrepreneurs but that they carry out agent's work.

the average (27.7%), while the positive answers of skilled workers amount to a considerably higher-than-average value (36.9%). Therefore, the latter constitute one of the most strongly business-minded groups. In Poland, by contrast, there is no significant difference between the various groups of workers, as all answered in the affirmative, some 10% higher than the average. In Serbia, the rate of industrial workers positively inclined towards entrepreneurship was slightly below average, and that of workers in the services was high above-average (average: 16.3%, workers in trade: 30%, workers in services: 31%). In Bulgaria, there are far more potential entrepreneurs among workers than the average (average: 27.4%, workers: 32.3%). Farmers and craftsmen exceeded the average figure by far, with 50% "yes" answers, though their overall weight within the sample (1.6%) is not significant. Finally, in Russia, entrepreneurial ambition among workers is below average (average: 30.4%, manual workers: 20.3%), the only group ranking lower being that of the old-age pensioners. At the same time, the positive responses among workers in the services were above average, as in Serbia.

Differences in the way the private economic sector is emerging in Russia on the one hand, and in Hungary and Poland on the other, are reflected in the entrepreneurial ambitions of managers. In Russia, potential entrepreneurs among middle-level leaders are below average, since the private economy is predominated by the former nomenclature. In Hungary and Poland, however, "yes" answers among middle-level managers were some 10% higher than the average, indicating that in these countries this group is most open to entrepreneurship. Yet in all five countries, those who had been in managerial positions replied "yes" to entrepreneurship in proportions high above the average.

Table 5: Entrepreneurial inclinations of those who have held managerial posts (%)

| | "Yes, I would like to be an entrepreneur" Have you ever been in managerial position? | |
	Yes	No
Bulgaria	34.7	25.5
Hungary	37.3	25.5
Poland	27.3	30.5
Russia	36.2	26.1
Serbia	24.8	14.0

Source: OMNIBUS survey, 1993.

Another important aspect is the far above-average rate of professional soldiers in Russia that are inclined towards private business (43% of them answering in the affirmative). In Russia and Bulgaria, where the surveys also contain figures for students, nearly half of the respondents still pursuing their studies would like to set up on their own.

The rate of the jobless saying "yes" to entrepreneurship was highest in Poland (half of them), but even in Serbia and Bulgaria positive responses were slightly above average. Entrepreneurial inclination and a fear of losing one's job were not correlated similarly in the five countries in question. In Poland and Hungary, the danger of losing one's job clearly boosted the willingness to become an entrepreneur. Those worried most gave the highest share of "yes" answers, about 10% above the average. In Serbia, Bulgaria and Russia worries about losing one's job definitely decreased entrepreneurial ambitions. Those afraid of unemployment displayed a below-average willingness to set up a business. This is so in spite of the fact that the Polish and Bulgarian unemployment figures are nearly identical, with 15.7% of the active population being out of work in Poland, and 16.4% in Bulgaria.

3.6   Entrepreneurial inclination with regard to parental entrepreneurship

The relation between the entrepreneurial past of parents or grandparents and entrepreneurial inclinations of those questioned was quite ambiguous. In all five countries those whose parents or grandparents used to be entrepreneurs were clearly more inclined to answer "yes" to the question "Would you like to be an entrepreneur". The smaller the share of those whose parents or grandparents were entrepreneurs, the stronger the ambitions to become independent earners. In Hungary, 23.4% of the sample had entrepreneurs as parents or grandparents but only 27.5% of them displayed an inclination to become entrepreneurs. The lowest rate of respondents with parental entrepreneurship in the family background can be found in Russia (12.7%), but half of the group would like to launch a business of their own.

Table 6: Entrepreneurial inclinations by family background (%)

	Percentage of respondents with entrepreneur-parents or grandparents	Percentage of respondents saying "yes" to entrepreneurship
Hungary	23.4	28.1
Serbia	17.0	30.1
Poland	15.8	36.3
Bulgaria	13.3	50.0
Russia	12.7	50.3

Source: OMNIBUS survey, 1993.

The figures suggest that it is not the preservation and bequest of entrepreneurial traditions and skills through which entrepreneur-parents or grandparents influence the entrepreneurial inclinations of their descendants. Thus, the figures above contradict the theory of interrupted embourgeoisement. It is more likely that in countries where entrepreneurship was heavily obstructed people gain moral support and encouragement from the fact that their parents or grandparents were entrepreneurs, and conversely, the easier it is to become an independent employer, the smaller the significance of the fact that a predecessor was an artisan, retail trader, or industrialist. In Russia, the entrepreneurs had certainly been in the worst situation during the 1990s, therefore their group coherence is the greatest and hence they stimulate their descendants very strongly.

3.7   Entrepreneurial inclinations by 'connections', part-time job and mobility

Connections increase entrepreneurial inclinations. Our data suggest that those who have a larger number of friends are more inclined to start a private business themselves. Though 'good connections' play an important part in well-established market economies, it is presumed that in ex-socialist countries their role is even more significant. In the centrally planned communist system, the economy of scarcity and the importance of ideological reliability made strong ties to those in power the most important social capital. Also during the extensive institutional transformation (changing banking system, emergence of new economic institutions) personal connections, for example to those in charge of privatisation, were inevitably stronger and more important than official contacts with frequently changing, often reshuffled institutions.

Table 7: Entrepreneurial inclinations by the number of friends (%)

| | "Yes, I would like to be an entrepreneur" Number of friends | | | |
	1-5	6-10	11-30	31 or more
Bulgaria	21.7	32.0	40.5	25.9
Hungary	27.9	33.7	42.7	24.1
Poland	28.3	26.5	35.8	31.4
Serbia	13.0	17.7	16.1	19.5

Source: OMNIBUS survey, 1993.

Entrepreneurial aspirations were also reinforced by an income from work other than permanent employment (table 8). In Bulgaria and Russia, nearly half of those earning an extra income would like to be entrepreneurs, in the other three countries it is 'only' a third. In this regard, the Serbian figures are closer to the Russian and Bulgarian figures again. People who are willing to move to another place in hope of a better job also have stronger aspirations to become an entrepreneur.

Table 8: Entrepreneurial inclinations by other sources of a second income (%)

| | "Yes, I would like to be an entrepreneur" Income from other sources than the main job? | |
	yes	no
Bulgaria	47.6	25.2
Hungary	35.7	25.2
Poland	32.7	29.0
Russia	45.5	27.4
Serbia	33.7	12.5

Source: OMNIBUS survey, 1993.

4   Conclusion

Logistic regression equations enable us to compare the explanatory force of the models and groups of variables by countries. In this respect, table 9 displays significant, at times strong correlations with most of the selected factors. In the Russian, Bulgarian and Polish cases the models account for some one-fourth or one-third of entrepreneurial inclinations. In the Yugoslavian and Hungarian cases, their predictions are even more precise, explaining over half the cases. It is not presumed that in every case cultural and national traditions underlie the differences. It is far more likely that the deviations are correlated to differences in the current economic situation and in public mentality (value systems). It is also probable that - if the income relations and the occupational factors, which are not analysed here because of multicollinearity, were also considered - further similarities would be exposed and the divergences between the countries would decrease.

The regression models confirm the assumption that entrepreneurial inclinations are most powerfully explained by demographic features and personal dispositions. Age has a clearly negative effect in every country, while a readiness to invest displays a very strong and unanimously positive correlation. Those who are inclined toward entrepreneurship are much more likely to believe that profit must be (re)invested, while the others would mostly spend it on consumption. The willingness to be mobile has a clearly positive correlation: Those who are open to mobility, who would leave their home for a better living, have something in common with the entrepreneur that strives to explore unknown territory and undertakes risks. Involvement in the second economy also has far-reaching implications. It is an important explanatory factor in the entrepreneurial inclinations of Russian, Bulgarian, and Yugoslav respondents. However, in the Hungarian and Polish contexts, where it used to play a very important role, it no longer seems to be a decisive prerequisite for becoming an independent earner.

Table 9: Logistic Regression Models of Entrepreneurial Inclination

| | Inclination to become an entrepreneur | | | | |
variable	Hungary	Russia	Bulgaria	Serbia	Poland
Gender	.7720	.4227	.5865	-	-
1 = male	(21.4233)	(8.4226)	(16.7787)		
0 = female					
Age between 18	-.0724	-.9656	.8428	-.1774	.0240
and 80 years	(131.8900)	(36.9224)	(86.8437)	(5.5348)	(27.8021)
Place of Residence	-.4706	+	-.3007	+	
1 = village	(7.5524)		(3.6657)		–
0 = city					
Fear of un-	-	-	-	-	.2809
employment					(2.8567)
1 = yes; 0 = no					
Use of profit	.5564	1.1590	.9228	1.6405	.4535
	(10.6051)	(58.1206)	(17.3920)	(5.7537)	(10.8311)
Parents	-	.7129	-	-	.5933
1 = yes; 0 = no		(12.8743)			(9.9906)
Manager	.8569	.4862	-	-	-
1 = yes; 0 = no	(16.2915)	(10.2786)			
Move	+	.5773	.2896	.9544	.7541
1 = yes		(13.4094)	(3.7006)	(4.3498)	(27.2878)
0 = no, hard to say					
Education	.7841	-	.4705	-	-
1 = higher educ.	(5.2700)		(5.3186)		
0 = else					
Second economy	-	.6907	.8562	.7475	-
1 = yes ; 0 = no		(14.8576)	(17.3920)	(2.9785)	
Friends	-	+	.0130	-	.0040
How many friends			(12.6852)		(3.1827)
do you have ?					
Constant	2.6013	-1.9941	.0399	.1774	-2.6486
	(62.1504)	(133.7914)	(.0311)	(.0580)	(127.1350)
-2 LL	906.509	1171.043	1268.906	131.422	1271.578
Chi-square	262.702	198.359	266.547	26.224	135.828
Df	5	7	8	4	6
Potential entre-					
preneurship in %					
Correct					
Yes	58.13	37.54	34.97	61.70	26.18
No	82.58	89.21	92.06	71.43	89.14

+ = Not included in the model; -= Not included in the model because its effect is insignificant
Fear of unemployment = Question: Are you afraid of losing your job?
Use of profit = Question: What would you use the profit for? Answers: 1 = mainly for investment; 0 =
    mainly for consumption, hard to say
Parents =   Question: Is there anyone among your parents or grandparents who was an entrepreneur?
    Answers: 1 = mainly for investment; 0 = mainly for consumption, hard to say
Manager =   Question: Have you ever been in a leading post?
Move =   Question: Would you move to another settlement if you could get a better job?
Second economy = Question: Do or did you have additional economic activity beside your main job?
Potential entrepreneurship = Question: Would you like to be a private entrepreneur?
    Answers: 1 = yes; 0 = no, it depends
Source: OMNIBUS survey 1993; authors' calculation.

The final conclusion of our survey is that the social composition of *potential* entrepreneurs is highly similar in all the five countries concerned, whereas there are far greater differences in the composition of *actual* entrepreneurs. One may venture the assumption that in the future, the Polish, Russian, Bulgarian, Serbian and Hungarian private entrepreneurs may draw closer to each other in a sociological sense. The social composition of present-day entrepreneurs is more strongly influenced by the cultural traditions of the past and by the peculiarities of each country than will be that of future entrepreneurs.

References

BOLCIC, S. (1997): Entrepreneurial Inclination and New Entrepreneurs in Serbia in the Early 1990s. International Journal of Sociology 27 (4), 3-35

CONNOR, W. D. (1991): The Rocky Road: Entrepreneurship in the Soviet Economy, 1986-1989. In: BERGER, B. (ed.): The Culture of Entrepreneurship. CS Press, San Francisco

CSITE, A. (1998): Embourgeoisement Theories and Debates. Review of Sociology. Special Issue. Budapest

CZAKÓ, L., KUCZI, T., LENGYEL, GY., VAJDA, L. (1994): A kisvállalkozók társadalmi összetétele. (Social Composition of Small-scale Entrepreneurs). Hunagrian Central Statistical Office, Budapest

KELEMEN, K. (1999): Kisvállalkozások egy iparvárosban. (Small Entrepreneurs in an Industrial Town]. Szociológiai Szemle (1)

KOVÁCH, I. (1998): Postsocialism and Embourgeoisement. Review of Sociology. Special Issue. Budapest

KUCZI, T., LENGYEL, GY., NAGY, B., VAJDA, L. (1991): Entrepreneurs and Potential Entrepreneurs. Aula (2)

KUCZI T., VAJDA L. (1991): A vállalkozók társadalmi összetétele. (Social composition of entrepreneurs). Közgazdasági Szemle (1)

LAKI, M. (1998): Kisvállalkozás a szocializmus után. (Small Entrepreneurs after Socialism). Közgazdasági Szemle Alapítvány, Budapest

LENGYEL, GY. (1999) (ed.): Kisvállalkozások megszűnése, bővülése és kapcsolatrendszere. (Bankruptcy, Growth and Networks of Small Entrepreneurs). Műhelytanulmányok, Budapesti Közgazdaságtudományi Egyetem Szociológia és Szociálpolitika Tanszék, Budapest

LEVELEKI, M. (1994): Les Fleurs du Mal: As VIDEOTON Was Falling to Pieces More and More of them Come to Use..." Szociológiai Szemle (2), Review of the Hungarian Sociological Association.

RADAEV, V. (1997): Practicing and Potential Entrepreneurs in Russia. International Journal of Sociology 27 (3), 15-50

RÓNA-TAS, Á. (1994): The First Shall Be Last? Entrepreneurship and Communist Cadres in the Transition from Socialism. American Journal of Sociology, 100 (1).

RÓNA-TAS, Á. (1997): The Great Surprise of the Small Transformation. The University of Michigan Press, Ann Arbor

RÓNA-TAS, Á., LENGYEL, GY. (1997): Entrepreneurs and Entrepreneurial Inclination in Post Communist East-Central Europe. International Journal of Sociology 27 (3), 3-14

SZELÉNYI, I. (1998): Socialist Entrepreneurs. The University Press of Wisconsin, Madison

Spatial and Social Disparities of Employment and Income in Hungary in the 1990s

Peter Meusburger

1 Introduction[1]

It is well known that in communist countries 15-20% of employees in state companies were not needed. This phenomenon was called "unemployment behind the gates" (DÖVÉNYI 1995, 115). Since many state-owned companies were not competitive, it comes as no surprise that since the introduction of the market economy, the percentage of gainfully employed has decreased considerably. In the late 1980s, several authors predicted that female employment rates would be affected to a larger degree by the transition process than those of men. This pessimistic forecast was based on the assumption that the communist systems had achieved gender equality, and that this equality would be destroyed by the introduction of a market economy. However, gender equality under socialism was more of a myth than a reality. While it is true that the percentages of gainfully employed women and men differed only slightly in communist countries, the gender-specific segmentation of employees among the work force was extremely pronounced and spatially varied. This fact was relatively unknown in the West, mainly because data and research results on social inequalities in communist countries were not published until the late 1980s or appeared only at a high level of spatial aggregation which concealed the extent of inequalities. When the relevant data became accessible in the 1990s, it became clear that many scientists in western countries had been led to believe in an ideological construct.

This paper analyses some of the developments and newly created structures of the employment and income situation in Hungary during the 1990s, with special emphasis on gender-specific and regional disparities. It will discuss at which extent Hungary followed in its transformation a path different from that of other formerly communist countries. Because (un)employment statistics provide little indication on which part of the population is searching for employment, the paper will consider how large a percentage of those not gainfully employed are looking for employment.

---

[1]   This study was sponsored by the German Research Foundation (DFG) as part of Project ME 807/9-2

2   The development of employment rates in Hungary

Long before the change of the political system and the introduction of the market economy, Hungarian state-enterprises started to shed unnecessary employees on a massive scale. Between 1980 and 1990, 10.7% of all jobs in Hungary were eliminated. In 70 out of 181 labour market districts the loss of jobs in the 1980s comprised more than 15% (figure 1). The job loss figures were highest in economically underdeveloped districts such as Sajószentpéter (40.0%), Gyönk (31.6%), Mórahalom (30.1%), Biharkeresztes (29.1%), Szerenc (27.9%), Komádi (27.4%), Létavértes (26.2%), Vámospércs (26.1%), Tokaj (25.6%) and Nagykálló (25.2%). Between 1989 and 1992, Hungarian economy suffered an 18% decline in GDP, and industrial output dropped by almost 40% (CSÉFALVAY 2000, 4). This trend caused a second wave of job losses.

Figure 1:   The decrease of work places in Hungary between 1980 and 1990

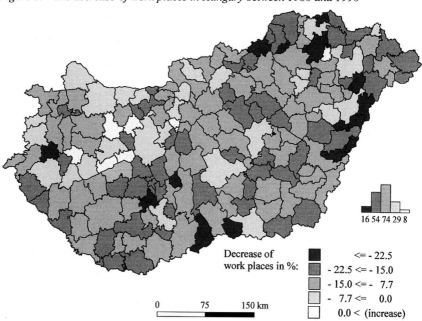

Source: Hungarian census 1980 and 1990; author's compilation from individual data sets.

The following factors have contributed to these changes of employment rates:

- change from 'unemployment behind the gates' to 'statistically documented unemployment';
- additional unemployment where state-enterprises and co-operatives collapsed and where the loss of jobs could not be substituted by the creation of new firms;

- large increase in students attending secondary schools, academies and universities, which delayed their entering the job market;
- change of the value system towards 'family values'; increase of women interrupting employment for a certain period of time after giving birth; growing proportion of women taking advantage of child and family allowances;
- decrease in employment of retirees; in the communist system, many retirees had to work to make a living, formal employment did not lead to a reduction of pensions;
- increase of Hungarians who work abroad;
- increase of Hungarians who commute daily to Austria and thus are listed in the statistics of inhabitants, but not in those about the Hungarian work force.

Even though women were less affected by job losses and unemployment than men until the mid-1990s, changes in employment among women are especially interesting for several reasons. While the employment rates of Hungarian men decreased during the 1980s relatively evenly for all age groups, among women a reduction in employment rates occurred predominantly among those younger than 37 years. Hungarian women aged 37 through 54 actually were able to increase their employment rates during the 1980s. During the early phase of the structural adjustment in the 1980s, women experienced mostly a lengthening and, for women with university degrees, also an intensification of the child-rearing phase (family phase).[2] Whereas men displayed a *general* decrease in employment, women have experienced a decrease and *structural change* in age-specific employment (figure 2).

Contrary to the expectations of many researchers, it were not the lowly skilled women who in the 1980s took advantage of taking a child-rearing phase but rather university graduates. This trend toward an interruption of employment at the birth of a child was partially a reaction to changes in the government regulations regarding maternity leave (child allowances, family support). On the other hand, there are many indications that this trend was a reaction to the anti-family values of the communist system. However, only highly qualified women - women with above-average incomes - could afford to interrupt their gainful employment for a certain period of time. Only between 1990 and 1996 did the employment rates of women decrease in all age groups.

---

[2] The small children or family phase is a temporary interruption of gainful employment for reasons of family (e.g. the birth of a child).

Figure 2:   Changes of age-specific employment rates in Hungary between 1980 and 1996

Men

Employment rate in %

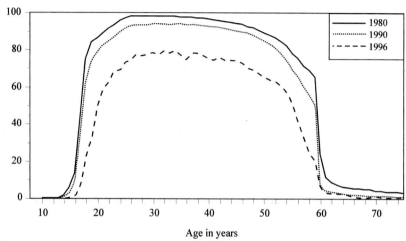

Women

Employment rate in %

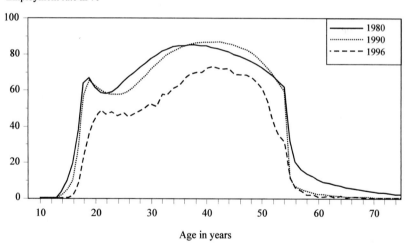

Source: Hungarian census of 1980 and 1990; Hungarian microcensus 1996; author's
compilation from individual data sets. The curve of 1996 was smoothed out by
three years averages.

2.1   The effects of educational achievement on labour force participation of Hungarian women

Social inequalities in work force participation could not be discussed openly in communist states. After 1989, it could be proved, however, that in 1980 age-specific employment rates of women varied strongly with their level of education and with the city-size class of their place of residence, a fact that contradicted the ideological claim of social equality. The higher the level of educational attainment of women the higher their employment rates. With regard to employment rates, gender equality was only to be found within the highest levels of education. The lower the educational attainment, the larger the gender gap with regard to gainful employment.

To achieve the ideologically important goal of full employment, state-companies in the communist system more or less were forced to employ lowly skilled people (even illiterates) which they did not need. The employment of unneeded unskilled labour was often the condition for the assignment of highly qualified experts by the centralised planning authorities. The introduction of market economy first of all demanded skills, knowledge and experience. Therefor, it was no surprise that employees without an eight year primary education were the most vulnerable, that they became the first to be dismissed and that their labour force participation decreased dramatically long before the official introduction of the market economy. The lower the level of educational achievement, the earlier and larger the exclusion from the labour force. The societal transition 'from ideology to meritocracy' or from the socialist goal of full employment to the efficiency and competitiveness demanded by the market economy has had drastic consequences for the employees with low skills and low educational attainment. Between 1990 and 1996, the correlation between level of education and labour force participation in all age groups younger than 50 has increased significantly (tables 1 and 2). The lower the level of educational achievement, the lower the labour force participation.

People who did not attend school or did not complete eight years of primary schooling were the most severely affected by the changes of the employment situation in the 1990s. Among this group, women had even smaller chances of being employed than men. Least affected were the employment rates of university graduates aged 30-49, the age group which represented the majority of new leaders in economy and society.

Figure 3:   *The development of age-specific employment rates of Hungarian women 1980-
1996 according to their level of educational achievement*

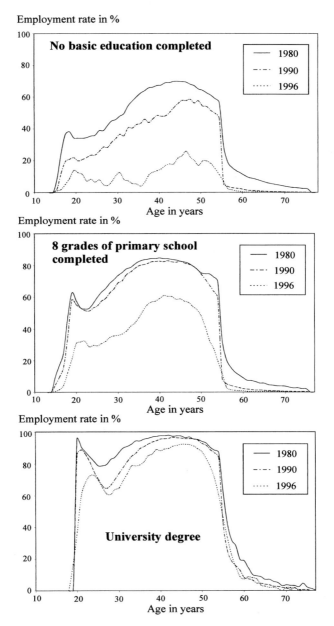

Source: Hungarian census of 1980 and 1990, Hungarian microcensus 1996; author's
compilation from individual data sets. The curve of 1996 was smoothed out by
three years averages.

Table 1: Labour force participation rates of Hungarian women in 1990 and 1996 with regard to their level of education and age

| | Age groups | | | | | | | |
| | 20-29 years | | 30-39 years | | 40-49 years | | >= 50 years | |
Level of education	1990	1996	1990	1996	1990	1996	1990	1996
No primary school completed	13.0	7.8	21.5	8.4	24.2	19.6	5.8	0.8
Primary school	53.8	31.9	75.5	48.1	78.7	58.1	13.5	8.3
Vocational or technical school	63.6	49.2	83.1	60.7	88.5	68.6	*	49.3
Secondary school	62.4	50.8	87.0	70.3	92.5	79.5	34.7	20.3
University or academy	71.3	66.0	86.7	79.5	95.9	90.8	37.9	36.5

*   1990 not available in this age group

Source: Hungarian census 1990; Hungarian microcensus 1996; author's compilation from individual data sets.

Table 2: Labour force participation of Hungarian men in 1990 and 1996 with regard to their level of education and age

| | Age groups | | | | | | | |
| | 20-29 years | | 30-39 years | | 40-49 years | | >=50 years | |
Level of education	1990	1996	1990	1996	1990	1996	1990	1996
No primary school completed	21.3	21.1	29.4	22.4	40.4	25.9	19.3	2.6
Primary school	86.7	55.6	88.4	56.2	86.3	58.2	32.1	18.1
Vocational school or technical school	96.3	75.3	95.6	79.1	92.8	71.7	*	59.9
Secondary school	77.7	63.8	96.6	87.2	94.6	81.9	41.5	30.3
University or academy	96.1	86.4	98.3	96.1	97.1	92.0	47.3	38.7

*   1990 not available in this age group

Source: Hungarian census 1990, Hungarian microcensus 1996; author's compilation from individual data sets.

2.2   Spatial disparities of employment rates

The slogan 'place matters' was equally true under communism and during the transformation period. Women living in villages and small towns with less than 5 000 inhabitants showed much lower employment rates in all three years under scrutiny than women living in Budapest (figure 4). With regard to the proportion of the gainfully employed, the central Budapest region and Western Transdanubia belonged to the winners, they had the smallest decrease of employment rates in the 1990s and thus further strengthened their position in the Hungarian labour market, whereas the north-east of Hungary experienced the largest losses of gainful employment (figures 5-7).

180   P. Meusburger

*Figure 4:   The development of age-specific employment rates of Hungarian women
between 1980 and 1996 according to the size their place of residence*

<= 5,000 inhabitants

Employment rates in %

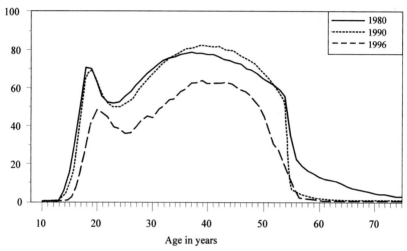

Budapest

Employment rates in %

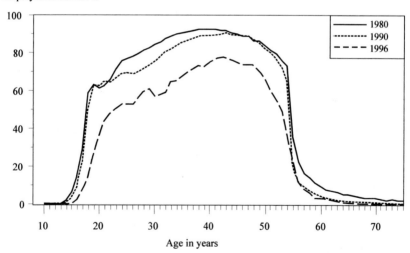

Source: Hungarian census of 1980 and 1990; Hungarian microcensus 1996; author's
compilation from individual data sets.

Figure 5: The rates of the gainfully employed among the 15-60 year old population, 1980 and 1990

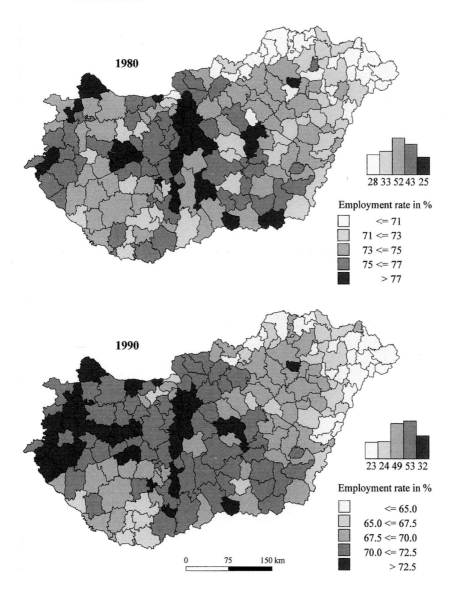

Source: Hungarian census of 1980 and 1990; author's compilation from individual data sets.

Figure 6:   The rates of the gainfully employed among the population of 15 years and older in the years 1980, 1990 and 1996

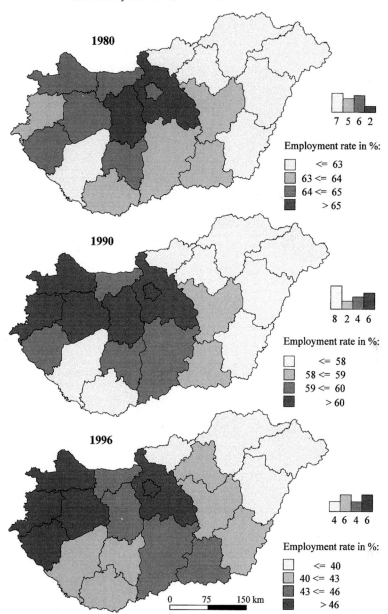

Source: Hungarian census of 1980 and 1990; Hungarian microcensus 1996; author's compilation from individual data sets.

Regional disparities of employment rates between the eastern and western parts of the country, as well as between the capital and the rural areas, which already existed in the socialist planned economy, have increased significantly during the 1990s.[3] In Budapest and in the three provinces at the western border, the loss of employment was very small. In these areas, there actually was a shortage of labour in the late 1990s. Contrasting this, the employment rates in the eastern and south-eastern provinces decreased by more than 20% between 1990 and 1996 (figure 7). The location quotient shows again, that the general decrease of employment rates did not change the *spatial* pattern of employment very much in the 1980s, and that the bifurcation between western and eastern Hungary happened mainly in the first half of the 1990s (figure 8).

Figure 7:   Decrease of employment rates between 1990 and 1996

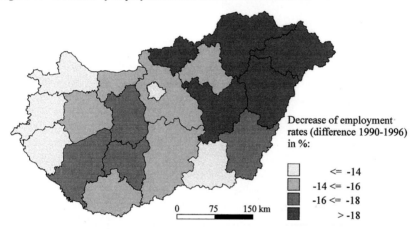

Source: Hungarian census of 1990; Hungarian microcensus 1996; author's compilation from individual data sets.

As men were more affected by the loss of jobs between 1980 and 1996 than women, the percentage of women in the work force[4] (i.e. without women on maternity leave) increased in all of Hungary from 43.4 to 47.1% between 1980 and 1996 (1980-1990: from 43.4% to 44.5%, 1990-1996: from 44.5% to 47.1%). However, one has to emphasise that the surprisingly positive employment situation for women only applies when the entire country of Hungary is considered. For some regions and social categories the situation is not so positive.

---

[3]   While the decrease in employment between 1980 and 1990 can be examined in great spatial detail based on the census results (MEUSBURGER 1995), the data of the microcensus allow only a differentiation by provinces.

[4]   The work force represents the people who are employed in Hungary. Thus, it includes the cross-border commuters to Hungary, but not the Hungarian cross-border commuters who work in neighbour states.

*Figure 8:   The extent of regional disparities of employment rates among the population of
15 years and older - measured by location quotient*

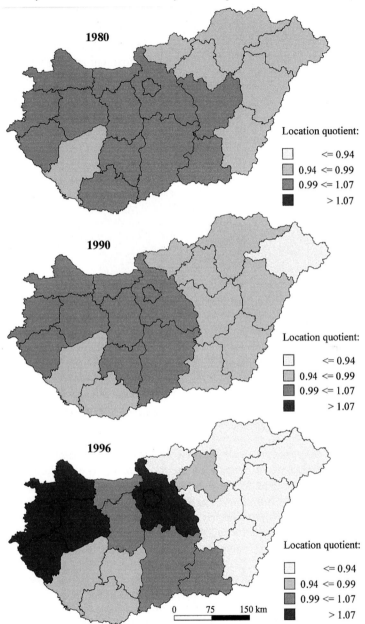

Source: Hungarian census of 1980 and 1990; Hungarian microcensus 1996; author's
compilation from individual data sets.

2.3   The effects of family size (number of children) on female employment rates

In many countries of Europe west of the iron curtain, the different models of family roles (PFAU-EFFINGER 1995; FASSMANN and MEUSBURGER 1997) and the number of school-age children strongly influenced the employment rates of women. Contrasting this, communist countries at least officially had only one family model, and the number of children for many reasons had no significant influence on the employment rates of women. First, there was a duty to work, people not gainfully employed were met with suspicion and were disadvantaged in many ways. Second, many social services, allowances and recreational facilities were available only through the workplace, so it was in a mother's interest to be gainfully employed. Third, the wages in the socialist countries were extremely low. The socialist countries represented the so-called "weak breadwinner" model (LEWIS 1992) in which the earnings of one person were by far not sufficient to sustain a family. As a result, the majority of women in the communist systems did not have the choice to interrupt their employment at the birth of a child or to take a longer family phase. Finally, most state-owned enterprises in the centralised planned economy had a large surplus of employees, and thus could tolerate long absences of women with small children rather generously.

Only toward the late 1980s, and especially in the years between 1990 and 1994, did the number of children become a factor that had a significant influence on the employment rates of women. However, the indicator 'number of children' hides other factors, such as age and level of education of these women. Because employment rates are computed for women older than 15 years, the categories 'women without children' and 'women with one child' include many young women who have not yet completed their education or training and have never been employed. The effect of the number of children thus can be assessed better by limiting the comparison to the three categories 'women with two or more children'. A comparison between women with two children and those with three or more children shows that the difference in the employment rate of these groups has increased between 1980 and 1996 from 7.6% to 36.3%. The rate of employment of women with more than three children dropped between 1980 and 1996 from 65.5% to 24.7%.

2.4   Why is the employment situation of Hungarian women much better than that of women in other former socialist countries?

In Hungary, the increasing percentage of gainfully employed women, the lower unemployment rates of women and the lower percentage of women seeking employment contradict all hypotheses and scenarios, which assumed that gender equality had been achieved in communist systems and which predicted that the introduction of market economy would first of all weaken the position of women on the labour market and again widen the gender gap in employment rates.

Table 3:   The change of female employment rates in Hungary between 1980 and 1996 with regard to the number of children per adult female

	1980	1990	1994	1996
No child	67.8	58.5	45.4	43.0
1 child	71.1	70.1	58.3	54.4
2 children	73.1	76.8	63.6	61.0
3 children	65.6	66.6	46.5	43.4
More than 3 children	65.5	48.6	27.7	24.7
Sample size	2 107 497	1 990 942	8 802	29 482

Source: Hungarian census of 1980 and 1990; Hungarian microcensus of 1994 and 1996; the sample size varies in the two microcensuses; author's compilation from individual data sets.

Why are women in other former socialist countries not in a similarly favourable situation? Whether the introduction of the market economy drove more women than men out of employment depended mostly on the question in which branches men and women were predominantly employed during the last two decades of the communist system. The starting conditions, in which Hungarian women found themselves toward the end of the 1980s, were entirely different from those of other communist countries. Because of the earlier liberalisation and opening of the system, which started already in the 1960s (see the other paper by Peter Meusburger), the development of the service sector (tourism, business services, restaurants, retail) was much more advanced in Hungary than in other socialist countries.

Table 4:   The development of the economic sectors in Hungary and the Czech Republic

	Employment in the three economic sectors in %				
	1991	1992	1993	1994	1995
Agriculture, Forestry, Fishery					
Hungary	7.8	6.5	5.8	6.0	6.2
Czech Republic	10.0	8.6	6.8	6.9	6.2
Industry and construction					
Hungary	31.6	29.6	27.9	27.4	27.7
Czech Republic	46.5	44.8	44.6	42.2	41.5
Tertiary Sector					
Hungary	60.6	63.9	66.3	66.6	66.1
Czech Republic	43.5	46.4	48.6	50.9	52.3

Source: WIIW Databook, http://www.wiiw.ac.at/C-03-2-0.html; http://www.wiiw.ac.at/H-03-1-0.html.

The earlier expansion of the tertiary sector led to an earlier move of Hungarian women from employment in industry and agriculture to jobs in the tertiary sector than in other socialist countries. Thus, in Hungary, many women in time had moved to that sector of the economy which experienced in the 1990s the largest expansion. A comparatively large proportion of Hungarian women have become self-employed. This happened not only in those services which require little qualifications (e.g. retail shops), but also in highly skilled professions. In other socialist countries, a large percentage of women still held low-paid industrial jobs with required low skills and were the first to be set off. The larger the proportion of jobs in industry and the smaller the share of jobs in the tertiary sector, the more vulnerable was the job situation in the 1990s.

Those who know the Hungarian society and history have little doubt that cultural differences regarding the societal status of women played an important role as well. Compared to other socialist countries, it was unthinkable for many Hungarians to allow women to work in steel mills, coal mines or in other dirty or hazardous occupations. This attitude was not a matter of class structure (bourgeois versus workers) but of history and culture. However, the influence of cultural values is difficult to prove with the data available for this study. Thus, early liberalisation, early introduction of elements of market economy, an early move to the service sector and cultural values influenced in Hungary the gender-specific segmentation of employment in a way that turned out to be beneficial for women after the market economy was introduced.

Presumably, fewer observers would have been surprised by the development of the job market situation for women if comprehensive and large-scale studies of employment of women and men had been allowed in the communist countries at an earlier stage. The studies that arrived at very negative predictions for the job market situation of women after the introduction of the market economy usually worked with small samples and qualitative methods. Dealing with 'ideologically charged' topics such as social inequality in communist countries, small case studies are prone to various kinds of fallacies and unrealistic social constructions. This objection is not intended to question the value of qualitative research on the micro scale, it is rather meant to point out that people interviewed in a totalitarian state gave rather the answers expected by the communist party than their own personal view. Census data disaggregated to small spatial units showed a quite different picture about equality in socialism and were therefore kept secret as long as possible.

Faced with the significant reduction in the percentage of the gainfully employed, several questions arise: What caused people to leave the labour market? How many of the unemployed really are looking for employment? And in which occupations and institutions the unemployed used to work?

3   The situation of people who are not gainfully employed

3.1   Reasons for not being employed

The Hungarian microcensus of 1994 and the Hungarian household and budget survey of 1995 make it possible to assess the situation of people not gainfully employed. The year 1994 is especially suitable for this analysis, because the period from 1993 to 1994 saw the low point of employment following the reduction of the workforce. Almost 62% of Hungarians above 15 years of age and not gainfully employed were retired. The second largest group are those who were searching for a job followed by those who have not finished their education yet.

The reasons of not being employed strongly differed by age group, gender, level of education and size of municipality, in which the unemployed lived. Among not gainfully employed *men* in the age group 20-34, 'notice' (being laid off by their employer) was the main reason for not being employed (73-80%), while for men in the age group 45-60, 'illness' and 'retirement' were given as the most important reasons for ending their employment (tables 6 and 7).

Among not gainfully employed *women* in the age group 20-35, 'raising children' was given as the most important reason for ending employment. For women aged 35-49, 'notice' was the most important reason for leaving employment, however, in all age groups much less women have been laid off by their employer than men. Women older than 50 most often cited 'retirement'. 'Illness' was given as a reason for lack of employment much more often by men older than 50 than by women of the same age group. This finding is confirmed by the significantly lower life expectancy for Hungarian men (KLINGER 1995).

Table 5: Activities and sources of income of Hungarians above 15 years of age without a gainful employment, 1995

	Men	Women
Student	15.2	11.3
Household	0.1	7.5
In search of an employment	18.6	8.5
Illness	0.8	0.4
Maternity leave	.	0.8
Various kinds of child support and allowances	0.1	8.7
Retired	62.1	60.1
Handicapped	0.6	0.4
Other reasons	1.9	1.9
No ambition to work	0.7	0.4

Source: Hungarian household and budget survey 1995; author's compilation from individual data sets.

Table 6: Reasons of Hungarian men for ending an income providing economic activity, 1994

Age group	Notice	Retirement	Illness	Other reasons	Sample size
20 – 24	73.3	2.5	3.1	21.1	318
25 – 29	80.1	2.9	5.4	11.6	276
30 – 34	77.8	4.7	12.1	5.4	297
35 – 39	63.3	10.6	20.4	5.7	417
40 – 44	55.5	14.3	27.3	2.9	532
45 – 49	38.4	23.8	34.7	3.1	453
50 – 54	27.8	30.7	39.1	2.4	616
55 – 59	13.2	60.4	25.4	1.0	948
60 and older	0.3	91.3	7.3	1.1	4 838
Total	20.2	61.8	14.6	3.4	8 795

Source: Hungarian microcensus 1994; author's compilation from individual data sets.

Table 7: Reasons of Hungarian women for ending an income providing economic activity, 1994

Age group	Notice	Child rearing	Retirement	Illness	Other reasons	Sample size
20 – 24	20.6	64.9	0.7	2.0	11.8	559
25 – 29	19.7	73.1	1.4	2.3	3.5	710
30 – 34	29.5	53.9	3.0	6.7	6.9	627
35 – 39	41.5	32.3	5.1	13.8	7.3	552
40 – 44	42.1	11.1	12.0	26.1	8.7	551
45 – 49	31.2	5.6	19.6	36.8	6.8	535
50 – 54	17.0	1.8	49.4	27.4	4.4	943
55 - 59	1.3	1.1	85.1	10.7	1.8	1 707
60 and older	0.3	1.4	89.0	6.2	3.1	5 545
total	11.1	13.8	59.9	10.8	4.4	11 837

Source: Hungarian microcensus 1994; author's compilation from individual data sets.

Whether raising children really is the most important reason for leaving employment or whether women used this as a pretext, because bringing up children conveys a higher societal status than being unemployed, cannot be answered unequivocally based on the data of the microcensus. Not gainfully employed women who graduated from vocational and technical schools said most often (46.9%) that they had terminated their employment to bring up their children. At the same time, women who graduated from vocational and technical schools had been laid off more often by their employers (24.8%) than women with other levels of education, because a disproportionately large number of women with this level of education were employed in the industrial sector. Among women between the ages of 21 and 40 years, 11.5% were not gainfully employed because they took care of the household, while 3.6% were on maternity leave. 37.3% of women in this age group said they were not searching for employment because they received child allowances or other forms of family related social benefits.

In contradiction to many expectations in the west, the women's level of educational achievement was positively related to the frequency of their answer 'raising children' as reason for giving up gainful employment. It were not the lowly skilled with the highest rates of unemployment and the lowest wages, but rather the women with the best perspectives on the labour market that gave up or interrupted their employment after birth of a child. Among those women who did not attend or complete eight years primary schooling, only 3% said they left their employment to bring up their children. Women who completed the compulsory schooling cited the reason 'raising children' in 10.6% of all cases. In contrast, 24.3% of not gainfully employed female secondary school graduates and 33.7% of female university graduates had given up their employment to raise their children. These results are underlined by the question in which period of time not gainfully employed women would be available, if they were offered a job. The higher the educational achievement of not gainfully employed women the less would be available on short notice (table 8).

Table 8: The period of time in which women without gainful employment would be available to the labour market with regard to their level of educational attainment, 1996

Available in	No basic education completed	Primary school	Vocational or technical school	University or Academy
Less than 1 week	71.6	67.2	66.1	58.7
1-2 weeks	6.6	10.2	12.2	8.2
3-4 weeks	3.8	4.1	3.7	4.3
More than 1 months	4.3	5.6	6.0	15.8
Do not know	13.7	13.0	12.0	13.0
Sample size	211	1 709	1 147	184

Source: Hungarian microcensus 1996; author's compilation from individual data sets.

There are several explanations for these results. First, these data indicate that the change of family values started in the upper classes with highest levels of educational achievement. However, it has to be kept in mind that women with a very low level of education are concentrated to a disproportionately large degree in peripheral rural areas. In small rural municipalities, the 'children phase' (maternity leave) was very pronounced even during socialist times (figure 4), thus there was no catching up necessary after the economic system changed. In addition, one has to consider that even during the final phase of socialism, the two lowest levels of education had the lowest employment rates.

Together with the fact that university graduates used the so-called 'maternity leave' to a larger degree than women with lower levels of education, and that less of them are looking for employment, these figures seem to show that the expected return to 'family values' at the end of communism is apparent mostly for well-educated women, who can afford to do so.

3.2   Central-peripheral disparities with regard to the proportion of people not gainfully employed

Various authors have stressed the fact that the authorities of central planning have deliberately discriminated against rural areas in resource allocation. Because of the decades-long insufficient allocation of infrastructure, qualified jobs, decision-making powers and financial resources in peripheral rural areas, especially villages with less than 5 000 inhabitants had a disadvantageous starting position and were least prepared to profit from the introduction of the market economy. Sorting the employment and unemployment structures by the size of the place of residence[5] shows that in municipalities with less than 5 000 inhabitants, many social and economic problems are much more pronounced. Municipalities of this size class feature the highest proportion of laid-off employees, of unemployed, of job seekers, of especially low qualified persons, of people who left employment because of illness and of people whose earnings put them in the lowest income segments.

Men, who lived in municipalities of the size class with up to 10 000 inhabitants, have left their 'active gainful employment' twice as often (22.8%) due to being laid off by their employers as those who lived in Budapest (11.9%). Furthermore, 'illness' was given as the reason for terminating employment almost twice as often (16.9%) in municipalities with less than 5 000 inhabitants compared to Budapest (9.7%). Women display similar trends, however, these are not as pronounced as they are among men (tables 9 and 10).

The motivation 'raising children', which was given by 13.8% of Hungarian women as the main reason for terminating employment, varied only slightly by size class of municipality. Only Budapest was with 8.2% significantly lower than the Hungarian average. This can be explained by the higher percentage of single women in large cities in general. Significant central-peripheral disparities were noticed in the percentage of women and men above 15 years of age who were not gainfully employed because the were attending a secondary school or university. In towns with more than 20 000 inhabitants, every fifth not gainfully employed man (over the age of 15 years) and every seventh not gainfully employed woman

---

[5]   Obviously, a differentiation by size classes of municipalities represents a model-like abstraction, which does not apply to all cases and which is modified by the well known west-east disparities in Hungary. Small villages which are located near the Austrian border or near a prospering agglomeration perform economically much better than the same size of villages near the Ukrainian border. Nevertheless, the variable size class represents better than any other available indicator the hierarchy of the urban system.

had not yet completed their education. By contrast, in municipalities with less than 5 000 inhabitants, this applied only to every tenth man and every fourteenth woman.

Table 9: Reasons of Hungarian men of 15 years and older for not being gainfully employed with regard to the size of their place of residence, 1995

City size: number of inhabitants	Study	In search of employment	Retired	Others
Up to 500	10.6	20.2	63.3	5.9
501 – 1 000	8.8	21.3	65.6	4.3
1 001 – 2 000	11.0	20.4	65.1	3.5
2 001 – 5 000	11.9	24.4	58.6	5.3
5 001 – 10 000	13.6	18.5	65.5	2.4
10 001 – 20 000	16.3	19.2	59.7	4.8
20 001 – 100 000	20.2	15.3	60.0	4.5
100 001 – 1 million	19.8	15.6	59.5	5.1
Budapest	20.7	11.1	66.0	2.2
Total	15.2	18.6	62.1	4.1

Source: Hungarian household and budget survey 1995; author's compilation from individual data sets.

Table 10: Reasons of Hungarian women of 15 years and older for not being gainfully employed with regard to the size of their place of residence, 1995

City size: number of inhabitants	Study	House-hold	In search of employment	Maternity leave, various kinds of child support and allowances	Retired	Others
Up to 500	7.2	13.3	7.9	7.5	61.8	2.3
501 – 1 000	8.1	8.8	9.0	9.2	61.3	3.6
1 001 – 2 000	8.8	9.9	9.5	8.8	59.9	3.1
2 001 – 5 000	10.0	11.2	8.0	9.8	58.3	2.7
5 001 – 10 000	10.3	9.1	8.8	8.5	60.6	2.7
10 001 – 20 000	11.0	7.1	8.5	10.5	59.2	3.7
20 001 – 100 000	14.6	4.8	8.2	11.2	57.5	3.7
100 001 – 1 million	13.8	2.8	7.8	8.5	63.4	3.7
Budapest	13.8	1.8	8.7	8.3	65.9	1.5
Total	11.3	7.5	8.5	9.5	60.1	3.1

Source: Hungarian household and budget survey 1995; author's compilation from individual data sets.

There are significant disparities between cities and rural areas in the percentage of women who see the main reason for a lack of gainful employment in the 'managing of the household', an answer that should not be confused with 'raising children'. The high percentages of women without gainful employment in rural areas have different consequences for family incomes than similar percentages in large cities. In rural areas, women have numerous opportunities to contribute to the family income.

3.3 The percentage of job seekers among those who are not gainfully employed

For several reasons, unemployment rates do not convey a clear image of the actual percentage of the population without gainful employment who are in fact looking for employment. On the one hand, people stop receiving unemployment benefits after a certain amount of time and thus do not count any longer in the unemployment statistics. Furthermore, unemployment rates give little indication of the so-called 'quiet reserve',[6] which might be relatively large among women, who gave 'managing the household' and 'raising children' as main reasons for lack of gainful employment. Among job seekers, one also has to distinguish between those who lost their job and those who have never been gainfully employed.

Among the most interesting results of the Household and Budget Survey in 1995 is the fact that only 8.5% of all not gainfully employed Hungarian women stated in 1995 that they were looking for employment (table 10). The percentage of men looking for a job was more than twice as high (18.6%). These results, as well as the lower unemployment rates of women, show once more that the transformation to a market economy has had significantly more negative consequences for the employment situation of men.

As expected, the percentage of job seekers varies with age, level of education and place of residence of the unemployed. In 1995, the highest percentage of job seekers was among the 21 to 40-year-old unemployed. In this group, 62.3% of men and 24.3% of women who were not gainfully employed were looking for a job in 1995. Considering the large central-peripheral disparities in employment rates of women, it is interesting to note that there were no such differences in the percentages of women looking for employment. The percentage of women without gainful employment, who received various forms of maternity and child allowances, differed only slightly between Budapest and peripheral rural regions. The relatively small percentage of women looking for jobs in rural areas may be caused by a disillusionment regarding employment opportunities, especially for women with a low level of education. Thus, it is likely that many women have given up their search for employment.

Contrasting this, very large central-peripheral differences were noted with regard to the proportion of men who are looking for employment. In municipalities with less than 5 000 inhabitants, every fourth to every fifth man without gainful employment was looking for a job, while in Budapest only every ninth was

---

[6]   The 'quiet reserve' comprises persons who, under certain circumstances, would be willing to work for money.

doing so. Among those women who had never before been gainfully employed, only 6.4% were searching for a job in 1995, while almost three times as many men (18.4%) of this group were doing so. Many Hungarians, who were looking for a job during the mid-1990s, already had found a new job in a private company after the changeover to the market economy but had lost this job. Obviously, employment at one of the many small private companies, which mushroomed after the system change carried the highest risk of becoming unemployed again. About 62% of men and 30% of women who had been employed by such an entrepreneur and were not gainfully employed in 1995, were again looking for a job. Also almost 15% of the former self-employed without employees were looking for a job again in 1995, indicating the high rate of failure of the companies they founded (tables 11 and 12).

Table 11: Formerly employed male population of Hungary without an income providing economic activity according to their latest employment, 1995

Latest employment	In search of employment	Retired	Others
Employee of a state enterprise or state institution	15.0	82.2	2.8
Employee of a co-operative or association	32.2	62.4	5.4
Employee of a private company	61.9	27.4	10.7
Member of an agrarian co-operative	6.1	92.9	1.0
Member of a non-agrarian co-operative	2.4	95.1	2.5
Entrepreneur without employees	15.7	75.3	9.0

Source: Hungarian household and budget survey 1995; author's compilation from individual data sets.

Table 12: Formerly employed female population above 15 years without an income providing economic activity according to their latest employment, 1995

Latest employment	In search of employment	Retired	Household, child care allowance, others
Employee of a state enterprise or state institution	8.1	75.4	14.5
Employee of a co-operative or society	13.9	63.1	19.9
Employee of a private company	30.2	24.0	39.6
Member of an agrarian co-operative	3.1	92.1	3.8
Member of a non-agrarian co-operative	2.6	87.1	7.8
Entrepreneur without employees	5.4	66.7	20.7

Source: Hungarian household and budget survey 1995; author's compilation from individual data sets.

*Figure 9:   Regional disparities in the proportion of people between 20 and 50 years of
age searching for a gainful employment, 1996*

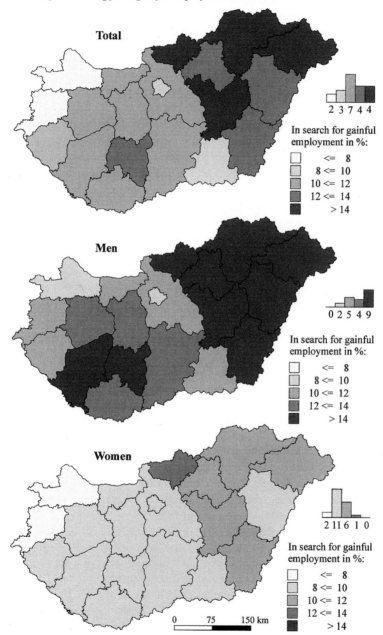

Source: Hungarian microcensus 1996; author's compilation from individual data sets.

In 1995, a surprisingly small percentage of men and women who had been members of an agricultural co-operative were looking for employment. Obviously, the members of agrarian co-operatives had aged so much at the time of the changeover to the market economy that most of them could retire.

Figure 9 shows that the regional disparities of the percentage of job-seekers displayed similar differences between western and eastern Hungary as those known from unemployment and other indicators of economic performance. In 1996, relatively few people among the population aged 20-50 were looking for employment in the economically prospering western border districts of Győr-Moson (7.8%), Vas (8.0%) and Zala (10.0%), as well as in Budapest (9.1%) and in the southern border province of Csongrád (9.5%). On the other hand, Nógrád (18.3%), Borsod-Abaúj (15.5%), Jász-Nagykun (14.5) and Szabolcs-Szatmár (14.2%) had the highest percentages of job seekers. These disparities between west and east mostly are caused by job-seeking men, who display both a higher level of job seekers as well as a larger range (lowest value 6.2, highest 25.8%) of spatial disparities than women (lowest value 6.2, highest 12.4%). It is interesting to note that those districts, which display the highest percentages of job-seeking (and unemployed) men, at the same time feature an especially high amount of gainfully employed women aged 20-50. This indicates again, that female employment rates are strongly influenced by the financial situation of the family and that the 'weak bread winner family' is much more frequent in the north-east of Hungary.

4   Regional and social differences in income

Only a few of the various types of income (gross income, income from main and secondary employment, income after taxes, income from retirement benefits, etc.), which were assessed in the Household and Budget Survey of 1995, will be examined here. The regional patterns of most types of income are very similar: All are characterised by clear west-east and central-peripheral differences (tables 14 and 15). The average total personal income of the female population was about 25% lower than that of males. Nevertheless, female entrepreneurs with up to 10 employees had a higher income than male entrepreneurs.

According to the microcensus of 1996, 3.4% of gainfully employed Hungarians (4.1% of men and 2.5% of women) had a *second occupation*. The proportion of persons with a second occupation increased with their level of educational attainment. Amongst persons with primary school certificate or less only 1.1% had a second job, among those with vocational or technical school the figure was 2.0%, among those with secondary school education the percentage was 3.6 and among the university graduates 11.0%. 50.7% of these declared themselves as entrepreneur (among them 92% without employees) and 9.3% were partners in a company.

Table 13:   Average total income of Hungarian active earners according to their main occupation and economic activity, 1995

Occupation, activity	Total	Average total income in HUF		
		Men	Women	Women compared to men men=100
Assisting family members, manual work	75 386	81 342	73 447	90.3
Assisting family members, non-manual work	79 500	-	79 500	-
Casual workers	111 921	110 044	117 506	106.8
Unskilled workers	217 027	229 188	204 586	89.3
Entrepreneurs without employees	248 693	249 158	247 580	99.4
Entrepreneurs, manual work	249 483	263 059	212 507	80.8
Skilled workers	269 507	305 003	243 096	79.7
Members of an agrarian cooperative	310 661	325 557	274 326	84.3
Artisans	354 169	384 147	270 873	70.5
Office workers	387 461	417 581	384 728	92.1
Employees in cooperatives	392 601	427 488	345 156	80.7
Employees in private companies	403 861	455 832	366 199	80.3
Entrepreneurs, non-manual work	433 807	405 314	467 365	115.3
Entrepreneurs with 1-10 employees	439 778	425 164	478 051	112.4
Employees with secondary school level	442 440	512 784	416 783	81.3
Employees with university degree	538 708	700 236	510 663	72.9
Executives, lower management	638 016	668 448	588 910	88.1
Executives, upper management	910 931	964 905	818 628	84.8

Source: Hungarian household and budget survey 1995; author's compilation from individual data sets.

4.1   Income disparities between western and eastern Hungary

Spatial differences in income mostly are a result of the regional patterns of jobs for various professions and skill levels, of the spatial distribution of jobs in joint ventures and affiliations of multinational enterprises, of the positive income effects of the Austrian border, of differences in export activities and of disparities in economic performance. Figure 10 illustrates the spatial disparities of the proportion of employees working in joint ventures, which generally pay significantly higher salaries as equivalent domestic firms (see chapter by Gábor Hunya).

Figure 10: Proportion of the workforce employed in joint ventures, 1996

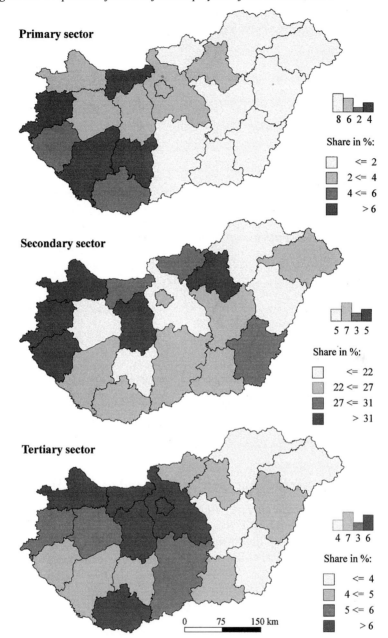

Source: Hungarian microcensus 1996; author's compilation from individual data sets.

Figure 11: Regional differences of income, 1995

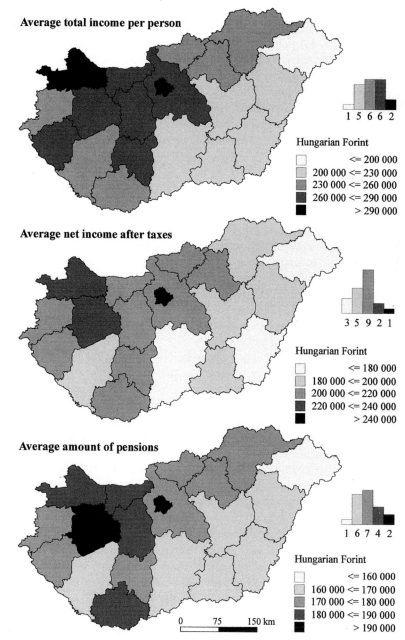

Average total income per person

1 5 6 6 2

Hungarian Forint

☐	<= 200 000
▨	200 000 <= 230 000
▨	230 000 <= 260 000
▨	260 000 <= 290 000
■	> 290 000

Average net income after taxes

3 5 9 2 1

Hungarian Forint

☐	<= 180 000
▨	180 000 <= 200 000
▨	200 000 <= 220 000
▨	220 000 <= 240 000
■	> 240 000

Average amount of pensions

1 6 7 4 2

Hungarian Forint

☐	<= 160 000
▨	160 000 <= 170 000
▨	170 000 <= 180 000
▨	180 000 <= 190 000
■	> 190 000

0   75   150 km

Source: Hungarian household and budget survey 1995; author's compilation from
          individual data sets.

The average *income after taxe*s shows significant spatial disparities. Apart from Budapest (HUF 272 711), the western Hungarian provinces of Győr-Sopron (232 208), Veszprém (225 344) and Komárom (218 462) have the highest net incomes per person. At the other end of the scale were, as expected from the other indicators, Szabolcs-Szatmár (169 531), Békés (178 305) and Bács-Kiskun (179 035). Thus, the average income after taxes in Budapest was 61% higher and that of Győr-Sopron was 37% higher than that of Szabolcs-Szatmár. Even within the same profession, wages and salaries usually are higher in the capital region of Budapest and along the strong economic growth axis between Budapest and Székesfehérvár, as well as the one between Győr and Moson, than in eastern and south-eastern Hungary

4.2   Income disparities between central and peripheral regions

Based on previous studies of the economic situation, most of the income dispari-ties between west and east were expected. However, little is known about income disparities between central and peripheral regions, and about the extent of poverty in rural regions. In Budapest, the average personal income was about 65-66% higher than that in municipality size classes with up to 5 000 inhabitants, and about 23% higher than that in municipality size classes with between 20 001 and 100 000 inhabitants (table 14). Among women, the range of the personal income disparities (Budapest compared to municipalities with less than 5 000 inhabitants) was significantly greater than among men.

While the average total income of Hungarians was 257 462 Forint, the percent-ages of people who earned twice as much as the average (more than 500 000 Forint), or more than four times as much than average are of special interest. 12.8% of male and 6.4% of female Hungarians earned a total income of at least twice the average income, while 1.8% of male and 0.6% of female Hungarians earned at least four times (more than 1 000 000 Forint) the average income.

A comparison of tables 14 and 15 illustrates the 'social gap', which opens be-tween small rural municipalities and large cities. In 1995, the percentage of people belonging to the lowest income group (below 100 000 Forint) in size classes with less than 5 000 inhabitants was more than twice as large as in Budapest. More women than men are in the lower income classes in all municipality size classes.

The range of central-peripheral disparities (between Budapest and municipali-ties with less than 5 000 inhabitants) of total personal income was larger among women than among men. Gender disparities were larger within the high income classes than among those belonging to the lowest income class. Among men liv-ing in Budapest, 23.2% had a total income that was twice as large as the Hungar-ian average. In the size classes with between 100 001 and 1 000 000 inhabitants, the percentage of this group was 20.6%, in the size classes with between 20 001 and 100 000 inhabitants 17.5% and in the size classes under 5 000 inhabitants, it still was 7.8%.

Table 14:   The proportion of people in the lowest income groups with regard to the size of their place of residence, 1995

Size of the city (commune) of residence	Proportion of persons receiving a total income of less than 100 000 Forint	
Number of inhabitants	men	women
up to 5 000	13.4	15.9
5 001-10 000	11.8	12.8
10 001-20 000	12.4	11.6
20 001-100 000	8.6	9.9
100 001 – 1 million	8.9	7.6
Budapest	5.7	7.7
Sample size	8 912	10 964

Source: Hungarian household and budget survey 1995; author's compilation from individual data sets. The term 'total income' comprises salary, child allowance, pension, unemployment fees and other kinds of income.

Table 15:   The proportion of people in the highest income groups with regard to the size of their place of residence, 1995

Size of the city of residence	Proportion of persons receiving a total income of			
	more than 500 000 Forint		more than 1 million Forint	
Number of inhabitants	men	women	Men	women
up to 5 000	7.8	2.9	0.5	0.1
5 001-10 000	9.0	2.9	0.6	0.1
10 001-20 000	11.4	5.1	1.0	0.0
20 001-100 000	17.5	8.5	3.6	0.8
100 001 – 1 million	20.6	11.1	2.8	1.2
Budapest	23.2	15.7	5.1	2.7

Source: Hungarian household and budget survey 1995; author's compilation from individual data sets. The sample comprises 8 912 men and 10 964 women receiving any kind of income (salary, social security, pension, unemployment fees).

5   Conclusion

In a period of transformation and dynamic growth, increasing spatial and social disparities are a normal phenomenon. Both the gap between the highest and lowest salaries and the disparities in economic performance between the dynamic growth poles and the underdeveloped peripheries have widened in the 1990s. Other spatial and social disparities created or enforced during the communist period of central planning, however, have decreased during the first years of market economy.

Whereas social and spatial disparities are unavoidable in any political system, close attention should be paid to unemployment and the proportion of people living below the poverty line. Due to the fact that the Hungarian economy and productivity was growing faster in the late 1990s than that of most EU-members, Hungary was very successful in reducing unemployment. With an unemployment rate of 7.0% in 1999, Hungary had a much better situation than Slovakia (16.0), Lithuania (13.3), France (12.1), Italy (11.7), Poland (10.6), Germany (8.9) or the Czech Republic (8.7). In May 2000 the period of eligibility for unemployment benefit was reduced to 9 months. As a result the unemployment rate will further decrease for statistical reasons and the number of long-term unemployed people who receive welfare from local governments will rise. The provision of regular welfare benefit is subject to a minimum of 30 days of communal work. The long-term unemployed doing communal work will receive a benefit equal to the minimum wage (ÉKES and KUPA/KREKÓ 2000, 38). The economic gap between Hungary and the EU-member states is narrowing in a quick pace. It is worth noting, however, that the GDP per capita in Hungary was only 53.3% of the EU-15 average in 2000 (CSÉFALVAY 2000, 5).

Poverty was tabooed in the communist period but became a big issue in the first years of the transformation process. Between 1990 and 1998, the number of financially endangered children has increased from 175 000 to 380 000 (ÉKES and FEKETE 2000, 65). In the mid 1990s 23% of the children in Hungary lived below the poverty line, that means they were raised in families with a household income less than half of the average income (in Germany this ratio was 7% and in Sweden 3%). 15% of the couples without children and 30% of single earners were poor at least once between 1992 and 1996 (ÉKES and KUPA/KREKÓ 2000, 40). In order to reduce poverty the Ministry of Finance is to raise the minimum wage to HUF 40 000 in the year 2001 and to HUF 50 000 in 2002. The raise to be introduced in 2001 will affect 750 000 people (ÉKES and FEKETE 2000, 64). Increasing the minimum wage will not only reduce poverty but may have a number of other impacts and side effects discussed by ÉKES and FEKETE (2000, 64-65). Even if net savings of poor families will not increase, their consumption will. Housewives who were not willing to work for the previous minimum wages might be inclined to return to the labour force again. The minimum wage in Hungary will be very attractive to Ukrainian and Romanian workers, which could further tempt entrepreneurs to use illegal workforce. A significant minimum wage increase will probably entail an increase of wages in general, as well as of pensions, health and maternity allowances. Whether increasing wages will lead to more unemployment or not, will partly depend on the development of wages in other Central and East European countries competing with Hungary for export markets and foreign direct investment.

References

CHAPMAN, J. G. (1979): Are Earnings More Equal under Socialism: The Soviet Case, with Some United States Comparisons. In: MORONEY, J. R. (ed.): Income Inequality: Trends and International Comparisons. Lexington Books, Lexington, 43-67

CSÉFALVAY, Z. (1995a): Ostmitteleuropa im Umbruch. In: MEUSBURGER, P., KLINGER, A. (eds): Vom Plan zum Markt: Eine Untersuchung am Beispiel Ungarns. Physica Verlag, Heidelberg, 19-28

CSÉFALVAY, Z. (1995b): Raum und Gesellschaft Ungarns in der Übergangsphase zur Marktwirtschaft. In: MEUSBURGER, P., KLINGER, A. (eds): Vom Plan zum Markt: Eine Untersuchung am Beispiel Ungarns. Physica Verlag, Heidelberg, 80-98

CSÉFALVAY, Z. (1997): Aufholen durch regionale Differenzierung? Von der Planwirtschaft zur Marktwirtschaft: Ostdeutschland und Ungarn im Vergleich.Steiner Verlag, Stuttgart (= Erdkundliches Wissen 122)

CSÉFALVAY, Z. (2000): Széchenyi Plan: The Background, Objectives and Funding. In: Economic Trends in Hungary 2, Ecostat, Budapest, 4-12

DÖVÉNYI, Z. (1995): Die strukturellen und territorialen Besonderheiten der Arbeitslosigkeit in Ungarn. In: MEUSBURGER, P., KLINGER, A. (eds): Vom Plan zum Markt: Eine Untersuchung am Beispiel Ungarns. Physica Verlag, Heidelberg, 114-129

ÉKES, I., FEKETE, G. (2000): The Impact of Increasing the Minimum Wage. In: Economic Trends in Hungary 2, Ecostat, Budapest, 64-65

ÉKES, I., KUPA/KREKÓ, I. (2000): On Key Social Trends. In: Economic Trends in Hungary 2, Ecostat, Budapest, 36-42

FASSMANN, H. (1995): Wegbegleiter nach Europa: Ökonomische Krisen und wachsende Arbeitslosigkeit in Ost-Mitteleuropa. In: MEUSBURGER, P., KLINGER, A. (eds): Vom Plan zum Markt: Eine Untersuchung am Beispiel Ungarns. Physica Verlag, Heidelberg, 1-18

FASSMANN, H., MEUSBURGER P. (1997): Arbeitsmarktgeographie. Teubner Verlag, Stuttgart

FOGARTY, M. P., RAPOPORT, R., RAPOPORT, R. N. (1971): Sex, Career and Family: Including an International Review of Women's Roles. Allen & Unwin, London

FONG, M., PAULL, G. (1992): Women and Employment: Eastern Europe. In: MOGHADAM, V. M. (ed.): Privatization and Democratization in Central and Eastern Europe and the Soviet Union: the Gender Dimension. World Institute for Development Economics Research of the United Nations University, Helsinki, 44-49

FREY, M. (1992): Nők és a munkanélküliség. (Women and Unemployment). Nőtudományi Füzetek 1, Magyar Nők Szövetsége, Budapest

204   P. Meusburger

GÖMÖRI, E. (1980): Special Protective Legislation and Equality of Employment Opportunity for Women in Hungary. International Labour Review 119, 67-77

HAJDÚ, Z. (2000): Transformation of the Hungarian Public Administration System after 1989. In: KOVÁCS, Z. (ed.): Hungary Towards the 21st Century: The Human Geography of Transition. Geographical Research Institute, Hungarian Academy of Sciences, Budapest (= Studies in Geography in Hungary 31), 99-115

HEGEDÜS, M. (2000): The Competitiveness of the Hungarian Economy. In: Economic Trends in Hungary 2, Ecostat, Budapest, 21-26

KISS, É. (2000): The Hungarian Industry in the Period of Transition. In: KOVÁCS, Z. (ed.): Hungary Towards the 21st Century: The Human Geography of Transition. Geographical Research Institute, Hungarian Academy of Sciences, Budapest (= Studies in Geography in Hungary 31), 227-240

KLINGER, A. (1995): Die demographische Lage von Ungarn zwischen 1960 und 1990 im europäischen Vergleich. In: MEUSBURGER, P., KLINGER, A. (eds): Vom Plan zum Markt: Eine Untersuchung am Beispiel Ungarns. Physica Verlag, Heidelberg, 29-61

KONCZ, K. (1994): A nők foglalkoztatása és a pályák elnőiesedése. (Women's Employment and Feminization of Occupation). Társadalmi Szemle 7-8, Budapest, 122-132

KOVÁCS, Z. (2000): Hungary at the Threshold of the New Millennium: The Human Geography of Transition. In: KOVÁCS, Z. (ed.): Hungary Towards the 21st Century: The Human Geography of Transition. Geographical Research Institute, Hungarian Academy of Sciences, Budapest (= Studies in Geography in Hungary 31), 11-27

LAPIDUS, G. (1976): Occupational Segregation and Public Policy: A Comparative Analysis of American and Soviet Patterns. In: BLAXALL, M. (ed.): Women and the Workplace: The Implications of Occupational Segregation. Univ. of Chicago Press, Chicago, 119-136

LEWIS, J. (1992): Gender and the Development of Welfare Regimes. Journal of European Social Policy 2, 159-173

MEUSBURGER, P. (1995): Zur Veränderung der Frauenerwerbstätigkeit in Ungarn beim Übergang von der sozialistischen Planwirtschaft zur Marktwirtschaft. In: MEUSBURGER, P., KLINGER, A. (eds): Vom Plan zum Markt: Eine Untersuchung am Beispiel Ungarns. Physica Verlag, Heidelberg, 130-181

MEUSBURGER, P. (1997): Spatial and Social Inequality in Communist Countries and in the First Period of the Transformation Process to a Market Economy: The Example of Hungary. Geographical Review of Japan 70 (Ser B.), 126-143

MEUSBURGER, P. (1998): Bildungsgeographie. Wissen und Ausbildung in der räumlichen Dimension. Spektrum Akademischer Verlag, Heidelberg

MOGHADAM, V. M. (1992): Gender and Restructuring: a Global Perspective. In: MOGHADAM, V. M. (ed.): Privatization and Democratization in Central and Eastern Europe and the Soviet Union: The Gender Dimension. World Institute for Development Economics Research of the United Nations University, Helsinki, 9-23

MORONEY, J. R. (1979): Do Women Earn Less under Capitalism? In: MORONEY, J. R. (ed.): Income Inequality. Trends and International Comparisons. Lexington Books, Lexington, 141-167

MÜLLER, W., WILLMS-HERGET, A., HANDL, J. (1983): Strukturwandel der Frauenarbeit 1880 - 1980. Campus Verlag, Frankfurt a.M

MYRDAL, A., KLEIN, V. (1960): Die Doppelrolle der Frauen in Familie und Beruf. Kiepenheuer & Witsch, Köln

NEMES-NAGY, J. (2000): Regional Inequalities in Hungary at the End of the Socio-Economic Transition. In: KOVÁCS, Z. (ed.): Hungary Towards the 21st Century: The Human Geography of Transition. Geographical Research Institute, Hungarian Academy of Sciences, Budapest (= Studies in Geography in Hungary 31), 87-98

OECD (ed.) (1988): Women's Economic Activity, Employment and Earnings: A Review of Recent Developments. In: OECD (ed.): Employment Outlook 1988. Paris, 129-172

OFER, G., VINOKUR, A. (1985): Work and Family Roles of Soviet Women: Historical Trends and Cross-Section Analysis. Journal of Labor Economics 3 (1), Supplement, 328-354

OPPENHEIMER, V. K. (1982): Work and the Family: A Study in Social Demography. Academic Press, New York, London

PALÁNKAI, T. (2000): Measuring Maturity for Integration. In: Economic Trends in Hungary 2, Ecostat, Budapest, 13-20

PAUKERT, L. (1984): The Employment and Unemployment of Women in OECD-Countries. Paris

PAUKERT, L. (1992): Women and Employment: Czechoslovakia. In: MOGHADAM, V. M. (ed.): Privatization and Democratization in Central and Eastern Europe and the Soviet Union: the Gender Dimension. World Institute for Development Economics Research of the United Nations University, Helsinki, 29-36

PERRONS, D. (1998): Maps of Meaning: Gender Inequality in the Regions of Western Europe. European Urban and Regional Studies 5, 13-25

PETÖ, I. (1990): Distinctive Features of the Hungarian Economy: Changes in its Workings from 1945 Onwards. In: New Tendencies in the Hungarian Economy: Studies on Hungarian State and Law 2, Budapest, 7-60

PFAU-EFFINGER, B. (1995): Erwerbsbeteiligung von Frauen im europäischen Vergleich. Am Beispiel von Finnland, den Niederlanden und Westdeutschland. Informationen zur Raumentwicklung (1/1995), 49-59

RECHNITZER, J. (2000): The Features of the Transition of Hungary's Regional System. Centre for Regional Studies of Hungarian Academy of Sciences, Pécs (= Discussion Paper 32)

SCHMIDT, M. G. (1993): Erwerbsbeteiligung von Frauen und Männern im Industrieländervergleich. Leske + Budrich, Opladen

SCHULTZ, T. P. (1990): Women's Changing Participation in the Labour Force: A World Review. Economic Development and Cultural Change 38, 457-488

SEMYONOV, M. (1980): The Social Context of Women's Labor Force Participation: A Comparative Analysis. American Journal of Sociology 86, 534-550

SHAFFER, H. G. (1981): Women in the Two Germanies. A Comparative Study of a Socialist and a Non-Socialist Society. Pergamon Press, New York, Oxford

SORRENTINO, C. (1983): International Comparisons of Labour Force Participation, 1960-81. Monthly Labour Review 106 (2), 23-36

SZABADY, E. (1969): Gainful Occupation and Motherhood Position of Women in Hungary. Population Review 13, 59-67

SZALAI, J. (1991): Some Aspects of the Changing Situation of Women in Hungary. Signs: Journal of Women in Culture and Society 1, 152-170

TESSARING, M. (1981): Qualifikation und Frauenerwerbstätigkeit. In: KLAUDER, W., KÜHLEWIND, G. (eds): Probleme der Messung und Vorausschätzung des Frauenerwerbspotentials. Beiträge zur Arbeitsmarkt- und Berufsforschung 56, 82-106

TIMÁR, J. (1993): The Changing Rights and Conditions of Women in Hungary. In: NASH, M. (ed.): From Dictatorship to Democracy: Women in Mediterranean, Central and Eastern Europe. Universidad de Barcelona, 181-194

TIMÁR, J. (2000): Geographical Aspects of Changing Conditions of Women in Post-Socialist Hungary. In: KOVÁCS, Z. (ed.): Hungary Towards the 21st Century: The Human Geography of Transition. Geographical Research Institute, Hungarian Academy of Sciences, Budapest (= Studies in Geography in Hungary 31), 151-167

TIMÁR, J., VELKEY, G. (1998): The Question Marks of the Relatively Favourable Rate of Female Unemployment in Hungary. In: BASSA, L., KERTÉSZ, L. (eds): Windows on Hungarian Geography. Geographical Research Institute, Hungarian Academy of Sciences, Budapest (= Studies in Geography in Hungary 28), 199-209

WEIL, G. (1992): Women and Employment: Hungary. In: MOGHADAM, V. M. (ed.): Privatization and Democratization in Central and Eastern Europe and the Soviet Union: the Gender Dimension. World Institute for Development Economics Research of the United Nations University, Helsinki, 24-29

Development and Spatial Disparities of Unemployment in Hungary

Zoltán Dövényi

1 Introduction and historical perspective[1]

Hungary's official introduction of unemployment benefits on January 1, 1989, just before the profound political transformation, marked the collapse of one of the foundations of the socialist ideology. Total employment was considered an important advantage of the socialist system when compared to capitalism. In early 1989, the socialist government gave up this important ideological argument by declaring the necessity of unemployment benefits.

The implementation of total employment, i.e. the elimination of unemployment, was more of a slogan than a reality during the long decades of socialism. Nevertheless, it is true that massive unemployment ceased to exist shortly after the communist take-over (1948). According to certain sources, the number of jobless people was over 400 000 in the summer of 1948 (BÁNFALVY 1989). The 1949 census registered about 126 000 as unemployed, indicating a 3% rate. Shortly thereafter, however, a hasty economic development policy soon put an end to unemployment, although it did not mean that people looking for jobs always managed to find the right jobs. Behind what was frequently referred to as a 'labour shortage' was substantial 'hidden' unemployment. This phenomenon particularly affected the less developed regions, where creating jobs for female labour was especially problematic. It is not accidental that the intensive involvement of manpower during the 1970s and 1980s led to serious tensions in the labour market in those regions where the highest unemployment rates are found today.

Unemployed people were not registered during the socialist decades. Therefore, there are only population estimates for people seeking job opportunities at that time, and these estimates differ from each other. According to some estimates, a 1 to 3% unemployment rate existed during the whole period (FERGE 1988). Other authors claim that the national unemployment average was 70 000 to 80 000 (1 to 1.5%) in the mid-1980s (FÓTI and ILLÉS 1992). Some estimate that only 0.1% of the active population was jobless through the 1980s (TIMÁR 1991, NAGY 1991). The above data, however, by no means refers to the unemployment 'inside the factory gates'. A well-known obstacle to the functions of the centrally planned economies was that the companies and firms were interested in accumu-

---

[1]   This study was sponsored by the National Scientific Research Fund (OTKA) project no. T 16757.

lating raw material, real asset, and manpower reserves. These hidden unemployment rates were estimated at an average of 20% in socialist states (FASSMANN 1992) and a 15% average in Hungary (VOGLER-LUDWIG et al. 1990). These figures are supported by calculations from the 1960s: during preparations for the 1968 economic reform, it was concluded that the reform would have led to a 10% unemployment rate (FÓTI and ILLÉS 1992). This estimate was not published at the time, and the government did not take responsibility for the conflicts that might have arisen in the case of official unemployment. This question of official unemployment remained a timebomb in the Hungarian economy, although the moment of its emergence (e.g. 1968 economic reform) was under question.

2   Evolution of unemployment and its structural characteristics

Unemployment has been officially recognised in Hungary for more than ten years (January 1, 1989). This short period can be subdivided into four phases:

1. *'Latent' unemployment.* This phase lasted from early 1989 until the autumn of 1990, with a slow rise in the number of registered unemployed people. The unemployment rate was insignificant by the end of this phase (1%). At that time, there was optimism that massive unemployment might be avoided in the transition from the centrally planned economy to the market economy in Hungary. In the beginning, this illusion was validated by the fact that the number of non-occupied jobs exceeded the number of registered unemployed people.
2. *'Running' unemployment.* This two-and-a-half year period was one of the most difficult phases in the history of Hungarian unemployment. Between the autumn of 1990 and early 1993, the number of registered unemployed rose from 50 000 to more than 700 000, i.e. the rate increased from 1% to more than 13%. This massive unemployment was a distinct component of the so-called 'transformation decline' (KORNAI 1993) and was a characteristic of the serious troubles of the Hungarian economy during this phase. At this time, unemployment seemed unmanageable, and there were fears that unemployment would reach one million.
3. *Decreasing and stagnating unemployment.* Following the peak in February 1993, surprisingly, the number of unemployed people started to decrease. This was the beginning of a slowly declining unemployment rate. The process, however, was interrupted by slight increases. The number of registered unemployed first dropped below 600 000 in April 1994, and the figure then fell to less than 500 000 by May 1995. Until early 1998, no important change occurred; the number of those seeking jobs fluctuated around 10%. It seemed probable that unemployment would be stabilised at this level.

4. *Improvement (?) of the situation.* The recovery of the Hungarian economy
    also had a positive effect on the labour market. Between January and October
    1998, the number of registered unemployed decreased by 90 000. Less than
    400 000, or 8.8%, were unemployed; this was better than the European Union
    average. Although a typical seasonal increase occurred in early 1999, the
    number of registered unemployed barely exceeded 400 000, or 9.4% (figure
    1).

Based on the current economic situation in the late 1990s, no increase in the
unemployment rate is expected. There is a chance that the rate will remain below
9% and the number of unemployed below 400 000.

Figure 1:   Unemployment in Hungary, 1990-1999

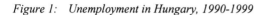

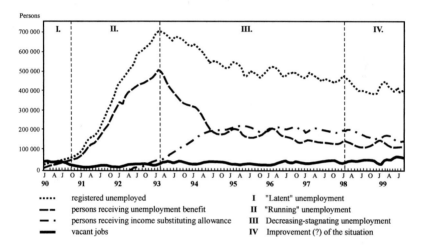

Source: National Labour Centre; author's design.

Over the past decade, there have been changes in the unemployment rate and in
the social composition of those seeking work. How the position of the social
groups in the labour market has been modified over time can be shown by looking
at development of the *educational level* within the registered unemployed (figure
2): During the first Hungarian unemployment phase, the main losers were those
with the highest and the lowest education. The fact that those with the least
education were disadvantaged is not surprising since the newly forming market
economy did not require the unskilled, virtually illiterate population segment. At
the same time, nearly one-tenth of unemployed people had a university degree.
This was only partly the outcome of the economic transformation; a substantial
role also belonged to those institutions that were overmanned during the socialist
era. As time passed, participation of both groups in unemployment statistics was
reduced to one-third of those who were registered in the early 1990s. Despite

210   Z. Dövényi

these apparently similar changes, however, there were basic differences: those with a higher education improved their situation dramatically due to a promotion of their position in the labour market, while most of the unskilled people either left the labour market or were forced out of it.

Figure 2:   Distribution of the unemployed by educational level, 1990-1999

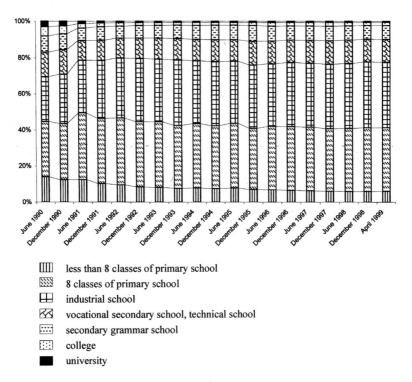

  ▥ less than 8 classes of primary school
  ▧ 8 classes of primary school
  ⊞ industrial school
  ▨ vocational secondary school, technical school
  ⸭ secondary grammar school
  ⸭ college
  ■ university

Source: National Labour Centre; author's design.

In the past decade, an overwhelming majority of the unemployed had a primary or industrial school education. Together they accounted for more than 70% of all active earners. The percentage of skilled workers in this category is especially striking: the ratio of skilled workers among the unemployed exceeds that of the total population significantly. This might be explained by the fact that the skills of workers trained during the socialist period were insufficient for the new labour market. About one-fifth of the unemployed in Hungary have completed secondary education, which is a high proportion when compared with the country's general educational level. The Hungarian labour market is under-utilising the potentialities provided by a relatively high general education and training (figure 2).

During the 1990s, the distribution of active earners and the unemployed showed a growing divergence by educational level. According to 1996 microcensus data, little more than 1% of active earners did not finish eight years of primary school, but their share within the registered unemployed population reached approximately 7%. An even greater discrepancy can be observed on the other side: almost 17% of the active earners had a university degree, but they comprised only 3% of the total unemployed population. Within the group of the unemployed lower educational levels dominate, while fewer people hold secondary and higher education degrees (figure 3).

Figure 3:   Active earners and unemployed by educational level in spring 1996 (%)

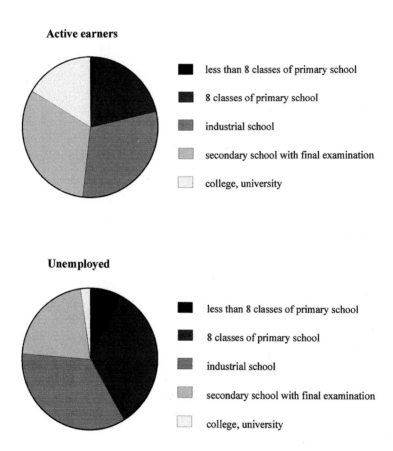

Source: Hungarian microcensus 1996; author's design.

There is a striking difference between job seekers and active earners, when looking at the *occupational structure*. A distinct difference can be observed between white collar and manual workers. According to 1990 census data, two-thirds of

active earners belonged to the manual worker category, but their ratio within the
registered unemployed group was permanently above 80% since the beginning of
the decade (figure 4).

Figure 4:   Registered unemployed by qualification categories, 1990-1999 (%)

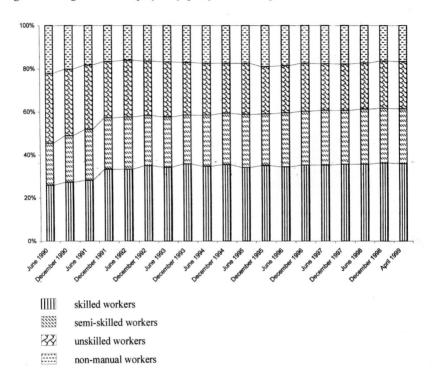

||||	skilled workers
▨▨▨	semi-skilled workers
☒☒	unskilled workers
::::::	non-manual workers

Source: National Labour Centre; author's design.

Unskilled workers were highly over-represented as part of the unemployed: according to the 1990 census, they comprised only 7.5% of the population, but they
accounted for one-third of the registered unemployed in the summer of that same
year. This percentage declined but has never reached below 20%.

Initially one-fourth of the registered unemployed were semi-skilled workers,
and the number of unemployed within this category has been steadily increasing,
though at a slow rate. More considerable changes were observed among skilled
workers. At first, their proportion was much lower among all categories of the
unemployed than within active earners (25.9% and 33.2%, respectively). During
the 'running' employment stage, a large number of skilled workers were made re-
dundant, their proportion among all unemployed people reached 35% by the end
of 1992. Since then, this figure has been fluctuating between 35 and 36% (figure
4). Skilled workers seem to be the main losers in the recent economic transforma-

tion. A favourable change could only be expected if the majority of skilled workers is trained according to the requirements of the market economy.

An unusual feature of unemployment in Hungary is that no specific *age group*
is particularly affected. Unemployment proportions among the different age
groups are mainly stable over time. However, there is a trend of decreasing unemployment within the age group below 20 years of age. At the same time, there is a
trend of increasing unemployment among those over 45 years old (figure 5).

Figure 5:   Registered unemployed by age groups, 1990-1999 (%)

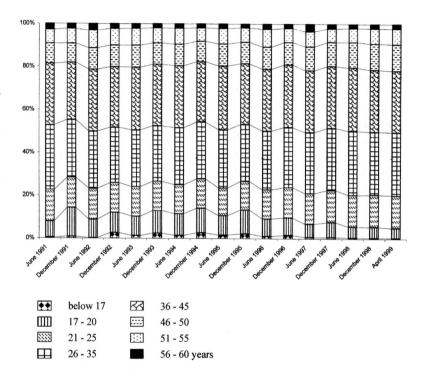

Symbol	Age group	Symbol	Age group
below 17	below 17		36 - 45
	17 - 20		46 - 50
	21 - 25		51 - 55
	26 - 35	■	56 - 60 years

Source: National Labour Centre; author's design.

The above phenomenon is reflected by a balanced pattern of unemployed people
in each group. According to the 1996 microcensus, the highest percentage of the
unemployed was in the age group twenty years and under, where nearly one-third
of the gainfully employed population was not working. This figure seems high,
but this age group plays a minor role in the labour market, since four-fifths of
them are students. In the long run, however, a group of people without a primary
school education or with only rudimentary school experiences is forming and is in
danger of becoming permanently jobless because of little demand for them on the
labour market.

An unusual feature of the Hungarian labour market is a lower female unemployment rate when compared to males while in most countries the reverse is true. The discrepancy was particularly large at the beginning of the 1990s, when the share of women among unemployed persons was only 40%. Since that time, the gap was decreasing. In spring 1999, the share of unemployed woman amounted to 45-47%. This phenomenon can be explained mainly by the fact that massive layoffs first affected economic sectors with an overwhelming male population (iron and steel production, mining, farming). Greater stability or even an increase in the number of jobs was typical in female occupations, e.g. services (FREY 1994). Unemployment increasingly affected manual workers, where the proportion of males substantially exceeded that of females. In contrast, the percentage of women was high among white-collar jobs, where layoffs were considerably fewer. A typical unemployed Hungarian male will often have finished vocational or industrial school, while a typical female often has a certificate from a secondary grammar school. Nevertheless, there is a higher unemployment rate among females with a university degree (LAKATOS 1997). In April 1999, 40% of unemployed women were previously blue-collar workers while almost three-fourths were non-manual labourers.

Male and female workers often differ in their strategies to fight unemployment: men who left the legitimate labour market usually found work on the black or grey markets; whereas women tend to remain inactive in the labour market (DEMKÓ 1999). As a result of these processes, the unemployment rates among females are lower than among males. This is confirmed by various surveys, even though it is rather diverse factual data. According to the spring 1996 microcensus figures, 13.7% of the male population of working age (15-59 years) were unemployed, while only 10.2% of the females of working age (15-54 years) were unemployed. A labour force survey carried out by the Hungarian Central Statistical Office in late 1998 showed that 7.6% of men and 6.3% of women qualified as unemployed (according to ILO standards).

Unfortunately, because it was not registered in the beginning, no complete data series are available on the *distribution of unemployed people by national economic sectors*. As a result, the only evidence that a high number of employees left the farming sector as well as the heavy industry and building industry sectors is a decrease in the number of workplaces during the years following the change of power. Accordingly, the distribution of the registered unemployed and active earners differed significantly between the different sectors in the first months of 1996. This was especially characteristic of two national economic sectors: the ratio of those who lost jobs in the agricultural sector was twice as high among all active earners; in the producing services sector, the ratio was the opposite (figure 6). Since then, the distribution of the unemployed and of active earners in national economic sectors has converged somewhat, but there are still considerable discrepancies. It should be emphasised that the tertiary sector accounted for about 50% of both unemployed and active earners.

Figure 6:   Active earners and unemployed by economic sectors in spring of 1996 (%)

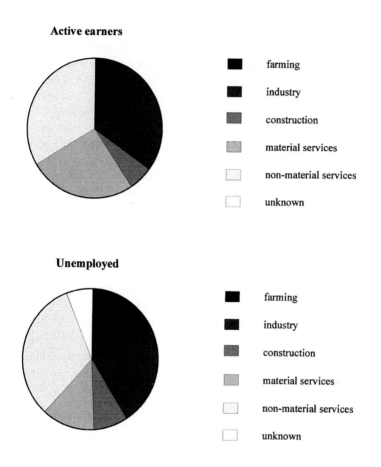

Active earners

- farming
- industry
- construction
- material services
- non-material services
- unknown

Unemployed

- farming
- industry
- construction
- material services
- non-material services
- unknown

Source: Hungarian microcensus 1996; author's design.

A grave problem with unemployment in Hungary is that many unemployed people do not even have the chance to return to the legitimate labour market. Unskilled workers or those with non-marketable skills form the bulk of the 'frustrated unemployed' group; they are forced to earn their living by non-legal work and/or to sustain themselves through social benefits. According to KERTESI (1994), the unemployment rate among the Roma was close to 50% in 1993, although exact data estimates and calculations are lacking.

Another serious problem is the increasing proportion of re-entrants among the unemployed. In 1995, 62.1% of the unemployed were re-entering the workforce, but this rose to 75.8% by 1998 (DEMKÓ 1999). This shows that a certain proportion of people became unemployed repeatedly, producing a layer of the workforce who are unable to find employment because of several disadvantages. This core group only partially figures in lists of the unemployed because after a time, they stop looking for work, become non-legal employees or inactive. Other indicators confirm that it will be rather difficult to relieve unemployment in Hungary. In the spring of 1999, people receiving unemployment benefits were not able to find work for an average of 280 days. Since only one-third of the registered unemployed receive benefits, this data does not reflect the real numbers. A better indication of prolonged unemployment is that two-thirds of the 400 000 registered unemployed were first registered as unemployed more than 720 days ago.

Unemployment hits large families in particular: there was at least one unemployed person in each twentieth family consisting of two members and in each sixth family having five or more members (LAKATOS 1997). This relationship was confirmed by the 1996 microcensus: as the number of household members increased, the ratio of households with at least one unemployed person also increased. Unemployment was only 6.6% for two-member households, 18.2% for four-member households, and 43.1% or more for eight-member or more households. This phenomenon was already revealed by the 1990 national census but with much lower values.

3   Spatial unemployment disparities

Unlike structural characteristics, spatial unemployment disparities were fairly stable in Hungary. Differences between macroregions had been established by 1990; since then, unemployment rates have changed but their spatial distribution has remained largely the same. Despite decreasing trends, unemployment rates continue to vary substantially by region. In July 1999, the unemployment rates were 4.5% in Budapest and its environs, 5.6% in western Transdanubia (along the Austrian border), 16.8% in north Hungary and 15.8% in the northern part of the Great Plain.[2]

The relationship of an above average GDP per person is generally coupled with unemployment rates which are below average. The most obvious example of this is Budapest, where the unemployment rate was half of the national average while GDP per head was twice the national average. The advantageous position of counties in north-western Hungary (Győr-Moson-Sopron, Vas) reflects a low unemployment rate and a relatively high GDP. In the northern and north-eastern parts of the country (Nógrád, Borsod-Abaúj-Zemplén, Szabolcs-Szatmár-Bereg), low GDP values were accompanied by the highest unemployment rates (figure 7).

---

[2]   See National Labour Centre report on changes in the labour market (July 1999).

Figure 7: Relationship between GDP and unemployment levels in Hungary by counties, 1996 (national average=100)

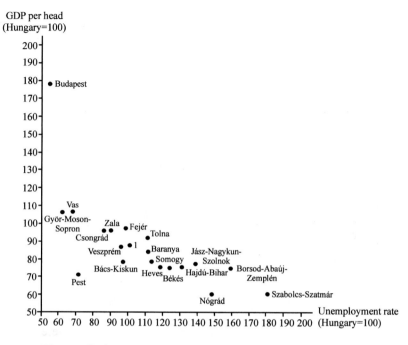

1 Esztergom-Komárom

Source: Hungarian Central Statistical Office (KSH) and National Labour Centre; author's design.

Spatial unemployment disparities, however, cannot be recognised on the county level, but can be observed among microregions. These differences reflect the economic efficiency of a given region and its ability to adapt to crisis situations. In favourable areas, the ability to attract innovative technologies and working capital means that unemployment always remains well below the national average. Even at the nadir of the economic crisis (spring-summer 1993), this spatial relationship occurred in some areas. At that time, unemployment rates of less than 10% comprised only one-sixth of labour market districts and were found primarily in Budapest and its surrounding areas and the north-western border region (figure 8). This distribution is far from incidental. These areas were in a favourable geographical and economic environment at the start of the era of political change, and therefore these areas attracted the most foreign investment. This foreign investments were accompanied by relatively positive processes on the labour market, and as a consequence of this, the danger of massive unemployment never arose in these areas. This was not the case in most parts of the country. Consequently, the overall labour market faced a disastrous situation in the summer of 1993. In districts with unemployment rates well above the national average, this was due to several factors. There are traditional backward regions in Hungary where deterioration has existed for about one hundred years. Certain border zones usually belong to these areas, as reflected by unemployment figures (DÖVÉNYI 1995).

Figure 8:   Ratio of unemployed within population of working age, by labour market districts, June 1993 (%)

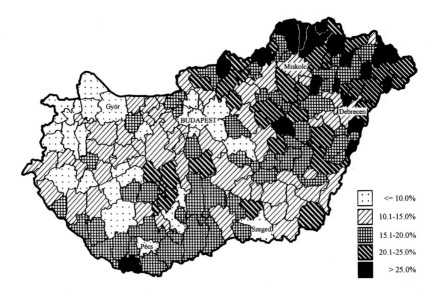

Source: National Labour Centre; author's design.

The fact that the percentage of unemployed people typical of a large part of the Great Plain is much higher than the national average can be attributed to two factors. First, the disintegration of COMECON caused the Great Plain to lose its earlier stable markets. Second, the intensive industrialisation of the macroregion which was mostly pursued in the socialist era (and already outdated at that time) caused industrial plants to go bankrupt following the change of power. As a result, many workers lost their jobs. A third type of region with a high unemployment rate comprised the declining industrial and mining areas. State-subsidised heavy-industrial plants and mines from the socialist period were not competitive in market environments and soon went bankrupt.

Basic changes in the Hungarian economy during 1992-93 brought about a nation-wide employment crises. When this crisis situation was over, the labour market improved considerably. By the summer of 1998, the percentage of unemployed people dropped below 10%. This improvement affected most of the labour market districts, and the picture became much brighter than five years earlier. In the capital city, its environs and in several districts of north-western Transdanubia, the unemployment rate was below 5%, and in some places there was even a workforce demand. The northern part of Transdanubia, the greater area surrounding Budapest, and the western margin of the Great Plain showed large contiguous areas with an unemployment rate below the national average.

Figure 9:   *Ratio of unemployed within population of working age, by labour market districts, June 1998 (%)*

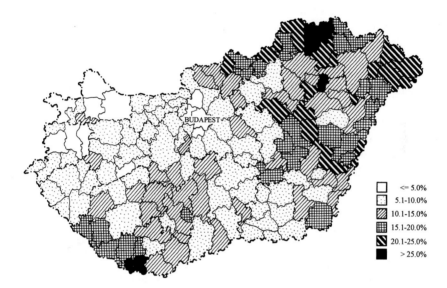

Source: National Labour Centre; author's design.

At the same time, unemployment rates significantly exceeded the national average in two of the country's regions. When comparing maps of unemployment from June 1993 and June 1998, a north-west - south-east divide of high and low unemployment rates becomes evident. This line, which stretches approximately between Balassagyarmat and Mezőhegyes, represents the boundary to the problem regions in the northern and eastern parts of Hungary and a divide between those regions that have benefited from the economic upswing and those who have not (figure 9). However, the situation was not generally favourable west of the divide since unemployment rates surpassed the national average over extensive areas in south Transdanubia in 1998. The same situation was true in the border zones; these areas are traditionally affected by structural economic problems.

Together with absolute unemployment rates, changes in the relative position of individual labour market districts between 1993 and 1998 should be examined when judging unemployment shifts within microregions. Using national unemployment rate indices five categories of positive and negative anomalies were established. Labour market districts could have theoretically fallen into 25 classes. 'Blank' classes comprised almost one-third of the total, i.e. eight classes did not have any labour market districts. This is understandable because the existence of all classes would have indicated extreme changes that rarely occur on the labour market for a five-year time period. Based on the changes in the relative position of individual labour market districts between 1993 and 1998, Hungary can be subdivided into three distinct sub-regions (figure 10). The line between Balassagyarmat and Mezőhegyes is once again a sharp divide in terms of unemployment rate modifications. Districts with unemployment rates exceeding the national average in 1993 and 1998 were found east of the divide. The serious unemployment situation is also indicated by the fact that 40 out of the 42 worst-positioned districts were found in this first sub-region. All of these aspects indicate that this sub-region is expected to face prolonged and serious unemployment. The situation in the south Transdanubian labour market and in some adjacent areas of the Great Plain was somewhat better but still perceptibly worse than the national average. Areas in the border zone with traditionally small and underdeveloped villages were in a particularly bad position. In this second sub-region, however, there are some chances for recovery and improvement. The labour market in the rest of the country was dominated by favourable processes between 1993 and 1998. Unemployment rates in this third sub-region were already lower than the national average in the summer of 1993, and this trend continued five years later. The same positive labour market trends have been visible in the national capital and its environs. They have been even more intense in the north-western part of the country (figure 10).

It is difficult to predict future trends with regard to spatial unemployment disparities. The gap tends to widen between the most advanced sub-regions (Budapest and surroundings and western Transdanubia) and the most backward sub-regions (northern parts of the country and the Great Plain). Studies of the settlements, however, point to a slight reduction in disparities. The Hoover index, which expresses spatial disparities, reached its maximum in the early 1990s. Since then, its value has decreased considerably (NEMES-NAGY 1999). The Hoover index indicates that the system of spatial disparities is controlled primarily by processes and shifts taking place in the countryside. Meanwhile, the capital's unemployment figures have been stable throughout the 1990s (NEMES-NAGY 1999). Other data refer to the differentiation of unemployment in Hungary by the settlement hierarchy. This type of distribution points to the towns' more favourable position and the villages' less advantageous position.

The overall picture can be interpreted more subtly than the simple relationship shown above. Areas with a dense, hierarchical urban network and elements of high potential are favourably positioned from an employment perspective. In contrast, areas lacking higher-ranking urban centres and those where the urban network is comprised of small, underdeveloped towns are in a more disadvantageous situation. The favourable situation of some urban settlements is connected with the situation in their rural environs in so far as many people who lost their job in the town are registered as unemployed in the rural areas in which they live (e.g. commuters). Big enterprises closed down factory branches in the countryside as a survival strategy. This aspect combined with other circumstances (e.g. larger and more adaptive labour market, skilled and more qualified workforce) has led to a generally lower level of unemployment in towns than in villages. This should be noted because the Hungarian public tends to regard unemployment as an urban phenomenon although the primary losers of the labour market transformation are people living in the rural areas of Hungary.

Figure 10: Changes in rates of unemployment between 1993 and 1998

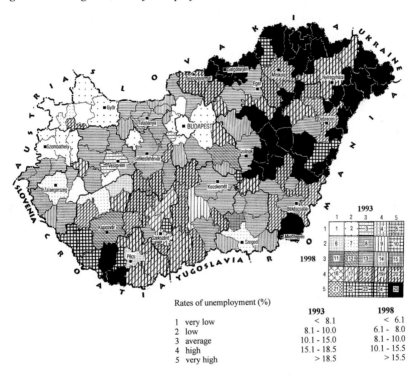

Rates of unemployment (%)

		1993	1998
1	very low	< 8.1	< 6.1
2	low	8.1 - 10.0	6.1 - 8.0
3	average	10.1 - 15.0	8.1 - 10.0
4	high	15.1 - 18.5	10.1 - 15.5
5	very high	> 18.5	> 15.5

Source: National Labour Centre; author's design.

Although the urban settlements cannot be considered as the obvious winners, they paid a much lower price. However, urban settlements should not be treated as a single category, because the larger towns have relatively less unemployment than the smaller towns. This difference came to light in the early 1990s, and according to a 1995 survey, almost 40% of the active earners lived in county seats (including

Budapest)[3] while only 25% of the unemployed were residents in small towns. The share of active earners was somewhat higher than that of the unemployed population even in towns with more than 20 000 inhabitants. Smaller-sized towns showed more rural features, i.e. the percentage of those seeking employment was higher than the proportion of active earners.

There was a reverse pattern for villages since the proportion of the unemployed exceeded the number of active earners, but the villages, however, did not form a uniform category either. A divide stretched along the settlements with approximately 5 000 inhabitants. The unemployment in the largest villages remained the same as that in the small towns, while the villages with a population of less than 5 000 were in a worse position.

The spatial unemployment pattern that characterises the hierarchy of settlements corresponds to disparities in other economic and social spheres. This can be shown by comparisons of several key values obtained during the 1995 Household and Budget Survey and by the 1996 microcensus (income conditions, household equipment, population's education level).[4]

4   Future perspectives

A promising future was expected when the unemployment rate stabilised below 10%.[5] There has been permanent economic growth since 1996 and the number of workplaces has been increasing since 1997. As a result of these favourable trends, unemployment will hopefully cease to be one of the most important issues for the Hungarian economy. Such general expectations are supported by the fact that in early 2000, the unemployment policy of the Hungarian government became more restrictive. For example, unemployment benefits are now available only for 9 months instead of 12 months. But it seems as if the slogan 'work instead of relief' is more appropriate in the highly developed areas of the country. In less developed regions with a high unemployment rate, this stricter policy might lead to a further aggravation of social tensions.

When extrapolating the unemployment trends identified in this article, additional spatial differentiation of unemployment rates can be expected. This mainly results from the fact that in regions where unemployment rates have been high from the very beginning, no significant improvement is in sight. Especially in areas characterised by small villages, unemployment might be the last issue that demonstrates the disintegration of local communities. Simultaneously, unem-

---

[3]   1992 data.
[4]   Data evaluation was performed by Robert Famulok (Institute of Geography, University of Heidelberg); his efforts are highly appreciated.
[5]   This figure was calculated by using the number of registered unemployed, while the ILO standard remains at 7%.

ployment reduction can be predicted for more developed regions where even a labour shortage might emerge. Such areas already exist today (e.g. Budapest and its surroundings, Győr, Székesfehérvár, Bicske, Szentgotthárd), but this phenomenon will probably become more important in the years to come because the population's spatial mobility dropped to a low level in Hungary during the 1990s. There is no sign of migration from areas with a manpower surplus to those with a labour shortage in the near future.

Prolonged unemployment will probably be the most severe problem. The number of long-term unemployed is estimated at 150 000-200 000, and this group has little hope for future employment due to their low qualifications. Especially for those who have been unemployed and out of the labour market for a longer period of time, a closer interrelationship between poverty and unemployment will be an inevitable outcome of the years to come (TÓTH 1999).

References

BÁNFALVY, Cs. (1989): A munkanélküliség. (Unemployment). Budapest

DEMKÓ, O. (1999): Munkaerőpiaci folyamatok 1998-ban. (Labour Market Report, 1998). OMMK, Budapest

DÖVÉNYI, Z. (1995): Die strukturellen und territorialen Besonderheiten der Arbeitslosigkeit in Ungarn. In: MEUSBURGER, P., KLINGER A. (eds): Vom Plan zum Markt: Eine Untersuchung am Beispiel Ungarns. Physica Verlag, Heidelberg, 114–129

FASSMANN, H. (1992): Phänomene der Transformation: Ökonomische Restrukturierung und Arbeitslosigkeit in Ost-Mitteleuropa. Petermanns Geographische Mitteilungen 136 (1), 49–59

FERGE, Zs. (1988): Teljes foglalkoztatás – foglalkoztatáspolitika – munkanélküliség. (Full Employment – Employment Policy – Unemployment). Valóság 6, 19–31

FÓTI, J., ILLÉS, S. (1992): A munkanélküliség demográfiai vonatkozásai. (Demography-related Issues of Unemployment). Budapest (= A Népességtudományi Kutatóintézet Kutatási Jelentései 43)

FREY, M. (1994): Az állam szerepe a foglalkozáspolitikában és a munkaerőpiaci programok menedzselésében Magyarországon. (The Role of the State in Employment Policy and in Managing Labour Market Related Programmes in Hungary). In: Foglalkoztatáspolitika és munkaerőpiaci programok Közép-és Kelet-Európában. Budapest

KERTESI, G. (1994): Cigányok a munkaerőpiacon. (Gypsies in the Labour Market). Közigazgatási Szemle XLI.11, 991-1023

KORNAI, J. (1993): Transzformációs visszaesés: Egy általános jelenség vizsgálata a magyar fejlődés példáján. (Decline in the Course of Tranformation: Studies on a General Phenomenon: The Case of the Hungarian Development). Közgazdasági Szemle XL (7-8), 569–599

LAKATOS, J. (1997): Foglalkoztatottság - munkanélküliség, 1992–1996. (Employment - unemployment, 1992–1996). Gazdaság és Statisztika 9 (48-6), 18–31

NAGY, Gy. (1991): "Aki segélyért jön és tudok neki munkahelyet találni, azt kiejtem...". ("Who Comes to Me for Benefit, and I can Find Him/Her a Place to Work is to be Selected Out..."). Mozgó Világ 1, 96–103

NEMES-NAGY J. (1999): Elágazó növekedési pályák az ezredvégi Magyarországon. (Branching Regional Growth in Hungary at the End of the Millennium). In: Helyek, terek, régiók. Budapest, 65–86

TIMÁR J. (1991): Foglalkoztatás egy új Rubicon közelében. (Employment Relations in a New Stage of Development). Társadalmi Szemle 5, 38–47

TÓTH, J. Gy. (1999): Mik a legfrissebb adatok a magyarországi szegénységről? (Latest Facts and Figures on the Poverty in Hungary). Magyar Tudomány 12, 1488–1490

VOGLER-LUDWIG et al. (1990): Verdeckte Arbeitslosigkeit in der DDR. München (= Ifo Studien zur Arbeitsmarktforschung 5)

Poverty Dynamics in Hungary during the Transformation

Zsolt Spéder

1   Introduction[1]

In western social science poverty research has long been a well-established field of research. However, in the former socialist countries, not only poverty research, but even the word 'poverty' was excluded from the scientific and public debate (ANDORKA 1995). Of course it does not mean that there was no research carried out on the most disadvantageous strata of the society (see e.g. BÖHM 1981; BOKOR 1985), but it had to be referred to as 'multiple deprivation research', and the results were not well-known in scientific, or in public debate. The first offi- cially acknowledged poverty research took place at the end of the 1980s (KSH 1991). The transformation brought about a complete change in the demand for research activities and, in the case of poverty research, the situation was reversed. "How many poor people are there?", "Which are the most risky social groups?", "How big is the increase in the poverty of the population?" Such questions became the most important ones for policy-makers and civil servants as well as for both affected and unaffected citizens. The growing interest in the poverty issue was clear and understandable, but could not be wholly fulfilled. The emerging picture was blurred and diverse. Nearly all discussions were highly loaded by politics, most of the participants did not differentiate between poverty and impov- erishment (the same word in Hungarian), and it was not unambiguously clear that poverty is not an objective concept and its meaning depends on the chosen approach. Of course the lack of research tradition, and the often unreliable data have also contributed to the confusion. Different institutions, using different data and employing different poverty concepts have projected dissimilar poverty rates in public and scientific discussion.

---

[1]   A version of this paper has been published in the journal *Economics of Transition* 6 (1/1999), 1-21.
   This research on poverty was initiated by my teacher Professor Rudolf Andorka. The first version of this paper was conceptualised and elaborated by us both and presented by myself at the EBRD conference on '*Inequality and Poverty in Transition Economies*', 23-24 May, 1997 in London. This longer and modified version was produced alone following the death of Rudolf Andorka in June 1997. Most of the ideas originated in our joint work and were recognised by both of us. I am very much indebted to Rudolf Andorka, and with this paper I would like to pay homage to him.

The Hungarian Household Panel Study Project (see appendices and Tóth 1995) was initiated at that time (1991), and one of our aims was to gain data as reliable as possible about the income, and also about the poverty situation of the population. It was just as important from the angle of the income development of the households and individuals. Discussing the different poverty concepts we have come to the conclusion that to reach a comprehensive understanding of poverty, it would be advantageous to use more poverty concepts side by side. This is not so important for defining which social group is most at risk, but it is indispensable to understanding, at least roughly, the extent of poverty and the dynamics of poverty positions.

Before discussing the applied poverty concepts, the meanings of the usage, and the results of our investigations, we should briefly summarise some well-known macroeconomic developments.

2 Macroeconomic indices and change in the inequality of incomes

The development of poverty has two basic macroeconomic sources: it depends on economic performance, and on change in income inequalities. In Hungary, there is a common understanding that growing poverty is a consequence of both types of development. So basic figures should be shown by all means. GDP fell by about 21% between 1989 and 1993 (table 1). According to macro-statistical data the fall of the per capita real income of the population was only about 11% up to 1993. The fall of real consumption by the population is somewhat less. After 1993, the development of indicators has fluctuated. The 'Bokros package', introduced in 1995, caused a real drop in households' living conditions (see later). We should mention that the rather unfavourable change had already begun in the 1980s: GDP was almost stagnant and, as a result, the real income of the population hardly increased. The index of real wages within it has been falling constantly since 1978. However, after 1989 far more unfavourable processes unfolded.

We should not forget that there are some uncertainties in the figures mentioned. It should be noted that the real development of economic performance is not unambiguous, because we have no information about the development of the unregistered economy. It is known that its contribution to economic performance was quite big in the former socialist system, and the increase of this part of the economy cannot be excluded now either (see TÁRKI and GKI 1994, 112).

It is nowadays a commonplace, that income inequality has risen enormously. Here it is important to mention that the increase began in the early 1980s (ANDORKA 1989). We do not know too much about the time around the systemic change, but it is generally thought that the increase accelerated around 1990 (KOLOSI and RÓBERT 1992). As mentioned above, since 1992 the Hungarian

Household Panel Project (HHP) reported repeatedly about the size of this rise (KOLOSI et. al. 1996; TÓTH 1996). The different inequality dimensions are shown in a table calculated by the HHP data (table 2).

Table 1: The main macro-statistical indices of the Hungarian economy, 1989-1996

Year	GDP 1989= 100	Per capita real income of the population, 1989=100	Index of real wages 1989=100	Per capita real Consumption 1989=100	Consumer price index 1989=100
1989	100	100	100	100	100
1990	96	98	96	97	129
1991	85	96	90	92	174
1992	81	94	89	91	214
1993	79	89	85	92	262
1994	82	92	91	93	312
1995	84	87	81	91	402
1996	85	88	78	89	497

Source: Hungarian Central Statistical Office (KSH), Income Surveys.

Table 2: Different income inequality measures in Hungary between 1992-1996

Year	Gini*	Robin Hood**	Percentile** 90/10	Decile-rates ***		
				Upper/ lower	lower/ average	Upper/ average
1992	0.295	19.8	3.38	6.676	0.364	2.430
1993	0.277	19.7	3.40	6.327	0.378	2.392
1994	0.295	21.3	3.81	7.374	0.335	2.469
1995	0.316	21.7	3.84	7.261	0.360	2.613
1996	0.308	21.6	4.13	7.575	0.299	2.531

*   On the basis of equivalent income (e=0.73) (see TÓTH 1996).
**   Individual deciles put in order on the basis of the per capita income, ordering, the author's calculation.
***   Household deciles made on the basis of per capita income (see KOLOSI et. al. 1996).

Source: Hungarian Household Panel.

3   The extent of poverty and its change

It is over the extent of poverty and its increase where disagreement is most significant. This is not surprising because of the several poverty concepts. Incidentally, one of the conclusions of our analysis may be that independent of the applied poverty concepts, poverty has increased during the period under survey (table 3). However, the dimensions of the increase depend on the applied concept.

Table 3: Poverty rates in Hungary between 1992 and 1996

Poverty concepts and poverty lines	1992	1993	1994	1995	1996
Relative income poverty*	10.1	10.4	11.6	12.4	14.0
Absolute income poverty	10.1	12.5	16.1	19.9	26.6
Subsistence minimum**	21.5	24.0	31.8	-	-
Minimum pension	5.2	5.4	6,7	5.1	6,2
Not enough money for clothing	67.1	69.3	69.7	71.1	75.2
Financial crisis in households					
Monthly	26.4	26.3	25.4	25.1	28.1
Never	38.7	39.0	41.1	40.9	31.2

*   50 % of mean income.
**   Household-specific poverty lines.

Source: Hungarian Household Panel.

As mentioned before, we do not want to discuss in detail the meaning of the different poverty concepts, but we would like to emphasise some important features of the longitudinal aspect, especially during transition. Transformation has reinforced the basic dilemma between the *absolute* and *relative* poverty approaches. According to the absolute concept, a household is poor if it has less income than a defined sum in real terms (e.g. subsistence minimum). Applying this approach here, its current relation to a previous condition of poverty would be emphasised. The relative poverty concept, however, stresses the consumption opportunities relative to the population norms. In other words, the poor have no opportunity to participate in generally accepted everyday life (consumption) (see ATKINSON 1989; RUGGLES 1994). During the period of lasting economic growth, most researchers made their choice between the two approaches mentioned here, opting for the relative poverty concept. That means that the income position of a particular household cannot be judged independently from the 'general' households (RUGGLES 1994, 40). According to this approach, the real income increase of a household only signifies an income position improvement if it is above the general increase. Otherwise the position is lowered. This can be applied to the estimation of poverty trends. We can speak about the growth of population in poverty, when households cannot increase their income according to the general rate, and so fall below the poverty threshold. Following this argument - and this can be seen as a paradox too - if there is an economic recession, there can only be an increase in poverty where a household's fall in income is greater than the general fall. Therefore citizens must participate not only in the general increase of welfare, but in its decrease too. It can also be argued that it is just at a time of recession that the absolute poverty concept should also be used. In other words, at a time of falling real incomes and when inequality does not decrease, the poor population will grow.

Turning to the data analysis, it can be said that during the period investigated, poverty increased by 1.5 times (table 3) based on the relative poverty concept. On the other hand, using the absolute concept the growth of the poor population was more than 2.5 times. For comparative purposes, both the well known '50% of mean equivalent income' poverty threshold, and the annual inflated 1992 poverty threshold were used.[2]

For assessing absolute poverty, the official subsistence minimum is much better, but suitable data were only available up to 1994.[3] The proportion of the population living below the subsistence minimum was about 50% higher in 1994 than in 1992. This rate can be compared with the figures of the 1980s. In the 1980s, the number and proportion of those living on incomes less than the subsistence level was estimated to be around one million, that is 10%. It should be noted that even in those days there were different estimates of the number of the poor. This also means that a rise of the poor population was already experienced around 1990 (see KOLOSI and RÓBERT 1992).

The official poverty threshold and the minimum pension in themselves do not reveal a clear increase in poverty in Hungary. These measures are the result of agreements between the different social forces, and most of the different social support programmes are linked to them (MONOSTORY 1997).

The several indicators of financial hardship in household economies can be characterised as subjective poverty indicators. Namely, these indicators show us the tension between resources and aspirations in families. In our table we listed two of them. The first one shows whether a household had financial difficulties at the end of the month or not and, if so, than how often. The second one reports the unsatisfied demands of the household in the case of clothing. We are aware that not only the poor but others can also experience financial hardship, but it is more probable that the income-poor will experience such situations, and also with greater frequency. Furthermore, we know that those households facing financial hardship worked out several methods of coping, e.g. intensifying household production and network support, changing the consumption structure, modifying aspirations.[4] So the above mentioned indicators should not be understood exclusively as poverty indicators, but I think it is not misleading to assume that with increasing poverty, the frequency of experiencing hardship in the household economy should rise. According to the answers to these questions, we see a rather moderate, but unambiguous, increase in the households' financial crisis.

Summing up: the increase in the extent of poverty is unquestionable, but the size of the increase depends on the applied poverty concepts. Probably the first strong expansion happened before the first wave of the Hungarian Household

---

[2] That is why in 1992 the poverty rate was the same in the case of both concepts.

[3] In 1995, the Hungarian Central Statistical Office started a new subsistence minimum counting, but it is not wholly comparable with the old one (see KSH 1991).

[4] For coping strategies of households during the transformation, see ROSE and HAERPFER 1993; SPÉDER 1996.

Panel Study (before 1992), the second major rise occurred in 1995. It was a year of accelerating impoverishment.

4   Poverty dynamics: permanent and temporary poor

One of the most remarkable and most disputed results of the familiar household panel surveys has been the differentiation between permanent and transitory poverty (DUNCAN 1984; HEADEY, HABICH and KRAUSE 1990; KRAUSE 1994). This showed that poverty in the industrial societies was mostly transitory, and people found poor in a given year would raise their income above the poverty threshold in the next one or several years. The main question was then whether the same, or a similar picture was also apparent in the relatively stable social groups of Hungary, at the time of transformation. In other words: could households living below the poverty line in one year, escape from poverty in the next year during the period of 'transformational recession' (KORNAI 1994). Based upon the growing extent of poverty, the likelihood of becoming poor seemed to be plausible, whereas a movement in the opposite direction was not so obvious.

Table 4:   Poverty dynamics in Hungary, N-times poor 1992-1996 (individuals, equivalent income)

N-times poor	Rate of individuals (50% of mean income)
Never poor	73.9
Once poor	13.9
Two times poor	5.9
Three times poor	2.4
Four times poor	1.8
Always (five times) poor	2.0
Total (%)	100

Source: Hungarian Household Panel.

One of the simplest methods of capturing the idea of poverty mobility is to construct the 'N-times poor' variable, to count how many years a household had income lower than the poverty threshold. We can sum up our results in two very general statements:

• First, during the five-year period studied, many more people had been at least once-poor (26.1%) than the annual cross-sectional view figures would suggest. In fact, the highest annual figure (in 1996) was only 14%. This difference can also help us to understand why so many individuals found themselves subjectively poor.

• Second, the proportion of permanently poor people is lower than expected. It can be estimated as somewhat less than 5% of the population. But if we relate the permanent poor to the annually poor population, and not to the once-poor, we come to a somewhat different conclusion. In 1996, about one-third of the poor were permanent poor, the other third transitory but more times poor, and the last third one-time poor.

This pattern immediately raises questions and doubts about the significance of the results, because it is contrary to our expectations. Hence we shall try to settle some of these doubts and questions in the following sections, as we attempt to capture the nature of poverty mobility during the transformation.

4.1   Poverty mobility using different poverty concepts

First, we are going to demonstrate that poverty mobility does not only exist on the basis of the relative poverty concept but also in regard to the absolute poverty approach. Unfortunately, as was mentioned before, we are unable to test the data for the whole period of the survey, because subsistence minimum numbers were only available for the first three years of the investigation. We found that there was also significant migration between poor and non-poor situations if the absolute poverty concept is applied (table 5), and that there are generally more permanent poor than temporary poor if we use the subsistence minimum concept. Using the household financial crises as poverty indices, we also found quite high mobility which is explored in more detail by SPÉDER (1996). Furthermore, we can compare the objective poverty concept, based on income, and the indicator of household financial crisis (table 6). This comparison shows that the longer poverty lasts, the more often the household suffers from expenditure problems.

Table 5: Poverty dynamics in Hungary 1992-1994 (two different poverty concepts)

N-times poor	Equivalent income		CSO subsistence minimum
	50% of the mean	60% of the mean	
Never	81.6	67.4	58.0
Once	11.7	17.0	16.8
Twice	3.7	8.6	15.1
3 times	3.1	7.1	10.2

Source: Hungarian Household Panel.

4.2   Poverty dynamics in comparative perspective

Let us compare Hungarian data with German data calculated in the same way. We have found small differences between the figures. To make the differences visible, we computed the so-called 'segregation-measure' to compare the poverty dynamics in the two societies (table 7). This measure shows the incidence of permanent poverty events and the sum of poverty events over a certain period of time. In other words, how sharp is the distinction between the permanent poor and the others? In Hungary, the segregation is less marked than in West Germany, but higher than in the former East Germany.[5] But if we count the East German poverty profile using the income distribution in the unified Germany, the segregation is the lowest in Hungary. As we know, expanding poverty induces growth in the proportion of temporary poverty. However, the rate of change can be high as well as low. The degree of segregation between temporary and lasting poverty is determined by the relation between these two complementary processes, the proportions of which are difficult to identify.

Table 6: Occurrence of the financial crisis at the end of the month and different longitudinal poverty positions in Hungary (households)

Poverty positions	Occurrence of financial crises in 1996				Total N=
	Monthly	Two-monthly	Some-times	Never	
Poverty position 1996					
Poor	59.4	13.4	15.0	12.1	218
Not poor	24.0	11.9	30.3	33.8	1 656
Poverty position in 1995-1996					
Both years poor	67.9	11.6	6.9	13.6	122
Fallen	42.4	18.4	27.3	17.1	82
Escaped	39.5	14.9	28.6	11.9	72
Never	23.6	12.0	30.3	34.0	1 517
Poverty position in 1992-1996					
Permanent*	(63.8)	(14.6)	-	(14.6)	69
Temporary**	45.3	17.5	21.2	16.0	137
Once poor	35.0	16.0	32.0	17.1	199
Never poor	22.5	11.9	30.0	35.6	1 325

*   Permanent poor: 5 times and 4 times poor.
**   Temporary poor: 3 times and 2 times poor.

Source: German Panel (GSOEP).

4.3   Distances in poverty mobility

It is very important to know how far those who escape poverty can get from their earlier income position. The other question is what was the earlier financial situation of those who have slipped into poverty? In other words, from which income classes did the new poor originate? To answer these questions we have defined income classes: we have marked the boundaries of the classes as a percentage of the average equivalent income. At first, let us examine how far those who escaped

---

[5]   In this case we handle East Germany as a single society.

from poverty could get (table 8). Many could only get as far as the next income category, but there are quite a few, whose nominal income increased notably, and who therefore succeeded in moving far from the poverty line. Except for the last year, this latter group was larger than the former one.

Considering the inflow side of the movements, we have found in every year that the proportion of the poor coming from income classes near the poverty-line is less than the proportion of those who used to belong to higher income classes (table 9). Namely, that primarily not only people who belong to income classes near or just above the poverty-line are slipping into poverty, but impoverishment also characterises the members of middle income classes. These two tables clearly show that during the transformation there were many quite large income movements and changes at the household level. It is very difficult to make statements about the trend of movements and changes over the given period of time. After all, we propose in our hypotheses that the trends changed around 1995. On the one hand, fewer people slipped into poverty from the higher income classes, though in that year there were plenty of 'new' poor. On the other hand, it seems that the 'escapees' from poverty were not able to get too far from the poverty-line in 1995. It means that the intensity of mobility as well as the distance of movement appear to have decreased. We found a similar tendency in the former East Germany for the period 1993-1994. After the next wave of our Panel Survey, of course, we hope to test the hypothesis mentioned above (KRAUSE and SPÉDER 1996).

Table 7: Poverty dynamics in Hungary, East Germany, and West Germany 1992-1995 (50% of mean equivalent income)

N-times poor	Hungary	East-Germany		West-Germany
		East-income	German-income	
Never	77.7	84.6	73.3	81.4
Once	13.0	8.3	12.3	8.6
Twice	4.3	3.9	5.8	3.9
3-times	2.7	2.1	5.1	2.8
4-times	2.3	1.1	3.5	3.3
Total	100	100	100	100
Segregation-measure*	44.5	39.9	55.0	56.8

* The proportion of permanent poverty (3 and 4-times poor) from all poverty events; Poverty event: poor in one year.

Source: KRAUSE and SPÉDER, 1996.

Table 8: Distances of the outflow shifts (Income position in t year of those, who were poor in the t-1 year)

Poor in t-1 year	Income position in t year (income classes in relation to the mean equivalent income)				Total N=
	under 50 (poor in t)	50-60	60-75	above 75	
1993	45.8	16.8	15.2	22.3	603
1994	51.7	34.3	12.9	13.3	492
1995	47.3	26.0	12.0	23.9	631
1996	68.5	20.0	8.2	9.9	523
The 1992 poor in 1996*	45.5	12.8	16.1	25.6	404

*   Income position of the 1992 poor in 1996.

Source: Hungarian Household Panel.

Table 9: Distances of the inflow shifts (Income position in t-1 year of those, who were poor in t year)

Poor in t year	Income position in t-1 year (income classes in relation to the mean equivalent income)				Total N=
	under 50 (poor)	50-60	60-75	above 75	
1993	49.7	19.0	16.4	15.1	555
1994	42.8	16.8	17.7	22.8	595
1995	45.3	27.4	11.8	15.3	658
1996	57.8	14.7	19.4	8.1	620
The 1996 poor in 1992*	32.8	12.7	19.0	35.7	565

*   Income position of the 1996 poor in 1992.

Source: Hungarian Household Panel.

The features of the Hungarian poverty mobility mentioned above have reinforced our hypothesis about its existence, and have given us an insight into the nature of the changes occurring during the transformation. That also implies, that 'transformation' took place not only at the institutional and macro levels, but at the level of everyday life too. The income mobility figures indicate a much more intensive and permanent change of position than was expected from the macroeconomic shifts. In the following, we clarify which social groups had the highest risk of becoming and staying poor.

5   Permanently poor social groups

In order to identify the poor social groups, we can use both the annual cross-sectional and the longitudinal approach. Here only the second is used, because the results of the cross-sectional analyses, as shown by our former analysis, are nearly the same. The directions of the social forces indicating poverty are quite similar in the case of permanent poverty and in the cross-sectional base.

Demographic poor are those whose poverty position depends heavily on life cycle and family types. We have already noticed the tendency during the late socialist period (in contrast to the early socialist period) that an increasing number of children may be classified as poor in addition to members of the older generations (tables 10 and 11) (ANDORKA 1990). Similarly, this tendency can be observed in the developed countries (RAINWATER 1988; SMEEDING 1988; SMEEDING and TORREY 1988). The tendency, however, seems to have grown stronger after the systemic change. It should also be mentioned that some families with children can probably escape from poverty, if the mother (and/or father) gets access to the labour market and gets a job when the children are older. Second, it is a new phenomenon that within the aged and retired population a differentiation has occurred. We are dealing with a stratified 'aged society'. The differences among the aged depend on the type of household, the kind of pension they have, their age, and the type of residence. Third, the effect of life cycle is heavily modified by the type of household. Living alone and bringing up children alone increase the risk of becoming poor.

A brief explanation is required to understand why the proportion of the poor is smaller among the elderly, and particularly among those aged 60-69 (see MILANOVICS 1995). The probable cause is that the majority of the elderly population have acquired the right to a relatively high pension, and though the real value of pensions has decreased, the income situation of the population of retirement age has deteriorated less than the national average. Furthermore, partly because the old age population is being constantly replaced demographically, older people on smaller pension die and are replaced by younger ones with higher pensions (the real value of whose pension will decrease in the coming years). Also pensioners were not threatened by unemployment, the major risk factor of the active age group. However, it should be stressed that the fact that the proportion of the poor is not particularly high among the elderly does not mean that there are not groups among the elderly who live in abject poverty, as do those on widow's and disability pension, and those who have retired early during the past few years (table 12). The elderly living alone (in single member households) may also belong to this group. A large part of the latter are old widows in rural regions.

Whatever definition of poverty is employed and whatever weighting or scale of equivalence is used, today the poverty of children in Hungary is a very conspicuous phenomenon. We agree with the statement of the Florence research centre of UNICEF (1993; 1995) that children are among the greatest losers of systemic change in East Central Europe (tables 10 and 11). It is considered to be highly problematic to the future of Hungarian society that a significant percentage of children are brought up under poor conditions at least for part of their childhood. This may be highly detrimental to many different aspects of their life, from nutrition to progress in school.

Table 10:   *Proportion of the poor by age groups in 1972, 1987 and 1992 (lowest decile, per capita income)*

Age groups	1972	1987	1992
0- 2	15.2	20.7	19.6
3- 6	15.3	17.0	19.2
7-14	15.1	15.1	15.6
15-19	6.3	7.2	13.9
20-24	5.0	6.0	11.9
25-29	5.1	12.9	12.3
30-34	7.9	11.5	9.3
35-39	6.7	9.7	9.7
40-44	4.4	6.4	7.2
45-49	3.2	4.7	8.0
50-54	3.6	5.4	5.5
55-59	6.2	6.0	5.5
60-69	15.1	9.8	5.8
70-	27.9	11.7	4.1
Total	10.0	10.0	10.0

Source: Hungarian Central Statistical Office (KSH), Income Studies, HHP first wave.

Table 11:   *Longitudinal poverty position between 1992 and 1996 by age groups in 1996*

Age group in 1996	How many times poor between 1992 and 1996 (50% poverty threshold)				Total N= (100%)
	Never	Transitory once	Transitory 2-3 times	Permanent 4-5- times	
3 – 6	71.5	18.9	6.5	3.1	123
6 –14	63.6	16.8	12.8	6.7	477
15 –19	66.9	16.8	11.5	4.9	389
20 –29	71.3	18.1	6.8	3.8	470
30 –39	72.9	16.2	7.1	3.8	577
40 –49	73.5	12.7	10.3	3.5	701
50 –59	82.7	7.9	6.5	2.9	505
60 –69	83.2	10.1	3.5	3.2	489
70 -	77.0	12.3	8.1	2.6	373
Total (%)	73.9	13.9	8.3	3.9	100
N=	3 034	571	339	159	4 103

Source: Hungarian Household Panel.

Table 12:   Longitudinal poverty position between 1992 and 1996 by the socio-economic position of the adults in 1996

Socio-economic position	How many times poor between 1992 and 1996 (50% poverty threshold)				Total N= (100%)
	Never	Transitory once	Transitory 2-3 times	Permanent 4-5- times	
Higher and medium manager	97.1	2.9	-	-	93
Professional	95.8	4.2	-	-	181
Clerical	89.2	9.8	0.5	0.6	215
Lower supervisor	91.0	9.0	-	-	90
Self-employed artisan, merchant	70.0	15.0	13.8	0.9	134
Skilled worker	81.3	12.4	5.2	1.0	357
Unskilled worker	72.6	18.1	8.3	1.1	337
Peasant, agricultural worker	(72.4)	(16.9)	(8.9)	(1.8)	70
Unemployed	57.4	21.1	14.5	7.0	119
Child care allowance and aid	50.5	34.0	11.2	4.3	124
Housework	(45.6)	(14.0)	(23.3)	(17.1)	63
Pensioner, working	(95.5)	(4.5)	-	-	68
Old age pensioner	85.2	9.8	4.0	1.0	814
Disability pensioner	64.0	13.5	13.4	9.1	248
Pensioner on widow's allowance	52.5	13.9	19.2	14.3	111
Student (above 16)	71.4	16.9	8.3	3.3	320
Others	45.6	23.7	17.5	13.2	169
Total (%)	75.4	13.4	7.7	3.5	100
N=	2 580	458	264	120	3 421

Source: Hungarian Household Panel.

The phenomenon called *new poverty* by us has emerged in close relation to the contraction of the labour market. The unemployed, the 'new entrants of the labour market' and the so-called 'inactive individuals' or 'adult dependants' are the most visible groups. However, the retired disabled, pensioners on widow's pensions and housewives could also be classified under this heading - in other words, all those who have no regular income-producing job, or who do not enjoy a pension paid after a career of more or less full employment (table 12). The 'new poor' come from among the 'traditionally poor' strata (see later) to a large extent, because the risk of unemployment is much higher among them. As is well known from other studies, the risk of becoming unemployed and being long-term unemployed is much higher than average among the low educated, the unskilled workers and peasants, and among the rural people. All this corroborates the commonly known fact that if somebody drops out of regular occupation this becomes one of the decisive factors of his life, causing poverty.

Table 13: Longitudinal poverty position between 1992 and 1996 by the different household characteristics in 1996

	How many times poor between 1992 and 1996 (50% poverty threshold)				Total N= (100%)
	Never	Transitory once	Transitory 2-3 times	Permanent 4-5- times	
Number of household members					
1	75.3	11.6	9.3	3.9	499
2	80.9	10.4	6.2	2.5	467
3	73.3	14.5	6.2	2.5	467
4	77.8	14.5	8.8	3.0	283
5	61.0	19.5	14.7	4.8	100
6 and more	(56.7)	(18.4)	(13.0)	(12.0)	49
Number of children in household					
No children	79.6	10.7	6.9	2.8	1 139
1 child	67.3	17.2	10.4	5.1	251
2 children	73.6	15.6	7.3	3.5	234
3 and more children	50.2	19.5	20.4	9.8	78
Age of the youngest children					
1-2	61.6	16.8	14.9	6.7	78
3-5	62.5	20.1	11.8	5.7	100
6-9	73.0	13.6	7.6	5.9	126
10-14	66.6	17.0	10.8	5.6	148
15-18	68.6	16.2	13.1	2.0	160

Source: Hungarian Household Panel.

There are some factors well known from sociological studies which caused significant income disadvantages in the post-socialist period as well as before, and were therefore accompanied by a far higher than average occurrence of poverty. This is why they are called *traditional* factors of poverty. Very low education makes it impossible to find a regular job, and was an underlying factor at a time of declining employment (table 14). This is the case with those belonging to the strata of unskilled workers and to that of the agricultural blue-collar workers. The type of residence is becoming an increasingly relevant factor. As before with badly paid jobs, unemployment - and consequently, poverty - is nowadays concentrated in villages, and the north-eastern part of Hungary (table 15).

Table 14: Longitudinal poverty position between 1992 and 1996 by education, ethnicity and gender

	How many times poor between 1992 and 1996 (50% poverty threshold)				Total N= (100%)
	Never	Transitory once	Transitory 2-3 times	Permanent 4-5- times	
School education					
Less than 8 class	42.7	13.0	18.8	25.5	409
Primary (8 class)	49.9	18.5	20.0	11.6	974
Apprentice school	57.2	20.5	16.6	5.7	851
Secondary school	73.4	15.9	9.5	1.2	802
University, college	91.9	6.9	1.0	0.2	383
Ethnicity					
Non-Roma	79.0	12.8	6.2	2.0	3 473
Roma	24.1	19.8	19.2	37.1	116
Gender					
Male	76.2	12.8	7.8	3.1	1 606
Female	74.6	13.8	7.7	3.8	1 817

Source: Hungarian Household Panel.

Table 15: Longitudinal poverty position between 1992 and 1996 by type of the residence, and regions in 1996

	How many times poor between 1992 and 1996 (50% poverty threshold)				Total N= (100%)
	Never	Transitory once	Transitory 2-3 times	Permanent 4-5- times	
Type of residence					
Village	66.1	16.6	11.5	5.8	1 619
Small town	78.5	12.4	4.8	4.2	1 135
County seat	70.1	14.9	14.1	0.9	589
Budapest	86.8	9.7	2.0	1.5	760
Regions					
Budapest	86.8	9.7	2.0	1.5	760
North-west Hungary	77.0	19.3	3.1	0.5	550
South -west Hungary	82.1	9.4	5.8	2.6	652
Middle and south-east Hungary	79.0	11.3	6.9	2.8	640
North-east Hungary	60.0	17.1	15.0	7.3	1 501
Total (%)	73.9	13.9	8.3	3.9	100
N=	3 034	571	339	159	4 103

Source: Hungarian Household Panel.

Last but not the least,[6] there is *poverty with an ethnic character* in today's Hungary. It was previously known that a much larger than average part of the estimated half a million people of Roma ethnicity (i.e. the gypsy population) is poor. However, the Hungarian Household Panel offers a mechanism for comparing the income relations of the Roma and non-Roma population. Despite all the reservations concerning the identification of the Roma group, the very fact that the 'over-representation' of the poor in every poverty threshold is the highest among the Roma people calls attention to the fact that they are the part of Hungarian society most endangered by poverty, and in all probability it is they who are the greatest losers of systemic change. It is clear too that in the case of the Roma population the already mentioned poverty risk factors are apparently cumulative (table 14).

Besides summing up the structural factors associated with poverty, I would like to underline a usually insufficiently stressed relationship. In our approach poverty is a household level category. Consequently the poverty situation of the individuals depends heavily on the position of other household members. Depending on the situation, the household can counterbalance an individual's hardship as well as accumulating the disadvantages. The individual poverty position is the result of personal careers and their accumulation in the household.

6   Processes behind temporary poverty

While the identification of the permanent poor population - as far as possible - has been successful, the components of temporary poverty could not be described very precisely. This is mainly because permanent poverty is the result of stable, structural processes, but temporary poverty is the outcome of accumulating macro- and micro-level changes. As already mentioned above, some macro changes, namely declining real income, growing income inequality and the parallel changes at the household level (e.g. becoming unemployed) explain the shifts toward poverty. But what are the relevant factors of escaping poverty? In order to answer this question we have examined the effects of certain life cycle events on one hand, and the income component effect on the other.

Although not too many demographic events have occurred in our sample, we have established some hypotheses based on them (table 16). Changes in the constitution of the household (changes in the household size in both directions) have an impact on the welfare position of the family. The influence of marriage is not clear as other poverty events cluster around it. We have had no opportunity to analyse whether marriage is a 'means' (as in the USA) by which a poor woman can escape from poverty (DUNCAN 1984), or whether marriage causes poverty. If

---

[6]   In the HHP sample there are no homeless and people living in institutions. They could be seen as a fifth type of poor.

we take into consideration that the first child usually comes just after marriage, then possibly not marriage, but the closely linked event of having a child is responsible for the fall into poverty. As we saw, childbirth is clearly an event of higher poverty risk to the family. And of course that is still the case if the child is the third, fourth, etc. On the other hand, if the children are old enough to begin to work, and there are enough job opportunities, then the family can escape from poverty.

Dissolution of the household, namely divorce and becoming widowed, increases the probability of impoverishment too. The first implies temporary poverty, while the latter brings on permanent poverty (table 16). Among those who divorced, the rate of temporary poverty is higher than among those remaining married. Among those divorced after 1992 there is quite a high proportion of once poor. Concerning widowhood, it can be assumed that the loss of a partner, especially of those living not far from the poverty line, or experiencing temporary poverty, quite probably causes long-term, permanent poverty.

Table 16:   Longitudinal poverty position between 1992 and 1996 by the different household characteristics in 1996

	How many times poor between 1992 and 1996 (50% poverty threshold)				Total N=100%
	Never	Transitory once	Transitory 2-3 times	Permanent 4-5- times	
Change in the household size between 1992 and 1996					
Decrease	70.3	15.6	9.9	4.3	470
No change	79.2	10.9	7.2	2.8	1 096
Increase	65.8	17.1	9.7	7.4	135
The date of marriage					
No marriage	73.6	14.2	8.1	3.6	634
Marriage before 1992	76.6	12.7	7.2	3.5	2 518
Marriage after 1992	70.4	16.5	9.9	3.1	127
The date of divorce					
Never married	73.6	14.6	8.1	3.6	634
Married never divorced	77.1	12.5	6.9	3.4	2 191
Divorced before 1992	72.8	13.9	9.4	4.0	382
Divorced after 1992	(70.0)	(17.9)	(9.6)	(1.8)	72
Widowhood					
Never married	73.6	14.6	8.1	3.6	634
Not widowed after 1992	76.4	13.0	7.3	3.3	2 517
Widowed after 1992	75.5	9.6	8.4	6.5	128
Total	75.8	13.2	7.7	3.5	100

Source: Hungarian Household Panel.

Elsewhere we have already argued that economic activity status-change, even if delayed, foreshadows slipping into, as well as escaping from poverty. Taking into consideration the last two waves of our survey (1995, 1996), we analysed the income component effect of the four poverty situations.[7] Therefore, we differentiated between five types of income, namely

a)   earnings (main job only);
b)   social insurance benefits;
c)   public social transfers;
d)   income from odd/informal, etc., jobs;
e)   joint income of the household (table 17).[8]

In the examined poverty categories, the income component effects were in accordance with the expected results. During the two years investigated there was a big difference in the income structure of the group whose members were never poor, and the group of the poor in both years. But there is a common characteristic too. There was very little change in the income composition of both mentioned groups. Among the non-poor, the most important income types are the salary/wage and benefit(s) from Social Security, while among the poor (poor in both years) these are Social Security benefits and social assistance. There is another difference between both groups concerning the importance of secondary incomes: in case of non-poor households this sum constitutes a significant part of the total income. But let us see those whose position has been changed in the given years, 1995 and 1996. We have found considerable restructuring in the income composition of the two groups. In the group of 'escapees' the proportion of salary/wage and the joint household income (e.g. incomes from small-scale production) increased. At the same time the proportion of Social Security incomes and social benefits decreased. Among those who slipped into poverty, it happened the other way round.

   In order to offer a clearer explanation, we have analysed the rise of mean income of the different income types and poverty categories (table 18). Among those escaping poverty the amount of social insurance benefits and social assistance did not decrease. On the other hand, they have experienced a really vague rise in earnings and joint household income. With further analysis we found that the rise in earnings is the result of both increasing individual earnings and the larger number of earners in the household (positive change in economic activity). Regarding the 'slippers' we see that the social benefits and public assistance could not counterbalance the fall of earnings and joint household income.

   Summing up the results of the analysis of the income components, we can assume that wages and joint incomes of the household members are mainly responsible for the dynamics of poverty. Social assistance and social security

---

[7]   The poverty mobility pattern, using the 50% poverty line, during the investigated two years is the following: 82.8% never poor; 7.9% both years poor; 5.7% not poor in 1995 and poor in 1996 (falling into poverty) and 3.6% poor in 1995, not poor in 1996 (escapee from poverty).
[8]   For the different kinds of income and income composition, see TóTH 1994.

benefits remained stable, and can not counterbalance the reduction of labour income.

Table 17:   Income composition in 1995 and 1996 of groups living in different poverty situation*

Types of income	Income distribution of groups living in different two years poverty situations							
	never poor		poor 1995, not poor in 1996		Not poor in 1995, poor in 1996		poor in both year	
	1995	1996	1995	1996	1995	1996	1995	1996
Earnings	52.3	52.0	27.1	34.6	47.5	34.8	15.5	17.1
Social insur-ance benefits	23.5	25.2	43.9	34.7	18.4	29.3	42.8	43.7
Public social transfers	6.1	6.0	20.7	14.9	18.4	30.7	33.0	33.2
Odd/informal jobs, etc.**	5.2	4.6	3.2	2.6	2.3	1.2	1.9	0.5
Joint income of the HH***	7.1	6.0	3.6	10.8	9.2	3.5	5.7	4.6
Others	5.8	6.2	1.5	2.4	4.2	0.5	1.1	0.9
Total (%)	100	100	100	100	100	100	100	100
(in Forint)	255 583	294 650	101 907	166 109	165 941	112 752	88 334	94 642

*   Yearly equivalent income.
**   Secondary jobs: all types of secondary incomes, not small-scale production.
***   Household incomes: mainly agricultural small-scale production.

Source: Hungarian Household Panel.

Table 18:   The percentage change of the different income types in the four poverty positions concerning the year 1994 and 1995 *

	Income distribution in 1995 and 1996 of groups living in different poverty situations in two years			
	never poor	poor 1995, not poor in 1996	not poor in 1995, poor in 1996	Poor in both years
Earnings	113	227	47	117
Social insurance benefit	122	139	101	107
Public social assistance	111	127	107	107
Household's joint income	92	536	24	86
Total income	114	176	64	106

*   Average equivalent income of 1996 in percentage of the 1995 income.

Source: Hungarian Household Panel.

7 Conclusion

In conlusion, we would like to emphasise some of our major findings. First of all, it is unquestionable that poverty rose during the transformation period, and that it became an undeniable social problem in Hungary. However, the extent of poverty depends to a large extent on the researcher's definition. Examining poverty from the longitudinal side, we demonstrated the existence and dynamics of permanent and transitory poverty at the time of transformation by using different absolute and relative approaches to poverty such as the subsistence minimum, the mean equivalent income, and subjective poverty indicators. Thus, we were able to iden-tify the most risky social groups living in permanent poverty. Particularly those living in north-east Hungary and/or in villages, those having low education, being excluded from the labour market, having more children, or not having a partner tend to face a very high risk of being permanently poor.

In order to say more about the transitory poor, we would need further analysis, but some findings may already be instructive. Besides the macro-social change, events in the life cycle of individual job careers, and their linkages in the house-hold seem to explain most of the dynamics in individual and household-based poverty positions.

Appendices

All of the data were taken from the series of surveys entitled Hungarian House-hold Panel (see TÓTH 1995), with the exception of macro-statistical data from the Hungarian Central Statistical Office given in table 1, and comparative data from the German Panel (GSOEP) in table 6. The Hungarian Household Panel follows the method of similar foreign longitudinal surveys (DUNCAN 1984), particularly that of the West German ones (HANEFELD 1987; RENDTEL and WAGNER 1991; ZAPF, SCHUPP, HABICH 1996). The latter were extended to the former East Germany immediately before reunification.

About 2 000 households were included in the original survey sample. They were randomly selected, but the sample was strongly concentrated territorially (originally not all counties were included). A questionnaire was filled in about the households and about every individual above the age of sixteen. About 4 500 in-dividuals above the age of sixteen live in these households. In addition, there were about 1 200 children in those households; their data figure in the household. It derives from the nature of panel surveys that these households are visited each year (in May) and are questioned. The questionnaires primarily contain data about incomes and employment behaviour, housing, households' coping strategies. Additional questions that are not asked with annual regularity deal with many other topics. Such questions enquire about satisfaction, symptoms of psychological problems, manifestations of anomie and alienation.

References

ANDORKA, R. (1989): Szegénység Magyarországon. (Poverty in Hungary). Társadalmi Szemle 44 (12)

ANDORKA, R. (1990): Public Debate on Poverty in Hungary since the 1960s. In: GOODWIN, C. D., NACHT, M. (ed.): Beyond Government. Westview, Boulder, 205-226

ANDORKA R., SPÉDER, ZS. (1996): Poverty in Hungary. Review of Sociology, Special Issue, 3-28

ATKINSON, A.B. (1989): Poverty and Social Security. Harvester, Hempstead

BOKOR, Á. (1985): Depriváció és szegénység. (Deprivation and Poverty). Társadalomtudományi Intézet, Budapest

BÖHM, A. (1981): A Magyar Szociológiai Társaság tudományos ülésszaka a "többszörösen hátrányos helyzetû rétegek vizsgálatáról". (Conference about the Multiple Deprived Social Positions, Organised by the Hungarian Sociological Association). Szociológia X (3-4), 279-232

DUNCAN, G. J. (1984): Years of Poverty, Years of Plenty. Institute for Social Research, University of Michigan, Ann Arbor

HANEFELD, U. (1987): Das sozio-ökonomische Panel: Grundlagen und Konzeption. Campus, Frankfurt am Main

HEADEY, B., HABICH, R., KRAUSE P. (1990): The Duration and Extent of Poverty: Is Germany a Two-Thirds-Society? WZB Paper, 90-103

KOLOSI, T., RÓBERT P. (1992): Munkaerőpiac és jövedelmek. (Labour Market and Incomes). In: ANDORKA, R. et. al. (eds): Társadalmi Riport 1992, 8-24

KOLOSI T., SZÍVÓS, P., BEDEKOVICS, I., TÓTH, I. J. (1996): Munkaerőpiac és jövedelmek. (Labour Market and Incomes). In: SÍK, E., TÓTH, I. G. (eds): Az ajtók záródnak!? Jelentés a Magyar Háztartás Panel V. hullámának eredményeiről. (The Doors are Closing!? Report on the Results of the 5th Wave of the Hungarian Household Panel). Department of Sociology, Budapest University of Economics, Budapest

KRAUSE, P. (1994): Armut im Wohlstand: Betroffenheit und Folgen. DIW Diskussionspapier No. 88., Berlin

KRAUSE, P., SPÉDER, ZS. (1996): Income Dynamics as an Indicator of Social Change: The Case of Hungary, East Germany and West Germany. Presentation at the Fourth International Social Science Methodology Conference, July 1-5, Essex

KSH (ed.) (1991): Létminimum 1989-1991. (Subsistence Minimum 1989-1991). Hungarian Central Statistical Office, Budapest

MILANOVIC, B. (1995): Poverty, Inequality and Social Policy in Transition Economies. Transition economics division, Policy Research Department, World Bank (= Research Paper Series 9)

MONOSTORI, J. (1997): Az önkormányzatok segélyezési gyakorlata. (The Social Assistance Practice of the Communities). Hungarian Central Statistical Office, Budapest

PARKER, J. L., SMEEDIG, T. B., TORREY,T. (1988): The Vulnerables. The Urban Institute Press, Washington D.C.

RAINWATER, L.. (1988): Inequalities in the Economic Well-Being of Children and Adults in Ten Nations. Luxembourg (= Luxembourg Income Study Working Paper 19)

ROSE, R., HAERPFER, C. (1993): Adapting to Transformation in Eastern Europe. New Democracies Barometer II. Center of the Public Policy Studies in Public Policy, Glasgow (= Working Paper 212)

RUGGLES, P. (1994): Drawing the Line. The Urban Institute Press, Washington D.C.

RENDTEL, U., WAGNER, G. (eds) (1991): Lebenslagen im Wandel: Zur Einkommensdynamik in Deutschland seit 1984. Campus, Frankfurt am Main

SÍK, E., TÓTH, I. G. (1996): Az ajtók záródnak!? Jelentés a Magyar Háztartás Panel V. hullámának eredményeiről. (The Doors Are Closing!? Report on the Results of the 5th Wave of the Hungarian Household Panel). Department of Sociology, Budapest University of Economics, Budapest

SMEEDING, T. M. (1988): Generations and the Distribution of Well-Being and Poverty: Cross National Evidence for Europe, Scandinavia and the Colonies. Luxembourg (= Luxembourg Income Study Working Paper 24)

SMEEDING, T. M., TORREY, B. B. (1988): Poor children in rich countries. Luxembourg (= Luxembourg Income Study Working Paper 16)

SPÉDER, ZS. (1996): Wenn man mit dem Einkommen nicht auskommt: Unfreiwillige Änderungen des Konsumverhaltens ungarischer Hausehalte während der Transformation. In: GLATZER, W. (ed.): Lebensverhältnisse in Osteuropa: Prekräre Entwicklungen und neue Konturen. Campus, Frankfurt am Main, 159-176

TÁRKI and GKI (eds) (1994): A magánszektor fejlődése Magyarországon: Kutatási beszámoló az ÁPV számára. (The Development of the Private Sector: Report for the National Property Rights Agency). Budapest

TÓTH, I.Gy. (1995): The Hungarian Household Panel: Aims and Methods. Innovation 8 (1), 106-122

TÓTH, I.Gy. (1996): A háztartások jövedelmi szerkezete: a munkaerőpiac és a szociálpolitika szerepe. (The Income Structure of the Households: The Role of the Labour Market and Social Policy). In: SÍK, E., TÓTH, I. G. (eds): Az ajtók záródnak!? Jelentés a Magyar Háztartás Panel V. hullámának eredményeiről. (The Doors are Closing!? Report on the Results of the 5th Wave of the Hungarian Household Panel). Department of Sociology, Budapest University of Economics, Budapest

TÓTH, I.Gy., ANDORKA, R., FÖRSTER, M., SPÉDER, ZS. (1994): Poverty, Inequalities and the Incidence of Social Transfers in Hungary, 1992-93. Paper prepared for the World Bank Budapest Office, Budapest

UNICEF International Child Development Centre (ed.) (1993): Public Policy and Social Conditions. (= Regional Monitoring Report 1)

UNICEF (ed.) (1994): Crisis in Mortality, Health and Nutrition. (= Economies in Transition Studies, Regional Monitoring Report 2)

UNICEF (ed.) (1995): Poverty, Children and Policy: Responses for a Brighter Future. (= Economies in Transition Studies, Regional Monitoring Report 2)

ZAPF, W., SCHUPP, J., HABICH, R. (eds) (1996): Lebenslagen in Wandel: Sozialberichterstattung im Längsschnitt. Campus, Frankfurt am Main

The Geography of Post-Communist Parliamentary Elections in Hungary

Zoltán Kovács

1 Introduction

Hungary has long been in the focus of attention of western media and political scientists as one of the front-runners of East European transformation. The reasons for this high-level interest are numerous, among others the geographical location of the country between East and West and its geopolitical importance plays a role just like the uniqueness of the Hungarian way of transformation.

In terms of the development of multi-party democracy, Hungary represents a specific case in Eastern Europe. Unlike the experience of many of its neighbours, the Hungarian communist dictatorship was transformed into a parliamentary democracy in a gradual and peaceful manner. Not only was the transition smooth, but all freely elected post-communist governments were able to complete their full terms with relatively few difficulties. Furthermore, all three post-communist elections (1990, 1994 and 1998) resulted in totally new structures of political power in the form of coalition governments which made possible rotation and, thus, the evolution of political parties.

This paper considers the electoral geography of Hungary since 1990. Electoral geography has been integral part of political geography in the West, but it was practically unknown in Eastern Europe. After World War II, under the conditions of a single party system there was neither the data nor the possibility to study regional variations of voting behaviour. Such efforts would also not have been tolerated by the communist regimes. The political changes of 1989/90 led to the formation of new multi-party democracies in Eastern Europe which consequently resulted in the mushrooming of electoral studies, many of them giving a geographical perspective (KOLOSSOV 1993; RIVERA 1996; KOWALSKI 1999). The present paper seeks to contribute to these new developments by comparing the results of three post-communist elections in a broad theoretical framework.

After a short overview of the post-1945 development of Hungary the new electoral system of 1990 is discussed. Following the presentation of the quantitative results of the first free elections, the process of cleavage formation in the country is analysed on the basis of the LIPSET-ROKKAN (1967) model. Geographical variations in voting behaviour and political-party support are then explained by historical factors as well as the present socio-economic structure of the country. Finally, the findings are integrated into a common theoretical framework.

2   Hungary's road to multi-party democracy

After World War II, Hungary set out on the road to a new democratic system under a coalition government of major democratic parties: the Smallholders Party representing the peasantry and small farmers, the two workers' parties, the Social Democratic Party and the Communist Party, and the National Peasant Party. However, as part of the Russian occupation zone, the Hungarian Communist Party was directed by Moscow.

The elections in 1945 were a setback for the Communists who received only 17% of the national vote behind the Social Democrats (17,5%) and the Smallholders (57%). Nevertheless the former 'grand-coalition' continued with considerable success in rebuilding the country's economy and restoring social life. Over the ensuing three years, in common with other East European countries, democratic parties were gradually eliminated by the Communists either by banning them (e.g. Smallholders), or by forcibly amalgamating them (e.g. Social Democrats). With political pressure and manipulation of votes the Communists managed to win the 1947 elections and the country became a single-party communist (Marxist-Leninist) system for the following four decades.

The events of 1988/89 leading to the collapse of communism were not a sharp break with the past in Hungary, as they were in many other communist countries. Rather they were vital to the process of continuity and change. In Hungary, claims towards western-style democracy and market economy had roots as far back as the 1956 anti-Communist revolution. Pressure for a more open society was steady in the 1970s and 1980s. In 1968, Hungary launched the New Economic Mechanism, a programme of reforms to replace the system of central planning by an economy reliant more upon decentralised management and the influence of market forces.[1] Despite the far-reaching economic liberalisation and reforms which made Hungary the most liberal part (or the 'Gayest Barrack') of the communist Eastern Bloc, hardly any political changes could take place until János Kádár came to power. Kádár and his entourage were finally swept from power by the new generation of party leaders inspired by Gorbachev at the communist party conference in May 1988. After that political groups of various types began mushrooming in the country.

When the collapse of the system became obvious at the end of 1980s, the communist party started 'roundtable discussions' with several opposition groups and organisations with a view to amending the Hungarian constitution and establishing a competitive multi-party system. The most active opposition groups participating in these talks were the Hungarian Democratic Forum, the Alliance of

---

[1]   The New Economic Mechanism of 1968 resulted in a greater availability of consumer goods and a gradual but steady rise in living standards. It became also easier for Hungarians to travel to the west.

Free Democrats, the Alliance of Young Democrats, the Christian Democrats, Social Democrats and the Independent Smallholders. During these negotiations a compromise was reached between the government and the opposition groups. This established legalised political parties. At the end of 1989 a new electoral law was passed by Parliament and the first free-election after 1947 were scheduled for March 1990 (KOVÁCS 1993).[2]

3   Electoral system and political parties in Hungary

The post-communist Hungarian election law (Act No. XXXIV. of 1989) is one of the most complicated in the world. The law reflects both the functional and practical electoral systems operating in Western Europe and can be seen as a compromise between the concept of strict geographical representation and proportional political party representation. The Hungarian electoral system, modelled on that of the Federal Republic of Germany, is a mixture of a single-member electoral district and proportional representation using two rounds of balloting. In practice two elections take place at the same time, each elector having two votes, one to cast for a specific local candidate and another to cast for a particular party (KOVÁCS 1993). These two elections are separate but linked. One link is that most, but not all candidates in the constituency elections represent particular political parties. The other, more important link is made by geography (DINGSDALE and KOVÁCS 1996; MARTIS et al. 1992).

The country is divided into 176 single-member electoral districts. The territories and boundaries of these districts are based on the geographical distribution of population to ensure broadly similar numbers of voters in each district. A candidate is elected from a single-member district if he or she receives a majority of the vote in the first round election. The election law requires that over a half of the electorate must vote in the first round for the election to be valid. If no candidate receives a majority in the first round, all candidates receiving 15% of the vote go on to the second-round election. If no candidate receives 15% of the vote the top three candidates compete for the second-round election. According to the election law, 25% of the electorate must vote in the second round for the election to be valid and the candidate with the highest number of votes is elected.

On the other hand, single-member districts 'nest' within counties. County boundaries are therefore incorporated into the system, making them discrete territories that become the units for the first tier of the two-tier proportional repre-

---

[2]   The first multi-candidate elections, within a one-party system, were already permitted in the 1980s. The ruling communist party supported a law passed in 1983 mandating contested elections in all parliamentary seats. In 1985, the first of these contested elections was held and a number of so-called independent candidates defeated official party candidates.

sentation system. Each party usually puts up a list of county candidates equal to the number of seats apportioned to that county. The county list is allowed to stand if the party is able to nominate a candidate (i.e. collects 750 'nomination coupons') in at least two districts in that county. The 'County List' of candidates elects members on the basis of votes cast for each party in the 20 counties of Hungary (the capital city Budapest is included in this system as a county). The second tier is the 'National List' of candidates who are chosen using votes cast for their party in the country as a whole, but which have not affected the result at the district or county level (i.e. 'scrap' votes). Political parties must initially organise county lists in at least seven counties to be eligible to participate in the national vote pool. Any party gaining 5% (in 1990 only 4%) or more of the total national vote has a right to representation in Parliament. The number of constituency members is fixed at 176 (45% of the Parliament), but the balance between candidates elected on the 'County List' or 'National List' can vary depending on the pattern of votes cast to make up the 386 members of Parliament.[3]

In 1990, there were 65 parties who met the legal registration deadline for participation in round one of the first free election. Four years later, on the eve of the 1994 general elections, there were already 136 officially registered political parties in Hungary. The proliferation of parties owed a great deal to the liberal requirements of the law (MARTIS et al. 1992). Few parties had any real organisation. Most of the smaller parties voice the opinions of specific interest groups such as peasants, smallholders, environmentalists, pensioners, entrepreneurs or the unemployed. This variety of 'niche' parties reflects the plurality of Hungarian society that has rapidly emerged since 1990 and is often a local manifestation of 'one issue' politics. Few of these parties could expect to play a role in the elections after 1990.

In the 1990 election, 19 parties had sufficient support to set up a county list and only 12 parties could set up a national list. In 1994 the same number of parties were able to set up a county list, however, the number of national lists increased to 15. Four years later, in 1998, only 15 parties gained sufficient support to set up a county list and 12 could participate in the national list competition. The decreasing number of parties both in the Parliament and in the contest reflects the gradual maturing of the party system and the development of the Hungarian electorate.

---

[3]   The number of mandates allocated on the 'National List' was 90 in 1990 elections, 85 in 1994 and 82 in 1998. This is a clear reflection of the gradual reduction of 'scrap' votes.

4   The seesaw of post-communist parliamentary elections in Hungary

In 1990, in the first free-election, the Hungarian Socialist Workers' (Communist) Party (MSZMP) which had ruled the country for 43 years was defeated. The party received only 3.68% of the votes and thus missed the 4% threshold which was necessary for parliamentary representation.[4] The conservative Hungarian Democratic Forum (MDF) won the election with 24.73% of the votes which secured 165 out of the 386 elected seats (43.7%) for the party in the Hungarian Parliament (table 1). Since the MDF did not win a clear majority it needed to form a coalition government with two smaller right-wing parties, the Independent Smallholders' Party (FKgP) and the Christian Democratic People's Party (KDNP) (figure 1). The biggest opposition party in the Parliament was the liberal Alliance of Free Democrats (SZDSZ) with 21.93%, backed by - at least ideologically - its smaller sister-party, the Alliance of Young Democrats (FIDESZ). The only left-wing party which received seats in the newly elected Parliament was the Hungarian Socialist Party (MSZP), the reform wing of the former Hungarian Socialist Workers' Party (MSZMP), with 10.89% of the votes.

Table 1:   The distribution of votes for party-lists in the post-communist elections (%)

	1990	1994	1998
Hungarian Democratic Forum (MDF)	24.73	11.74	-
Alliance of Free Democrats (SZDSZ)	21.93	19.74	7.57
Independent Smallholders' Party (FKGP)	11.73	8.82	13.45
Hungarian Socialist Party (MSZP)	10.89	32.99	32.92
Alliance of Young Democrats (FIDESZ)	8.95	7.02	-
Christian Democratic People's Party (KNDP)	6.46	7.03	-
Hungarian Truth and Life Party (MIEP)	-	-	5.47
Other parties	15.31	12.66	11.11
Total	100	100	100

Source: Author's compilation.

Four years later, in May 1994, the centre-right parties were defeated and the election returned to power the reform communists with a large majority. The 32.99% of the votes cast for the Hungarian Socialist Party (MSZP) meant that the Socialists gained 209 out of the 386 seats in Parliament and thus achieved an absolute majority. The second most successful party in the election was the Alliance of Free Democrats (SZDSZ) with 19.74% of the votes and 69 seats in the Parliament. These two parties formed a coalition after the 1994 election holding together a comfortable majority of 278 seats (72%) in Parliament.

---

[4]   The threshold was lifted to 5% in the next election which is most common in Europe.

Figure 1:   Distribution of seats in the Hungarian Parliament

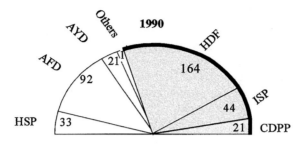

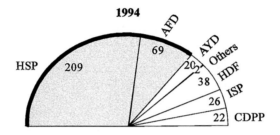

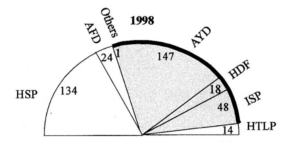

Coalition government

HSP   - Hungarian Socialist Party (MSZP)
AFD   - Alliance of Free Democrats (SZDSZ)
AYD   - Alliance of Young Democrats (FIDESZ)
Others - Independent or joint candidates
HDF   - Hungarian Democratic Forum (MDF)
ISP   - Independent Smallholders' Party (FKgP)
CDPP   - Christian Democratic People's Party (KDNP)
HTLP   - Hungarian Truth and Life Party (MIEP)

Source: Author's compilation.

The conservative parties lost support in all parts of the country. The former winner, the moderate centre-right Hungarian Democratic Forum (MDF) gained less than half of its 1990 vote and finished in the third place with 11.74%. The fourth biggest party, the Smallholders' Party (FKGP), received only 8.82% compared to 11.73% in 1990. Two parties, FIDESZ and KDNP, performed even less successfully and ended up as the smallest parliamentary parties with roughly 7% of the votes each. In spite of the fact that the threshold for parliamentary representation was raised from 4% in 1990 to 5% in 1994, the same six parties were able to achieve representation in both elections, but the balance of power shifted enormously.

In 1998, the Hungarian Socialist Party (MSZP) nearly repeated its 1994 performance and finished at the first place with 32.92% of the votes in the first round of the election. The second biggest party was the moderate conservative FIDESZ with 29.48%. Since other parties received significantly less votes than the two front-runners, the final decision remained for the second round of voting, the single candidate competition. Thanks to its clever campaign and skilful coalition tactics, FIDESZ was able to integrate all conservative votes in the second round and won 90 out of the 176 single constituency seats, compared to the Socialists' 54 seats. As a result the biggest party in Parliament became the FIDESZ with 147 seats on aggregate, as opposed to the Socialist Party (MSZP) with 134 seats. Thus, after four years of socialist-liberal government, the FIDESZ could form a conservative coalition government with the Independent Smallholders' Party (FKGP) and the remnants of the Hungarian Democratic Forum (MDF).

A new feature in the Hungarian Parliament after 1998 was the appearance of the nationalist/populist Hungarian Truth and Life Party (MIÉP), a former fraction of the Hungarian Democratic Forum, which gained 5.47% of the votes and became the smallest parliamentary party. On the other hand, due to inter-party rivalries and the subsequent split between different platforms of the party the Christian Democratic People's Party missed the 5% threshold and, after two parliamentary cycles, remained out of the Parliament.

5   Political cleavages and party system in Hungary

Usually an election does not hinge upon just one or two issues but covers a wide range of topics. In normally functioning multi-party systems each party represents a 'package' of responses to these issues. As such, they simplify the decision of the voter: instead of having to find out all the various positions each candidate takes on each issue, the voter simply has to find out which party the candidate represents (LIPSET and ROKKAN 1967).[5] The cleavage model elaborated by LIPSET and

---

[5]   In order that a party label may have this property of simplifying the voting decision, it

ROKKAN (1967) has had an outstanding impact on the development of electoral geography, but it has also encountered strong criticism. In the following section, an overview about the most important factors determining cleavage formation in the post-1989 Hungarian politics is given by using the Lipset-Rokkan model as a theoretical framework.

LIPSET and ROKKAN (1967) and later ROKKAN (1970) defined in their famous study four major conflicts which they termed the critical cleavages in the development of modern European states in the 20th century. Two of the cleavages (nationality and religion) are the product of the National Revolution, and two (economic sector and occupation) are the product of the Industrial Revolution. Much of the political history of European countries in the last 200 years reflects the interaction between these two revolutionary changes, which largely diffused out of France and Britain respectively (TAYLOR and JOHNSON 1979). These four basic issues have got eight alternatives which characterise western party systems: central versus peripheral cultures (nationality), state v. church (religion), industrial v. agrarian interest (economic sector), and workers v. employers (occupation). Changing electoral fortunes of individual parties represent shifts in the underlying political cleavage structure (RIVERA 1996).

5.1   Central versus peripheral cultures

As Hungary can be considered ethnically nearly homogenous (96% of the population is ethnic Hungarian) this issue has had somewhat less importance in the post-communist elections than in other post-communist East European countries (e.g. Czechoslovakia, or member states of the former Soviet Union) (ZARYCKI 1999). Ethnicity has been articulated in Hungarian politics in two particular ways. Firstly, the question of Hungarian ethnic minorities in the neighbouring states was repeatedly raised in debate, and secondly, the status and representation of ethnic minorities (including gypsies) within Hungary has been a factor (KOVÁCS and DINGSDALE 1998).

Excluding the Russian minorities in some successor states of the former Soviet Union, Hungarian minorities are the largest minority group in Europe, totalling 3.5 million (KOCSIS 1994).[6] The concern for Hungarian minorities in neighbouring countries and the promotion of their protection was intensively manifested primarily in the campaign of conservative parties, especially by the Hungarian Democratic Forum (MDF) in 1990, by the nationalist/populist Hungarian Truth and Life Party (MIÉP) in the 1994 and, to a lesser extent, the 1998 elections. These parties warned voters that a victory for 'cosmopolitan' socialist or liberal

---

is necessary for political parties to be relatively stable groups of politicians with relatively consistent responses to issues (see TAYLOR and JOHNSTON 1979).

[6]   There are more than 2 million Hungarians living in Romania, 700 000 in Slovakia, 400 000 in Serbia and the rest in other neighbouring countries (e.g. Ukraine, Croatia).

parties (MSZP and SZDSZ) would lead to a surrender of the interests of Hungarian minorities to the nationalist regimes of Meciar, Ilescu and Milosevic. However, nationalist campaign has not always proved to be successful in the Hungarian elections. In 1990, on the eve of the elections, the worst outbreak of anti-Hungarian violence was reported from Romania (Tirgu Mures). The Romanian clashes seemed to give a boost to the Hungarian Democratic Forum, which based its appeal partly on nationalist feelings. Four years later, in the 1994 elections, the newly established Hungarian Truth and Life Party (MIÉP) raised the question of ethnicity again with xenophobic overtones. The lack of appeal of the nationalist campaign was shown by the tiny 1.59% of the vote achieved by the MIÉP.

In the reaction of voters we could trace the internal socio-economic development of the country between 1990 and 1994. The first years of transition were difficult ones for the majority of Hungarian society. Most families were hard-hit by decreasing living standard and falling social security. Under these circumstances voters were much more oriented towards issues related to the internal economic affairs of the country than towards the question of ethnicity and the future of Hungarian minorities in the Carpathian Basin. Having learnt from the 1994 election, most parties tried to avoid the ethnic question and concentrate their campaign on political and economic issues in the 1998 election.

On the other hand, the question may also raise, how 'non-Hungarians' participate in the Hungarian elections.[7] Earlier investigations, made on the voting behaviour of 87 German, 44 Slovak, 42 south Slav and 8 Romanian settlements (settlements where Hungarians are in minority), demonstrate that ethnic groups vote in a fairly similar way to the population in general (KOVÁCS and DINGSDALE 1998). Most parties received around average votes in these settlements. The only exception was the MIÉP, which had little success among the ethnic groups both in 1994 and 1998.

The situation is somewhat different in the case of the biggest 'non-Hungarian' ethnic group, the gypsies.[8] After 1989 more than one hundred gypsy organisations emerged in Hungary. Yet there is no single gypsy organisation which would unify representation of this minority and unite gypsy voters.[9] Since gypsies have been discriminated in many different ways within society their voting patterns differ significantly from the national average and from other ethnic minorities as well. In gypsy settlements it is important to emphasise the role of community leaders in voting decisions. Bearing this in mind, it is easy to understand the success of the former communist party (MSZMP) in 1990, the Hungarian Democratic Forum

---

[7]   There are about 200 000 ethnic Germans, 125 000 South Slavs (mostly Croats), 110 000 Slovaks and 25 000 Romanians living in Hungary (KOCSIS 1994).

[8]   There are about 500 000 gypsies in Hungary, i.e. 5% of the population belong to this minority.

[9]   Such a party would be probably able to achieve the 5% parliamentary representation threshold.

(MDF) in 1994 and the Socialist Party (MSZP) in 1998. The tendency of the gypsy communities to support the 'once in power parties' is related to their political and economic insecurity and consequent dependence on the good will of the 'authorities'. As might be expected, the extreme right wing nationalist MIÉP received hardly any support in gypsy communities in 1994 and 1998.

5.2   Religious dimensions

Religion can be considered a relatively weak electoral factor in Hungary. The only church based religious party which won parliamentary representation in the 1990s was the Christian Democratic People's Party (KDNP). Both in 1990 and 1994 the KDNP ended as the smallest parliamentary party with only about 7% of the votes. In both elections votes cast for the KDNP followed basically the same geographical pattern, which in turn reflected the relative stability in the spatial distribution of religious votes. The party received most support in the northern and western regions of Hungary (Nógrád, Vas and Zala counties) where the influence of the Roman Catholic Church is most prominent.[10] In the eastern part of Hungary, where the Calvinist reformed church is dominant, the KDNP achieved negligible support, not even reaching the 5% threshold.

The traditional voting base of Christian Democrats is concentrated in the villages and in the over 60 years age group. Since other conservative parties, especially the Smallholders' Party (FKGP) and the Hungarian Democratic Forum (MDF), have also targeted the support of this social group, the KDNP has always had to struggle for survival. The relatively weak role of religion as a factor in voting can also be demonstrated by the fact that, as rivalries increased between the different platforms of the KDNP before the 1998 election, many voters turned away from the party and supported other right-wing parties (FKGP, MIÉP and especially FIDESZ). As a consequence, the KDNP gained only 2.31% of the votes in 1998 and fell out of the Parliament.

5.3   Industrial versus agrarian interests

On the basis of the 1990, 1994 and 1998 elections it seems that Hungary's dominant cleavage lines are primarily urban versus rural, with an overlapping subtext of nationalism and religion.[11] Figure 2 displays the distribution of vote by settlement for the major (i.e. parliamentary) parties. On the basis of the first three post-communist elections we can observe a strong rural/urban dichotomy, or even a Budapest/country town/village trichotomy in terms of voting behaviour.

---

[10]   The Roman Catholic Church is historically the major religion of Hungary. Although official church-attendance figures are not available, about two thirds of the population is estimated to belong to this religion, the rest being Protestant, mainly Calvinist.

[11]   According to RIVERA (1996) this coincides with cleavages in the inter-war period.

Figure 2:   Voting pattern of parliamentary parties by settlement types

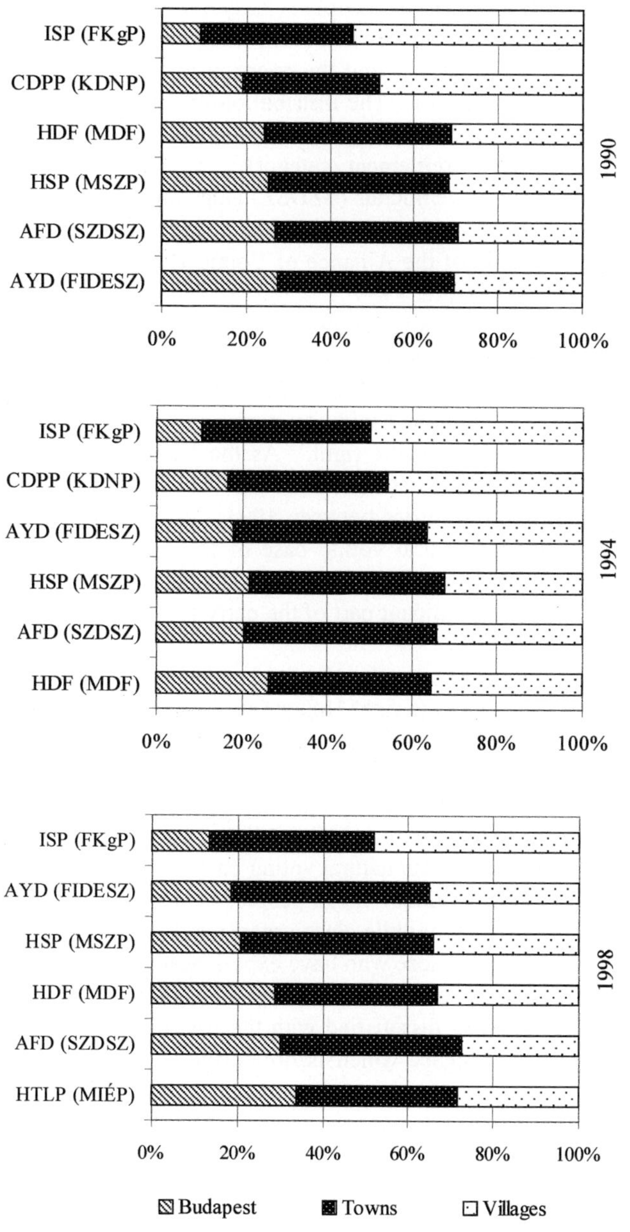

Source: Author's compilation.

The traditional voting base of the Smallholders Party (FKGP) and to a lesser extent the Christian Democratic People's Party (KDNP) lies predominantly in rural areas and small towns with agricultural character. The traditional peasant and religious values of these parties, and the question of agricultural production, are more relevant to village people. The distribution of votes cast for the Hungarian Socialist Party (MSZP) and the Hungarian Democratic Forum (MDF) is somewhat more balanced in these settlement categories, whereas the liberal, western-oriented Alliance of Free Democrats (SZDSZ) gains most of its votes in Budapest and other major towns. These parties have had a fairly stable voting pattern over time whereas in the case of the Alliance of Young Democrats (FIDESZ) we can observe a substantial shift after 1990.

The FIDESZ has its origins in the ideas and attitudes of the intellectuals' opposition movement during the second half of the 1980s. In the election of 1990 the FIDESZ gained support primarily from urban (Budapest) intellectuals. After the first election the party struggled with its image and self identity and gradually shifted from the centre-left to the right.[12] As the Hungarian Democratic Forum (MDF), the Christian Democrats (KDNP) and the Smallholders (FKGP) were weakened by intra-party rivalries between 1994 and 1998, FIDESZ managed to attract and integrate the natural voting base of these conservative parties. This change is well reflected by the distribution of FIDESZ votes by settlements in the 1998 election, when the dominant part of the party support came already from the countryside.

Interesting voting patterns can be observed at the two poles of the ideological axis, in the case of the extreme right Hungarian Truth and Life Party (MIÉP) and the extreme left Workers' Party (Munkáspárt). The former was supported mainly by urban voters, with the highest proportions in Budapest, whereas the working class oriented communist Munkáspárt had disproportionately large support in rural areas. The success of the nationalist/populist MIÉP in urban areas and in Budapest supports the idea that the natural voting base of the party consists of not so much the conservative, religious village people, as we could estimate at the first glance, but rather of lower middle-class urban inhabitants (civil servants, small traders and entrepreneurs), those who have experienced the most dramatic decline in their living standard and social status during the eight years of transformation.[13] These strata were the most dissatisfied with the first eight years of the new parliamentary system in Hungary which is obviously reflected in their votes. The

---

[12]   In 1990, FIDESZ viewed itself as a left-of-centre party with fairly similar programme to the Free Democrats (SZDSZ). Hence, many voters considered them as a 'youth organisation' of the bigger liberal party (KOVÁCS and DINGSDALE 1998).

[13]   In the 1998 election, MIÉP received 34% of its total votes in Budapest, where 18% of the total population of Hungary lives. This was clearly the highest rate among the parliamentary parties.

relatively good result of the extreme-left Munkáspárt in rural areas can be explained by nostalgia for the communist regime and its social security.

5.4   Worker-employer divide

Occupation and social status are also strong factors in Hungarian voting patterns. After the long years of social homogenisation under the communist regime it is therefore interesting to ponder the extent to which social class plays a role in voting behaviour in post-communist Hungary, and what sort of cleavages can be observed in this respect.

Given the lack of adequate polling data, indirect statistics for the analysis of the socio-demographic background of party support had to be used. Based upon the 1990 census, the results of the major - i.e. parliamentary - parties in the 1990 and 1994 elections were analysed in regard to social characteristics. Age structure and educational attainment were used as basic indicators for social class variables. The Hungarian settlements were classified into small groups according to the ratio of elderly as well as university and high school graduates. The distribution of votes cast for the major parties in the different settlement groups shows distinct social characteristics in the voters' support for the different parties (figure 3).[14]

The Christian Democrats (KDNP) and Smallholders (FKGP) enjoy generally high support among elderly and less educated people. On the contrary the left wing Hungarian Socialist Party (MSZP) and especially the urban based liberal Alliance of Free Democrats (SZDSZ) are more favoured by younger and better educated voters. The centre-right Hungarian Democratic Forum (MDF) and the Alliance of Young Democrats (FIDESZ) have a very balanced voters' support according to age and education variables, which confirms the centrist character of these parties. These findings coincide very much with the rural/urban dimension of political party support discussed above.

6   Geography of voting in Hungary

In this section, two particular aspects of voting will be discussed. These can be considered as the key geographical patterns of voting. First, the long-term regional variations of voting turnout are analysed, then the spatial distribution of political-party support is explained. For that purpose a series of thematic maps have been produced for each of the post-communist elections in the Institute of Geography at the Hungarian Academy of Sciences. The explanation of these pat-

---

[14]   These findings are also confirmed by exit-poll data which were first systematically collected in the 1998 election.

terns is based upon the historical traditions and the present human geographical structure of the country.

Figure 3:   Distribution of party support in settlements with different social characteristics 1994

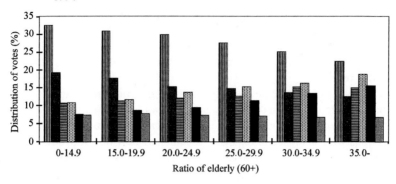

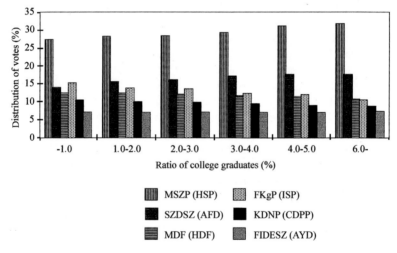

Source: Author's calculation.

6.1   Voter participation

In 1990, 65% of the electorate voted in the first round and 45% in the second. In 1994, 69% voted in the first round and 55% in the second. This overall increase in participation was a favourable feature and one would have expected the turnout to increase well above 70% in the 1998 election and thus Hungary to approach the level of western democracies (HAJDÚ 1992). The 56% turnout in the first round of the 1998 election was therefore a disappointingly bad result which was not coun-terbalanced by the astonishingly high participation (57%) in the second round.[15]

The geographical pattern of voter participation shows high stability when com-pared over time (figure 4). As in 1990, both the 1994 and 1998 elections recorded the highest turnouts in the north-western parts of the country. The difference be-tween eastern and western Hungary is about 20% on average, in two eastern counties (Szabolcs-Szatmár-Bereg and Hajdú-Bihar) participation rates remained below 50% making necessary a repeat of the first-round election. It seems likely that this pattern of political awareness is related to divergent socio-economic developments of these regions.

According to survey data the most important variables predicting voting par-ticipation in Hungary are educational background and the rural-urban composition of the district (KOVÁCS and DINGSDALE 1998).[16] Western Hungary, especially along the Budapest-Vienna corridor, is more urban, has a higher education level, and a more western orientation (TURNOCK 1989). Eastern Hungary is basically rural-agricultural and has a somewhat lower level of educational attainment. In Transdanubia (W-Hungary), higher turnout rates may also be accounted for by increasing economic prosperity after 1990 with increasing contact with the west marked by high level of foreign investment.

6.2   Geography of political party support

In Hungary, we can find significant spatial variability in the distribution of politi-cal-party support. The profound transformation of the country after 1990 has gen-erated large-scale regional restructuring and increasing regional disparities. Some regions were able to utilise their more favourable geographical location and better developed human and technical infrastructures to take advantage of the new situa-tion. Other less well-endowed regions were unable to keep pace with the rapid rate of change and fell into deep recession. In this context the post-communist elections represent important political barometers of the transition, reflecting the fortunes and misfortunes of regional performance in the dynamism of transition.

---

[15]   As explained elsewhere, it seems likely that the unexpected decline of voter participa-tion in the 1998 election was related to more than one factor (KOVÁCS 1999). Among the reasons were a general disappointment of people from politics, as well as the 'quiet campaign' of the ruling Socialist Party or the nice sunny weather on the day of elec-tions which attempted many voters for weekend activities rather than for voting.

[16]   This is also reflected by the differences in turnout by settlement types: Budapest has always an above average turnout, whereas voter participation is generally below aver-age in villages.

Figure 4:   Election turnout

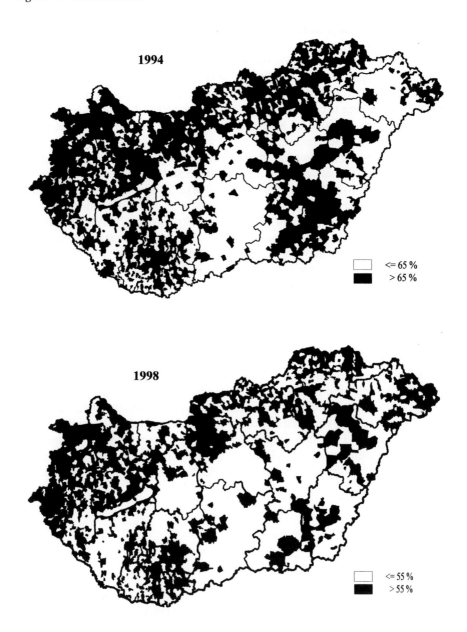

Source: Author's calculation.

In 1990, two parties, the moderate centre-right Hungarian Democratic Forum (MDF) and the liberal Alliance of Free Democrats (SZDSZ), won most of the votes in an election that resembled a two-horse race.[17] The Free Democrats with their more radical economic programme enjoyed the strongest support in Budapest and in the western, more developed regions of the country. On the other hand, the Hungarian Democratic Forum envisioned in its programme a gradual movement from a socialist economy to a free-market economy and denied the necessity of a more radical shock-therapy promoted by the SZDSZ. The MDF programme placed greater emphasis on traditional values and culture, such as the role of women in society, or the future of Hungarian minorities in neighbouring countries. As a consequence, the political-party support of MDF was geographically more balanced than the SZDSZ. The MDF did well in all regions but it won especially strong support in the eastern and south-eastern parts of the country, in regions where fear of the possible outcome of a financial shock-therapy was greatest.

Figure 5:   Changes of GDP in Hungary

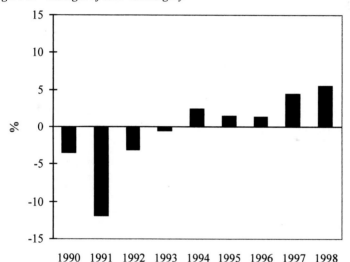

Source: Hungarian Central Statistical Office (KSH).

The 1994 election result was the outcome of the dissatisfaction of the electorate with the four-year performance of the MDF government. Many voters, disappointed by the politics of the centre-right coalition government were motivated to

---

[17]   In the second round of voting the Hungarian Democratic forum backed by the Smallholders and the Christian Democrats secured a convincing majority with 56% of the vote and won the election.

vote 'against' the MDF-led government rather than 'for' some clearly stated alternative. The party to replace the MDF was the biggest left wing party, the Hungarian Socialist Party (MSZP). The protest nature of voting in 1994 is also supported by the fact that the greatest support for the MSZP came from crisis-ridden eastern Hungary, where industrial decline and unemployment were especially grave. The MSZP came top in all 20 county lists in 1994 but the ratio of non-left-wing votes was generally higher in the west.

The four years of left-liberal MSZP-SZDSZ coalition government between 1994 and 1998 also proved to be difficult. An increasing budget deficit, high unemployment and economic stagnation were the heritage of the new regime. The new government implemented strict restrictions (the so-called Bokros package)[18] in the budget and financial policy in spring 1995. This was a near disaster for most people, especially the older or retired workers and public sector employees. Partly as a consequence of the harsh fiscal policy, the Hungarian economy started to recover gradually. By the 1998 elections, the economy showed an astonishing 5% growth, a performance experienced last time in Hungary in the mid-1970s. The ruling parties became increasingly optimistic about their chance of re-election (figure 5).

In 1998, the Alliance of Young Democrats (FIDESZ) was able to integrate most of the non-left-wing votes and won the election thanks to the single candidate competition. The party recorded its best results in north-western and north-eastern Hungary (figure 6). This seemingly contradictory result can be explained by two different factors. In western Hungary, the FIDESZ managed to replace the MDF and attract most of the former votes of the liberal SZDSZ. In eastern Hungary, people disappointed with the economic (and regional) policy of the MSZP-SZDSZ government turned also towards FIDESZ.

The spatial pattern of satisfaction (or dissatisfaction) with the economic policy of the earlier regimes can also be detected in the result of the communist Workers' Party (Munkáspárt). The geographical spread of votes for the Workers' Party shows a high level of stability (figure 7). The party is clearly the strongest in a north-south zone east of the Tisza River. This region has experienced the least benefit from the post-communist transformation. As a result, nostalgia for the communist regime is strongest in this part of the country.

---

[18]   Bokros was the shock-therapist minister of finance of the Horn cabinet.

Figure 6:   Voting results of the Alliance of Young Democrats (FIDESZ)

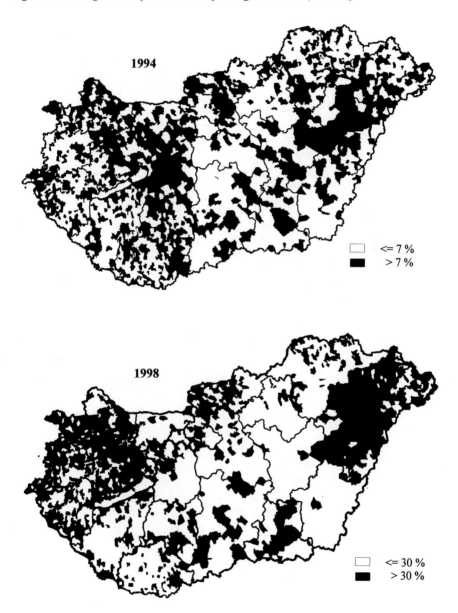

268   Z. Kovács

Figure 7:   Voting results of the Workers' Party (Munkáspárt)

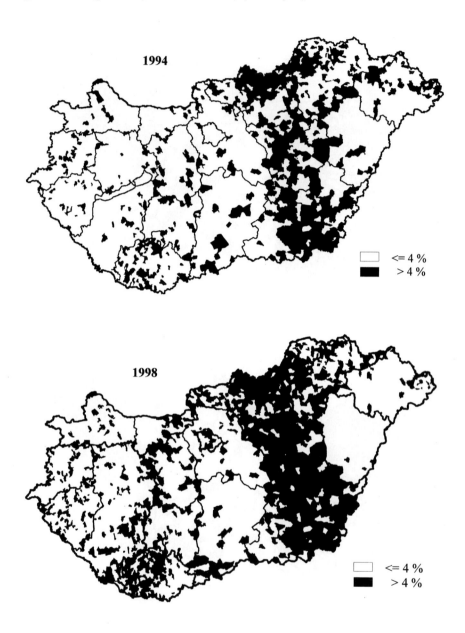

Source: Author's calculation.

7   Conclusion

The new party system of Hungary shows discontinuity rather than continuity with the democratic past of the 1930s and 1940s. In the 1990s, only two parties of the pre-communist period successfully revived: the Independent Smallholders' Party (FKgP) and the Christian Democratic People's Party (KDNP). None of the famous historical parties of the inter-war period regained their former political influence after the unfreezing of the party system except the Smallholders' Party. Despite the institutional discontinuity, however, we can observe a great deal of continuity in the underlying political constituencies in Hungary.

Accounts of Hungarian voting behaviour in the 1990s have stressed the development of growing spatial divides within the country, especially an east-west divide which reflected economic success and a consequent voting stability in the west and depression and instability in the east. As our geographical analysis reveals, in those districts in which foreign investment and economic dynamism had stimulated increased prosperity, voters tended to support parties of the centre (i.e. MDF, SZDSZ, FIDESZ) in all three elections. Where the economic changes had the most serious consequences, voters turned out strongly to register their desire for policy changes, and supported in 1994 the MSZP, in 1998 the FIDESZ.

Figure 8:   Political geographical profile of Hungary in the 1990s

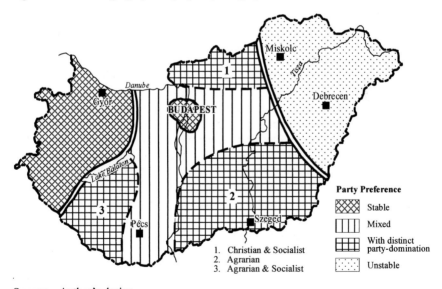

Source: Author's design.

The geographical analysis of the votes suggests that the victory of both the MSZP in 1994 and the FIDESZ in 1998 were the result of a protest from disillusioned economic 'losers' of the transformation. In this respect the factor of 'regional development' was decisive in the first post-communist elections and overshadowed other factors such as social class or religion. The mass of people who suffered from the dismantling of the socialist economy and the economic recession voted against the ruling parties both in 1994 and 1998.

On the basis of the results of the 1990, 1994 and 1998 parliamentary elections a map of political stability and awareness can be constructed for Hungary (figure 8). It reveals, that Hungary can be divided into three major regions with respect to political awareness and party preference. In this framework, western Hungary and Budapest can be classified as stable and politically more mature, with high turnout rates and ideologically consistent party preference. On the other hand, eastern Hungary can be classified as an unstable, politically less motivated and mature region, with generally low turnout rates and considerable swings in party preference. Between the two politically stable and unstable zones, a mixed central region can be distinguished, which can be further differentiated with some districts are moving towards particular party orientations and have greater stability whilst others tending in the opposite direction.

References

DINGSDALE, A., KOVÁCS, Z. (1996): A Return to Socialism: the Hungarian General Election of 1994. Geography 81 (3), 267-278

HAJDÚ, Z. (1992): A választási földrajz Nyugat-Európában. (Electoral Geography in Western Europe). INFO-Társadalomtudomány 22, 71-75

KITSCHELT, H. (1995): Party Systems in East Central Europe: Consolidation or Fluidity? Centre for the Study of Public Policy. University of Strathclyde, Glasgow

KOCSIS, K. (1994): Contribution to the Background of the Ethnic Conflicts in the Carpathian Basin. GeoJournal 32 (4), 425-433

KOLOSSOV, V. (1993): The Electoral Geography of the Former Soviet Union, 1989-91: Retrospective Comparisons and Theoretical Issues. In: O'LOUGHLIN, J., VAN DER WUSTEN, H. (eds): The New Political Geography of Eastern Europe. Belhaven Press, London, 189-215

KOVÁCS, Z. (1993): The Geography of Hungarian Parliamentary Elections 1990. In: O'LOUGHLIN, J., VAN DER WUSTEN, H. (eds): The New Political Geography of Eastern Europe. Belhaven Press, London, 255-273

KOVÁCS, Z. (1999): Geographical Patterns of the 1998 Hungarian Parliamentary Elections. In: DURÓ, A. (ed.): Spatial Research in Support of the European Integration. Centre for Regional Studies, Discussion Papers, Pécs, 97-112

KOVÁCS, Z., DINGSDALE, A. (1998): Whither East European democracies? The Geography of the 1994 Hungarian Parliamentary Election. Political Geography 17 (4), 437-458

KOWALSKI, M. (1999): Electoral Geography in Poland. In: DURÓ, A. (ed.): Spatial Research in Support of the European Integration. Centre for Regional Studies, Discussion Papers, Pécs, 87-95

LIPSET, S. M., ROKKAN, S. (1967): Cleavage Structures, Party Systems and Voter Alignments. In: LIPSET, S. M., ROKKAN, S. (eds): Party Systems and Voter Alignments. Free Press, New York, 3-64

MARTIS, K. C., KOVÁCS, Z., KOVÁCS, D., PÉTER, S. (1992): The geography of the 1990 Hungarian parliamentary elections. Political Geography 11, 3, 283-305

MÉSZÁROS, J., SZAKADÁT, I. (1995): Magyarország Politikai Atlasza (Political Atlas of Hungary). Konrad-Adenauer-Stiftung, Budapest

O'LOUGHLIN, J., VAN DER WUSTEN, H. (eds) (1993): The New Political Geography of Eastern Europe. Belhaven, London

RACZ, B., KUKORELLI, I. (1995): The 'Second-generation' Post-communist Elections in Hungary in 1994. Europe-Asia Studies 47 (2), 251-279

RIVERA, S. W. (1996): Historical Cleavages or Transition Mode? Influences on the Emerging Party Systems in Poland, Hungary and Czechoslovakia. Party Politics 2 (2), 177-208

ROKKAN, S. (1970): Citizens, Elections, Parties. McKay, New York

TAYLOR, P. J., JOHNSTON, R. (1979): Geography of Elections. Penguin, Harmondsworth

TURNOCK, D. (1989): The Human Geography of Eastern Europe. Routledge, London

ZARYCKI, T. (1999): Electoral Map of Central and Eastern Europe. In: DOMANSKI, R. (ed.): The Changing Map of Europe: the Trajectory Berlin-Poznan-Warsaw. Warsaw, 79-84

The New Role of Budapest
in the Central European City System

Zoltán Cséfalvay[1]

Since the iron curtain was lifted, the question of whether Budapest could become the centre of the region has often been raised. Realistic politicians and experts consider the Hungarian capital the future financial centre of Central and Eastern Europe, while those who are overly optimistic regard Budapest as the future economic organiser in the region. The arguments supporting these new roles can be divided into three groups:

- In Central and Eastern Europe an economic vacuum was created after the iron curtain was lifted.
- Budapest was a significant regional economic organiser in the last third of the 19th century.
- Globalisation processes in the first decades of the 21st century will make it possible for Budapest to assume an economic organising role in Central and Eastern Europe.

However, one criticism is that the first two of these three arguments cannot be justified. On the one hand, the former way of economic organising disappeared in Central and Eastern Europe after 1990, but this did not signify the beginning of a permanent vacuum. Quite the contrary, several cities - including Berlin, Vienna, Budapest and Prague - are competing with each other for the role as dominating control point, but the outcome cannot be predicted.

In former centuries Budapest never played a central part in Central and Eastern Europe. It was, however, one of the most quickly developing European cities in the second half of the 19th century. During the 'fin de siècle' - using BELUSZKY's (1998) proper phrase - Budapest was just a "world city with a small radius", which only had limited regional influence in Vienna's shadow, the capital of the Austro-Hungarian Empire.

Nevertheless, the third argument, the evaluation of the effects of globalisation processes, is not as obvious as the previous ones. Therefore, the main aim of this paper is to analyse the following questions: What effect does globalisation have on Budapest's development? What roles does or can the Hungarian capital fulfil in the European city system?

---

[1] This work was supported by the Research Support Scheme of the Open Society Support Foundation, grant No.: 566/1999.

1   Changes in geographical scales

A common theme in the debate on globalisation is that the role of national states has decreased radically. It seems that it is only an ideological question whether to consider this change a victory of the free market or, as popular best-sellers suggest (MARTIN and SCHUMAN 1996; KORTEN 1996), the initial step toward the world power of capitalist companies.

However, the national state and the world's regional distribution into national states is, historically speaking, quite a new phenomenon. Before the 16th century the history of mankind was not characterised by national states but by supranational empires which included several different nations (WALLERSTEIN 1986). Using a modern term, it was characterised by the domination of a global organisational level.

The 16th century marked a significant turning point in the withdrawal from a global organisational level in several respects. First, none of the empires could keep the world under control. Second, it became obvious that the emerging world economy would not be built on the great powers but on a system of competing national states. The supranational empires, for example, such as the Austro-Hungarian Empire or the Soviet Union, were marginalised by the 20th century and did not remain competitive on the periphery as did the national states. The third reason why the 16th century was significant is that Central and Eastern Europe and Hungary started to become less developed in this period. This is the era when the development of the region's cities came to a standstill and remained on a standard reminiscent of the Middle Ages.

Despite the decline of the supranational empires and the triumph of the national states, the global organisational level has never totally disappeared. Its growth is visible in the last third of the 20th century. The controlling and co-ordinating role of supranational empires was taken over by international organisations in the 1960s and by transnational business giants since the 1980s.

The precarious situation of the national states is further impeded by the fact that at the end of the millennium, they are not only being attacked from the global level 'above' but also from the cities and regions 'below'. The globalisation of the world economy is an economically and regionally divided process. This contradicts the common belief that it is the world's economic and cultural homogenisation. Globalisation is basically a polarisation process. Globalisation simultaneously involves and excludes different regions and narrow social layers from getting involved in the producing, capital, and information networks of the world (ALTVATER and MAHNKOPF 1996). In the globalising world, the transnational business giants, and each city and region, must fight each other for the most important resources of capital, knowledge, and economic and political power.

Agreement stops at this point. While there is unanimous agreement on the change of the regional scale, on the decreasing importance of the national state organisational level, and on the increasing significance of the global and regional level, opinions differ regarding the following two questions:

- How can the cities' competitiveness be measured?
- How will the city system change in the age of globalisation?

2   Some critical remarks on the competitiveness of cities

Those working on the research of competitiveness can agree on one thing: the competitiveness of national economies, regions and cities is extremely difficult to define (IRMEN and SINZ 1989; FLASSBECK 1992; SCHMIDT and SINZ 1993; SUNTUM 1986; TRABOLD 1995). The main problem is that the mainstream definitions are based on microeconomics. To be more precise, the criteria for measuring the competitiveness of companies are extended to non-economic formations like national states, regions and cities.

According to microeconomics, the competitiveness among companies has two main characteristic features. First, the competition among companies is a zero-sum gain. In other words, in company rivalry the only possible way to achieve an advantage or increased market shares is to undermine the performance of other companies. Second, competitiveness is shown by success on the market, so the only measure of competitiveness is company output. Applying these microeconomic criteria to cities, we can interpret the competitiveness of cities as an ability:

- to maintain rivalry with other cities,
- to increase output (how cities increase their economic achievement with real income),
- to export (how cities cover their import with export),
- to gain resources (how cities attract the resources of the global economy, mainly the working capital),
- to adapt (how cities adapt to the basic trends of the global economy, including privatisation, liberalisation and deregulation),
- to generate knowledge (how cities create and transform technological knowledge).

However, the variety of definitions is misleading since none of them takes the basic differences between the competitiveness of companies, national economies, regions and cities into consideration. As Krugman writes "competitiveness is a meaningless word when applied to national economies. The obsession with competitiveness is both wrong and dangerous" (KRUGMAN 1996, 22). According to Krugman, in the age of globalisation we can talk about competition only in relation to companies because this microeconomic term for competition cannot be applied to national states. Hallin and Malmberg also mention that "the widespread metaphor that regions 'compete for development' in much the same way as firms compete for market shares is basically misleading" (HALLIN and MALMBERG 1996, 324).

The competition among companies, regions and cities differ from each other in three basic ways.

1. While the competition of companies can be regarded as a zero-sum gain, regions and cities cannot gain market shares to the disadvantage of other regions and cities.
2. If a company is not competitive enough, it will go bankrupt and will be liquidated. However, a city cannot be liquidated only because it is not competitive enough or because the city management has carried out a wrong policy. Therefore, the consequences of the competitiveness among companies, regions and cities are also different.
3. The economic output, which is mainly substituted by economic indexes like per capita GDP or unemployment rate, indicates the regions' and cities' wealth and development accumulated throughout history rather than their ability to compete.

3   The competitiveness of European cities

Although applying microeconomic criteria to the competitiveness of cities is quite controversial, a wide range of regional research has dealt with the European city system in the past two decades. The findings of this research can be divided into three main theories (HALLIN and MALMBERG 1996):

1. The convergence theory is based on a neoliberal approach which states that the European city system is characterised by significant development differences. However, these differences continuously disappear and will be levelled as a result of the spread of expansive impulses and the strong interconnections between cities.
2. The divergence theory claims that the increasing specialisation of cities and the concentration of economic control functions point to the polarisation of the European city system.
3. Dunford represents the middle-of-the-road viewpoint in the two-theory debate. He talks about the temporal switch between convergence and divergence in the European city system. "The years after 1945 can be divided into two main phases: the years of high growth and income convergence lasted up to the mid-1970s and in the subsequent era growth was slower and convergence came to an end" (DUNFORD 1994, 99). According to Dunford, the reason for this change is that the 'golden age' of Fordism was taken over by postfordism. This was when the "neo-liberal programme of market integra-

tion and an intensification of competition for investment created a zero-sum gain in which the gains of winners are often at the expense of the losers" (DUNFORD 1994).

Nonetheless, despite these differing theories, the results of this research are methodologically one-sided because competitiveness is examined almost exclusively on the basis of a city's output (mainly by per capita GDP). Therefore, these theories do not describe competitiveness but instead describe regional inequalities. The term 'competitiveness' is politically neutral and fashionable and substitutes 'regional inequalities', which is politically stigmatised and therefore less fashionable. It should be noted that these theories do not really differ from each other in their contents but in the way they are presented.

Two different types of theories can be distinguished in the research that examines the transformation of the European city system during the globalisation of the 1990s.

1.   The theory of global city hierarchies claims that the city hierarchy within national states will loosen due to the decreasing significance of the national state organisational level and the increasing significance of the global level. This will be substituted by a new, global city hierarchy in the 21st century (KEEBLE 1991; KRÄTKE 1995; PALME 1995).
2.   In contrast, the proponents of regional patterns claim that in the new millennium the polarisation of cities will not be determined by their position occupied in a hierarchical order but by their geographical location (BRUNET 1989; GORZELAK 1996; KUNZMANN 1992).

Some critical remarks can be made in this case, too. The hierarchical city classifications overemphasise the role of hierarchy when the hierarchical distribution of cities takes place according to Fordist-type mass production and the dichotomy of Postfordist-type flexible production structures. However, the regional patterns overemphasise the importance of geographical location (IRMEN and SINZ 1989). Patterns such as the Western European development zone (see 'blue banana') or the Central European development zone (see 'Central-European boomerang') are publicised in city marketing strategies in the form of geodesign. A city with a favourable geographical position, however, will not necessarily become a regional centre. An advantageous geographical location is an opportunity which a city can make use of to develop into a real centre. If a city misses the opportunity, it will remain an insignificant mark on the map.

4   Budapest in the global city hierarchy

The city hierarchy issue and the term 'world city' were introduced in the mid-1960s by Peter HALL (1966). A city's role in the world is what makes it a 'world city'. This role can be primarily political, economic, financial, commercial, transportation or cultural. Neither the number of residents nor the favourable position occupied by a city in the national state city hierarchy guarantees a 'world city' status.

It is certainly important what functions are taken into consideration when defining 'world cities'. That is why the 'world city' concepts that appeared in the 1980s restrict the range of possible functions in comparison to Hall's rather broad criteria. Thus, Friedmann identifies 'world cities' as those that possess a leading or a commanding role in the world economy (FRIEDMANN and WOLFF. 1982). According to Friedmann, this commanding role is demonstrated by the phenomenon that transnational company headquarters tend to accumulate in a particular city. This supposition is logical since those decisions which have enormous influence on the world economy are made in these companies' headquarters.

FRIEDMANN (1995) suggests that the 'world cities' standing at the forefront of the world economy are divided into a specifically structured hierarchy. Thus he distinguishes primary and secondary 'world cities'. The territorial radiation of primary 'world cities' is felt in the whole world, while that of secondary 'world cities' is felt only in one macroregion. He claims that in Europe, London, Paris, Rotterdam, Frankfurt am Main and Zurich are primary 'world cities', whereas Brussels, Milan, Vienna and Madrid belong to secondary 'world cities'.

However, a metropolis can attract transnational company headquarters and fulfil a commanding role in the world economy only if it can also collect business-related services whose main customers are the transnational companies. Moreover, according to Saskia Sassen, the concentration of these enterprise-oriented services, such as company law advisors, chartered accountants, public relations offices or advertising companies, raises a city to 'global city' status (SASSEN 1991). A 'global city' is, however, much more than a pure concentration of services working for transnational companies. Due to the concentration of enterprise-oriented services and transnational company headquarters, these metropolises are simultaneously becoming communication centres. 'Global cities', using Reich's phrase, are turning into 'symbol-processing zones', where the representatives of occupations attempt to interpret the world's processes with symbols, pictures, words or writing (REICH 1997).

SASSEN (1991) claims that only London, Paris, New York and Tokyo are 'global cities' and they are the offspring of globalisation. The daily experience in the globalising world shows that information and resources can quickly become decentralised in time and space. However, it is obvious that because of this fast movement, there is a need for centralised information and power centres. A whole new structure will evolve from the information that was scattered all over the world. As SASSEN (1994, 15) writes: "Control functions and top managerial tasks concentrate according to territories in a way which is strongly similar to the regional spread of economic activities." Economic activities are decentralised territorially in the same proportion as the functions associated with management and economic organisation are centralised in 'global cities'.

It is not surprising that Budapest does not have a role in the global city hierarchy in either Sassen's or Friedmann's concepts. Most research on global city hierarchies was carried out before the iron curtain was lifted. Budapest and the other large Central and Eastern European cities were excluded.

Despite the fact that Budapest attracted more than a-third of the capital coming to the region after 1990, none of the transnational company headquarters have settled in the Hungarian capital, although the headquarters of less significant companies have been attracted to the city to a limited extent. Furthermore, it is unlikely that they will move to Budapest in the first decades of the 21st century. What can be predicted from the current trends is that the large Central and Eastern European cities may become logistic, distribution or organisation centres, but they will not have any co-ordinating or decision-making responsibilities. The other possible strategy, the formation of Budapest-centred, Hungarian or Central European transnational companies does not look very hopeful either, because the economic power and markets of Hungary and Central and Eastern Europe are too weak despite recent economic growth.

5   Budapest in regional pattern models

The best known and most popular regional pattern of the transformation of Central and Eastern Europe into a market economy was created by GORZELAK (1996). In Gorzelak's view the losing and winning regions follow the traditional east-west division in development typical of Central and Eastern Europe (see figure 1). The losing regions along the eastern borders of Poland, Slovakia and Hungary constitute a single continuous underdeveloped zone he calls the 'eastern wall'. This 'eastern wall' bears the characteristics of a peripheral position with a low development standard, lack of city centres, a high agrarian population, low-level infrastructure, weak integration of national and international markets, and weak capital attraction. In contrast, the winning regions are located mainly in the western parts of Central and Eastern Europe. A 'boomerang' shape appears when the winning cities are connected by a fictitious line, commencing from the port of Gdansk, proceeding westward through Poznan and Wroclaw to Prague and then following the curve of the boomerang to pass Brno, Vienna and Bratislava down to Budapest.

However, this 'boomerang' is not a consistent formation because the Budapest-Bratislava-Vienna triangle also has a special role. This is the region where the speediest transition to a market economy took place and which became a region of growth by the mid-1990's. The northern arm of the 'boomerang' is more of a fictitious supplement to this development zone and Warsaw has been left out due to its political distance.

Figure 1:   The Central and Eastern European 'boomerang'

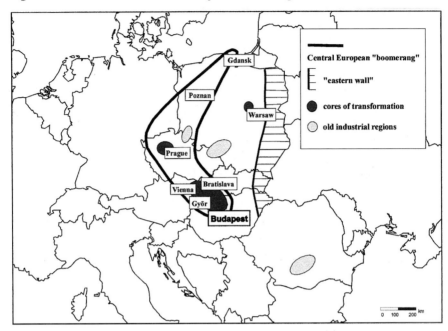

Source: GORZELAK 1996, 128.

Its shape suggests that this 'boomerang' wishes to be a counterpart to the well-known western 'blue banana' (i.e. the primary economic zone of European ur-banisation that stretches from south-eastern England through the German Ruhr area to the large agglomerations in southern Germany and northern Italy). This theory/model assumes that a similar development zone evolved in Central and Eastern Europe after 1990. Furthermore, Dostal's and Hampl's models, which are updated versions of the Gorzelak-model, claim that there is a secondary European development zone which connects Budapest, Vienna, Prague, Leipzig and Berlin and can also extend towards Hamburg and Copenhagen. DOSTAL and HAMPL (1996, 124) make it clear that the 'blue banana' is the predecessor of this secon-dary European development zone: "It is obvious that this geographic form is simi-lar to the primary economic zone of European urbanisation that stretches from Middle and Eastern England to Randstad (Holland), through Antwerpen and Brus-sels (Belgium), to the Rhine River, the Ruhr area and Frankfurt (Germany) and extending to the large agglomeration in southern Germany and northern Italy."

However, it should be mentioned that the two zones differ significantly from each other in several respects. The highly industrialised Western European 'blue banana' zone is not a recent development since the region has always been the innovative European driving force. It was the point where literacy evolved during the early Middle Ages, where the Reformation took hold in the 17th century and where industrialisation occurred during a later period. The cities of the 'Central European boomerang' did not invent but received many innovations from the 'blue banana', with a considerable time delay. The cities of the 'blue banana' are linked through myriad economic and social ties. However the economic and social relations of the cities in the 'boomerang' are rather weak despite their common historical development. This is due to the fact that the Central and Eastern European cities have never focused on each other. Instead, each of them has had a western orientation. Prague, for example, was linked to Berlin and Nuremberg, while Budapest was oriented towards on Vienna and Munich.

These regional patterns seem to provide more important roles for the Hungarian capital than for global city hierarchies. Budapest is a significant point in the secondary European development zone, it is the core of the 'boomerang'. Despite this fact, the 'boomerang' is no more than a geopolitical metaphor which tries to make the aspirations of the Central and Eastern European cities 'marketable' and popular. It is a future perspective which is far from the real situation at the dawn of the new millennium in Central and Eastern Europe. However, only a positive future perspective will lead to a positive future.

6   Budapest - The centre of Central and Eastern Europe?

In his book on world economy, WALLERSTEIN (1986, 521) summarises the lessons of history in the following way: "a capitalist world economy appreciates accumulated capital and human capital more than a raw labour force." Although his statement is based upon an analysis of the 16th century world economy, this sentence remains true 500 years later. In the globalising world of the new millennium, the competitive position of regions and metropolises depends mostly on how much capital and knowledge they were able to accumulate in previous centuries and to what extent they are able to transmit capital and knowledge today.

Certainly, the capital and knowledge transmitting role of metropolises ranges over many activities. Therefore, this paper will only analyse the activities demanding capital and knowledge transfer in Budapest and in large European cities using characteristic examples for each, including:

- on the basis of competition for premises in the banking sector, and
- on the basis of competition for international conferences.[2]

---

[2]   Large Western European cities were analysed based on the following criteria: more than 10 international conferences per year; banks belonging to the top 500 in Europe and having accumulated capital stock valued at more than 500 million dollars. Large Central and Eastern European cities were analysed without these minimum criteria.

6.1   Budapest as a financial centre

In the mid-1980s, HAMILTON (1986, 13) wrote "What is going on now is a revolution: a revolution in the way finance is organised, a revolution in the structure of banks and financial institutions and a revolution in the speed and manner in which money flows around the world." Although 'revolutionary' may be a bold description, Hamilton is right in the sense that in previous centuries, the role of politics or production was the determining factor for a metropolis' central position. Today, the determining factor is the position they occupy in the financial sphere. Since Hamilton's 'revolutionary statement' in the 1980s, an irreversible process of concentration has been going on in the banking sector which has clearly affected metropolises.

Public debates are dominated by fears of the overpowering influence of giant banks, but the underlying principle behind bank mergers is simple economics. The bigger the banks, the greater the possibilities to internationalise banking activities and the greater the advantage on global markets. In other words, the bigger and the stronger the bank, the wider its transfer activities and the greater the profits on the global markets. Therefore, if banks in a metropolis have more capital stock, they will play a more significant co-ordinating and decision-making role in the globalising world economy.

For historical reasons, European banks accumulate less capital stock when compared to their American or Japanese counterparts. In the future it will become important to accelerate further the capital stock concentration process so that European banks can compete with their rivals in the economic triangle, including the United States, Japan and Western Europe. Finally, the strengthening of the European Union and the planned introduction of a common currency has also accelerated the concentration processes.

The 500 largest European banks, published annually as a list in *The Banker* (a British financial magazine) were examined considering these concentration processes. According to *The Banker's* statistics in 1997, the 500 largest European banks owned 662 billion USD capital ('tier one capital'). The progress of the concentration process is shown by the fact that 60% of this accumulated capital is concentrated in German, French, British and Italian banks. In those countries where the accumulated capital of the largest banks exceeds 10 billion USD, 90% of the accumulated capital of European banks is concentrated in 11 countries. These countries are: Germany, France, Great Britain, Italy, Switzerland, Spain, Holland, Belgium, Austria, Sweden and Denmark.

A general picture is more precise because the banks' average capital stock differs significantly in the various countries. While the capital stock of France's largest commercial banks was an average of 4.4 billion USD in 1997, it was 3 billion USD in Holland, 2.6 billion USD in Great Britain, 1.3 billion USD in Germany, and only 1 billion USD in Italy. Therefore, historical influences such as the long-lasting formation of the Italian and the German national state and the political differences of today's world notably between a strongly centralised state (France) and a federalist state (Germany), are reflected in the banks' average capital stock.

It is a general trend that the banks' average capital stock is moving from Western Europe toward the eastern countries. In Central and Eastern Europe, the average amount of bank capital belongs to the European top 500 in each country and is well over the average European 1.4 billion USD. The capital bank stock in these countries is extremely low. This also applies to Hungary, where the largest banks possessed an average of 0.17 billion USD capital in 1997 and even lagged behind the regional rivals (Czech 0.75 billion/bank; Polish 0.29 billion USD/bank).

The extreme concentration in the banking sector was even more significant in the competition among large cities. In 1997, 37% of the accumulated capital stock of the 500 biggest European banks was concentrated in three financial centres, which includes Paris, London and Frankfurt am Main. The 15 leading European financial centres (Paris, London, Frankfurt am Main, Zurich, Amsterdam, Munich, Milan, Rome, Madrid, Brussels, Utrecht, Stockholm, Vienna, Dusseldorf and Basel) each have a total capital bank stock exceeding 10 billion USD, and possess two-thirds of the accumulated capital stock of Europe's 500 biggest banks.

The Hungarian capital occupied a very modest place in this European competition. In 1997, six Budapest-based banks were on the European top 500 list, but the total capital stock of these banks did not reach one billion US dollars.[3] As a result, the Hungarian capital only ranks 68th among the 231 most important European financial locations. Therefore, despite the fact that the banking sector has gone through significant developments in the past ten years (see chapter by Heike Jöns), Budapest has not been able to break out of its peripheral position and join the dominant European financial centres.

The Hungarian capital's lag is not only significant compared to the rest of Europe but also in comparison to Central and Eastern Europe (see figure 2). This is proven by the fact that the total capital stock of the largest Budapest-based banks is a small fragment of the largest southern German financial centre, Munich, and the largest north Italian centre, Milan. Even Verona, Bologna or Padua, the less significant north Italian regional centres, surpass Budapest. In terms of its bank capital stock, Budapest is in the same category as less significant regional centres such as Brescia in Northern Italy or Innsbruck in Austria. Budapest does not play a significant role in the Central and Eastern European region since the accumulated capital stock of large banks in Prague and Warsaw is well above that of the Hungarian capital (nearly three billion US dollars each).

---

[3]   OTP Bank (Rank: 325, Tier one capital: 243 million USD), Magyar Külkereskedelmi Bank (355, 210), CIB Bank (390, 181), Budapest Bank (417, 162), Postabank és Takarékpénztár (475, 110), Raiffeisen Unicbank (500, 87) - Source: The Banker 1997. Top 500 Europeans. September, 35-57.

Figure 2:   European banking centres in 1997: accumulated capital bank stock of banks belonging to the European top 500

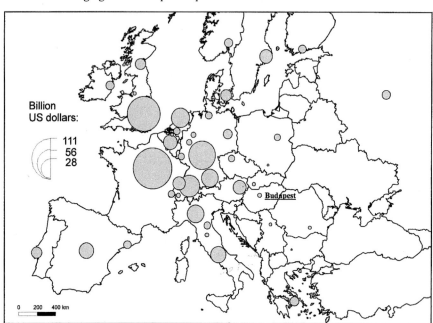

Source: The Banker 1997. Top 500 Europeans. September, 35-57.

Despite Budapest's aspiration to become the financial centre of the Central and Eastern European region, on the basis of its current position at the beginning of the 21st century, the Hungarian capital cannot yet function as the financial co-ordinating and decision-making centre. However, even today it is possible that Budapest will become the financial centre of south-eastern Europe. Moreover, as history and BEREND and RÁNKI's (1976) research indicates, at the end of the 19th century and between the two World Wars, Budapest was not an important finan-cial centre from a Central and Eastern European point of view but from a south-eastern European perspective instead. The city transmitted capital from Western Europe mainly toward south-eastern Europe.

The Hungarian capital may not be a financial decision-making centre in Central and Eastern Europe, but it may yet become an important innovative financial service centre. The rapid development of financial services in the past decade pro-vides a good basis for this. There is a connection between the internationalisation of financial services and the capital stock of the banking sector: the larger the capital bank stock in a financial centre, the wider and more international the fi-nancial service sector in this region. Therefore, the strengthening of the Hungarian banking sector is indispensable for Budapest to keep or further increase its current pivotal financial services position in Central and Eastern Europe.

6.2   Budapest as a communication centre

Capital, knowledge and information-demanding activities influence the central role of large cities in the globalising world economy. One indication of this fact is that conference tourism has been booming world-wide in the past decade. Consequently, the number of conferences organised in a city does not only show the tourist conditions of the particular city (accommodations, transportation infrastructure, attractive natural or artificial environment) but also to what extent it is able to transmit knowledge and information and to organise information and knowledge exchange.

According to the statistics of the Union of International Associations, which registers international conferences and congresses, 8 991 international conferences were organised throughout the world in 1996. Half of these conferences (5 146) were hosted in Europe.[4] One-third of the conferences held in Europe took place in France, Great Britain and Germany. Hungary was on 11th place on the list in 1996 with 161 conferences, well ahead of several highly developed countries.

While Hungary fell towards the middle of the European list, Budapest was at the top of the European list, a member of an exclusive club that organised more than 100 conferences (see figure 3). This leading group included the classic European political powers: Paris, Vienna, London and Brussels. However, the Hungarian capital comes right after this group; it was 7th place on the European ranking behind Genf and Copenhagen and was ahead of Amsterdam with 125 conferences in 1996. With this result, Budapest was also ahead of Berlin, Rome, Madrid and Stockholm. This is particularly impressive because Stockholm, for example, represented the huge Scandinavian region.

This perspective can be further enhanced by the fact that in Paris, 45% of all conferences were national conferences of international significance.[5] Based only on the conferences organised or supported by international associations, Budapest successfully rivalled Paris which was only sixth on the European list. Budapest competed with regionally important metropolises like Berlin or Madrid and with leading European co-ordinating and organising centres such as Brussels or Paris by successfully hosting international conferences that were significant in knowledge and information transfer.

---

[4]   These statistics take the large international conferences into consideration (minimum number of participants: 300; minimum proportion of foreigners: 40%; minimum number of participating countries: 5; minimum conference duration: 3 days).

[5]   The Union of International Associations differentiates two types of conferences: (1) international conferences organised or supported by international associations, and (2) national conferences of international significance with many foreign participants.

Figure 3:   Centres of International European conferences, 1996

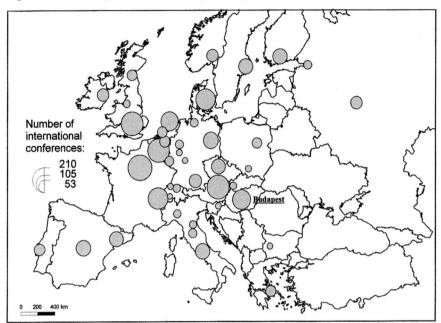

Source: UNION OF INTERNATIONAL ASSOCIATIONS 1996.

Budapest's excellent position is clear when compared to its own counterparts in Central and East European cities. Vienna was the only city ahead of the Hungarian capital. Budapest's leading role is shown by the fact that German cities like Berlin or Munich and north Italian cities like Milan and Turin lag far behind the Hungarian capital. Central and East European cities did not pose a serious threat to Budapest either because, except for Prague and Warsaw, the number of international conferences and congresses was insignificant. Thus, Budapest's integration into the European metropolises system is much more successful in knowledge-demanding transfer activities than in capital demanding ones.

7   What does the future have in store?

On the basis of global city hierarchies, regional patterns and Budapest's competitive position in capital concentration and knowledge transfer, the following question is still on the agenda at the beginning of the 21st century: how can the Hungarian capital become the centre of Central and Eastern Europe? The exact answer will probably emerge in future years, but different scenarios are presented below.

7.1   The scenario of west-east development corridors

The first future perspective is based on the supposition that even if the Central and East European countries join the European Union, they will not be able to break out of their current semi-peripheral position. Consequently, the Central and East European countries will remain 'extended workbenches' of West European companies with cheap but qualified labour force, and with cities that will remain the eastern bridgeheads to West European centres. The only difference is that when compared to today's zones, the corridors from west to east will strengthen the European regional pattern, which includes the Budapest-Vienna-Munich and the Budapest-Balaton-Venice-Milan axes from a Hungarian viewpoint. If the latter axis becomes stronger, Budapest may join the 'European sunbelt' in the Mediterranean, which is the most rapidly developing region in the European Union.

7.2   The scenario of consumer market 'blue banana' and production zone 'boomerang'

The second future perspective, called the consumer market 'blue banana' and the production zone 'boomerang', is based on the supposition that the semi-peripheral position of Central and Eastern Europe will significantly decrease by joining the European Union. Therefore, besides the industrialisation governed by multinational companies, which is known as postfordist reindustrialisation, local mid-sized enterprises will strengthen and the cities will become internationally acclaimed service sector centres. The most important consequence of this change will be a boom in economic relations between Central and East European cities. Therefore, the 'Central European boomerang' which only exists in theory today, will become an important industrial production zone. It will sell its products in the 'blue banana' zone, a significant European consumer market.

7.3   The scenario of Central and East European innovative zones

The third future perspective, the scenario of Central and East European innovative zones, claims that Central and Eastern Europe will integrate into the European regional pattern with independent and innovative centres. This will occur with the help of the new and growing, mid-sized entrepreneurial layer, the rebirth of craftsmanship and the development of knowledge- and research demanding industries. According to MATOLCSY and DICZHÁZI (1998), the biggest possibility for the emergence of such a zone is in the Budapest-Győr-Vienna-Brno-Bratislava pentagon. Arnold demonstrates that this is not only a dream but also a possible future perspective: "By 2020, the booming industrial locations in Germany, Austria, Switzerland, France and northern Italy will eventually waste away [...] The qualified labour force, engineers and scientists will study Russian, Czech or Hungarian and will move to the east. Their destination are the new Silicon Valleys near Bratislava, Győr, Sopron, Estonia, the western part of the Czech Republic and Poznan" (ARNOLD 1995, 147). This scenario, however, is regarded as a 'gloomy future' in Western Europe.

References

ALTVATER, E., MAHNKOPF, B. (1996): Grenzen der Globalisierung: Ökonomie, Ökologie und Politik in der Weltgesellschaft. Westfälisches Dampfboot, Münster

ARNOLD, H. (1995): Disparitäten in Europa: Die Regionalpolitik der Europäischen Union. Birkhäuser Verlag, Basel

BELUSZKY, P. (1998): Budapest - nemzetközi város: Történeti áttekintés. In: MTA (ed.): Budapest - nemzetközi város. MTA, Budapest, 27-46

BEREND, T. I., RÁNKI GY. (1976): Közép-Kelet-Európa gazdasági fejlődése. Közgazdasági és Jogi Könyvkiadó, Budapest

BRUNET, R. (ed.) (1989): Les villes europeennes. Reclus/Datar, Paris

DOSTAL, P., HAMPL, M. (1996): Transformation of East-Central Europe: General Principles under Differentiating Conditions. In: CARTER, F. W., JORDAN, P., REY, V. (eds): Central Europe after the Fall of the Iron Curtain: Geopolitical Perspectives, Spatial Patterns and Trends. Peter Lang, Frankfurt am Main (= Wiener Osteuropastudien 4), 113-128

DUNFORD, M. (1994): Winner and Losers: The New Map of Economic Inequality in the European Union. European Urban and Regional Studies 1 (2), 95-114

FLASSBECK, H. (1992): Theoretische Aspekte der Messung von Wettbewerbsfähigkeit. Vierteljahrshefte zur Wirtschaftsforschung (1-2), 5-26

FRIEDMANN, J. (1995): The world city hypothesis. In: KNOX, P., TAYLOR, P. (eds): World Cities in a World-System. Cambridge University Press, Cambridge, Mass., 317-317

FRIEDMANN, J., WOLFF, G. (1982): World City Formation. An Agenda for Research and Action. International Journal of Urban and Regional Research 3, 309-344

GORZELAK, G. (1996): The Regional Dimension of Transformation in Central Europe. Regional Policy and Development 10, Regional Studies Association, Jessica Kingsley, London

HALL, P. (1966): The World Cities. Weidenfeld and Nicholson, London

HALLIN, G., MALMBERG, A. (1996): Attraction, Competition and Regional Development in Europe. European Urban and Regional Studies 3 (4), 323-337

HAMILTON, A. (1986): The Financial Revolution. Penguin, Harmondsworth

IRMEN, E., SINZ, M. (1989): Zur Wettbewerbsfähigkeit der Regionen in der Europäischen Gemeinschaft. Informationen zur Raumentwicklung (8-9), 589-602

KEEBLE, D. (1991): Core-Periphery Disparities and Regional Restructuring in the European Community of the 1990s. In: BLOTEVOGEL, H. H. (ed.): Europäische Regionen im Wandel: Strukturelle Erneuerung, Raumordnung und Regionalpolitik in der Europaischen Union. Dortmunder Vertrieb für Bau- und Planungsliteratur, Dortmund (= Duisburger Geographische Arbeiten 9), 49-58

KRUGMAN, P. (1994): Competitiveness: A Dangerous Obsession. Foreign Affairs 73 (2), 28-44

KRÄTKE, S. (1995): Stadt - Raum - Ökonomie: Einführung in aktuelle Problemfrelder der Stadtökonomie und Wirtschaftsgeographie. Birkhäuser Verlag, Basel (= Stadtforschung aktuell 53)

KRUGMAN, P. (1996): Pop Internationalism. MIT Press, Cambridge, Mass.

KUNZMANN, K. R. (1992): Zur Entwicklung der Stadtsysteme in Europa. Mitteilungen der Österreichischen Geographischen Gesellschaft 134, 25-50

KORTEN, C. D. (1996): Tőkés társaságok világuralma. Kapu Könyvkiadó, Budapest

MARTIN, H.-P., SCHUMAN, H. (1966): Die Globalisierungsfalle: Der Angriff auf Demokratie und Wohlstand. Rowohlt, Reinbek bei Hamburg

MATOLCSY, GY., DICZHÁZI B. (1998): A növekedés regionális forrásai. Társadalmi Szemle 1, 21-36

PALME, G. (1995a): Struktur und Entwicklung österreichischer Wirtschaftsregionen. Mitteilungen der Österreichischen Geographischen Gesellschaft 137, 394-416

REICH, B. R. (1997): Die neue Weltwirtschaft. Fischer Taschenbuch Verlag, Frankfurt am Main

SASSEN, S. (1991): The Global City. Princeton University Press, Princeton, New Jersey

SASSEN, S. (1994): Metropolen des Weltmarkts. Die neue Rolle der Global Cities. Campus, Frankfurt am Main

SCHMIDT, V., SINZ, M. (1993): Gibt es den Norden des Südens? Aspekte regionaler Wettbewerbsfähigkeit in der Europäischen Gemeinschaft. Informationen zur Raumentwicklung (9-10), 593-618

SUNTUM, U. van (1986): Internationale Wettbewerbsfähigkeit einer Volkswirtschaft: Ein sinnvolles wirtschaftspolitisches Ziel? Zeitschrift für Wirtschafts- und Sozialwissenschaften 106, 495-507

TRABOLD, H. (1995): Die internationale Wettbewerbsfähigkeit einer Volkswirtschaft. Vierteljahrshefte zur Wirtschaftsforschung 64 (2), 169-185

UNION OF INTERNATIONAL ASSOCIATIONS 1996. International Meetings: Some figures – 1996. UIA, Brussels

WALLERSTEIN, I. (1986): Das moderne Weltsystem: Die Anfänge kapitalistischer Landwirtschaft und die europäische Weltökonomie im 16. Jahrhundert. Syndikat, Frankfurt am Main

Recent Differentiation Processes in Budapest's Suburban Belt

Éva Izsák and Ferenc Probáld

1   Introduction: Budapest, the agglomeration's core

Academic interest in the development of Budapest, a city whose history goes back to prehistoric times, has been kindled by its importance within the Hungarian urban network and by its potential emerging future role in the web of prime European cities. Several disciplines, including many geographical studies, papers and monographs, deal with this issue, and some of them are written in German or English (e.g. FRIEDRICHS 1985; ENYEDI and SZIRMAI 1992; LICHTENBERGER et al. 1994; BERÉNYI and DÖVÉNYI 1996; KOVÁCS and WIESSNER 1997; BURDACK and HERBERT 1998). The major characteristics of the Hungarian capital that are relevant to the topic of this paper can be summarised as follows:

Although Budapest as it is known today first came into existence in 1873 through the unification of three towns - Buda, Pest and Óbuda - its functional role as capital dates back to the thirteenth century. A series of favourable physical and socio-geographical factors played a part in the city's growth and in the enrichment of its central functions. Their significance, however, has changed from time to time in accordance with the demands of differing historical conditions (PROBÁLD 1997; table 1).

In 1873, Budapest occupied an area of 194km^2 and had 296 000 inhabitants. By 1910, the city's population had increased to almost one million. After World War I, Hungary lost 71.5% of its former territory; this left Budapest in an uncontested primary position at the top of the Hungarian urban hierarchy. The annexation of the surrounding suburban settlements in 1950 increased the capital's municipal area to 525 km^2, while the collectivisation of agriculture and the socialist industrialisation policy induced a migration wave within the country, thus further increasing the area's population. At the time of the 1980 census, the capital's population peaked at 2.06 million.

Budapest's industrial function gradually decreased during the past decades since it entered the post-industrial development stage after 1970. Along with the growing predominance of the tertiary and quaternary sectors, the concentration of economic management functions was further enhanced by the transition to a market economy. Budapest became the main beneficiary of the political and economic changeover from the former socialist system by attracting the bulk of the foreign capital influx. The city's valuable human capital and innovative potential must also be emphasised as a basis for future development (table 2).

Table 1: Key factors contributing to Budapest's development

Ancient times	Middle Ages	19th century	20th century
Local building material (limestone, wood)	Crossing-place of the Danube River (ferry and bridge)	Rapid development of cultural, political and economic central functions	Location advantage provided by the agglomeration of various manufacturing industries
Abundant water sources (thermal springs)	Crossing point of major trade routes	Centre of the emerging road and railway network	Early development of tertiary sector and infrastructure
Fortress at the border of the Roman Empire	Location at the boundary of plains and hills	Mass labour force immigration	Concentration of research and higher educational institutions
	Topography suitable to fortification	Cultural melting pot	Concentration of highly skilled labour force
	Central location in the Hungarian Kingdom embracing the entire Carpathian Basin	Expanding consumer market	Flexibility and ability to adapt and develop innovations
		Large scale construction projects	
		Coal mines near the city	

Source: Authors' classification.

Table 2: The Hungarian primate city: Budapest's proportion compared to the country total (1995)

Population	18.8%	Foreign direct investment (stock)	45.0%
GDP	33.4%	Number of joint ventures	48.6%
Industrial employment	18.0%	Employment in business services	49.7%
Tertiary employment	25.8%	Employment in financial services	40.5%
R&D expenditure	64.6%	Number of enterprises	44.4%
R&D employment	55.5%	College and university students	40.4%
Public administration employment	63.3%		

Source: HORVÁTH and ILLÉS (1997); Hungarian Central Statistical Office (KSH); Statistical Yearbook of Budapest, 1996.

Out of the seven Hungarian regions, the central region embracing Budapest and its agglomeration and all of Pest county stands out because of its exceptionally large population and economic potential both domestically and internationally (table 3).

Table 3: The primacy of the Budapest region: an international comparison (1994)

Name of the central region (CR)	Share of the CR in the active labour force	Share of the CR in the labour force employed by the modern (advanced) service sector
Île de France	22%	33%
NW-Italy (Piemonte, Liguria, Lombardia)	29%	33%
W-Netherlands	48%	56%
SE-England	32%	43%
Lisbon and Tejo Valley	34%	57%
Budapest (incl. Pest county)	32%	63%

Source: HORVÁTH and ILLÉS (1997).

Regional development policies have unsuccessfully aimed at reducing the pre-dominance of this region for a long time. These policies have often disregarded the fact that Budapest's comparative advantage extends beyond the national boundaries. This refers to its eventual 'gateway' role within the European urban network as well as to its East-Central European sphere of influence (see chapter by Zoltán Cséfalvay). These advantages result primarily from the major clustering of powerful resources, and to the size and favourable location of the Hungarian capital city.

2   The development and main features of the agglomeration

Several stages can be distinguished in the evolution of the agglomeration belt surrounding Budapest's central city (ENYEDI and SZIRMAI 1992). The first two stages lasted from 1870 until 1914 and were characterised by the industrialisation of the neighbouring settlements and by the transformation of villages into satellite towns. There was a dynamic population increase due to in-migration. From the turn of the century onwards, improvements in mass transportation (rail and tram construction, bus service) allowed commuting from the agglomeration belt into the central city to increase considerably. The agglomeration continued to sprawl and enlarge its sphere of influence between the two World Wars. Satellite and resort towns developed to the north of the city along the Danube River; these were suburbs inhabited mostly by the 'white collar' middle class. The idea of the administrative unification of the capital and its neighbouring settlements was first suggested at the end of the 1930s. This was based on the recognition of the agglomeration's functional interconnectedness (as mentioned in the 1937 Urban Planning Act, and the 1938 and 1941 Ministerial Orders about the delineation of

Budapest's agglomeration belt). However, it wasn't until 1950 that the 23 suburban settlements neighbouring to Budapest, which had already been amalgamated with the central city, were annexed to the capital.

The fourth stage in the agglomeration's development was characterised by a policy of forced industrialisation and by massive in-migration to the capital. Apart from a chronic housing shortage, moving to Budapest was further impeded by administrative restrictions. This resulted in a new wave of commuting from the rapidly growing settlements outside the confines of the capital's new municipal boundary. The government began curtailing industrial development in Budapest in the 1960s and established directives to remove certain industries from the city. This turned the suburban belt into an attractive location for industry. Although restrictions for industrial development were extended during the 1970s to include the whole agglomeration, construction and enlargement of a few mammoth industrial plants continued unimpeded. The largest oil refinery and thermal power plant in the country was built in Százhalombatta and a truck factory was built in Csepel. Important new industrial centres emerged outside the accepted official boundary of the agglomeration in Vác and Gödöllö. Since the 1980s, however, the number of industrial jobs has begun to decline even in the agglomeration zone, indicating the beginning of the post-industrial development stage.

The official delineation of the agglomeration's outer boundary in 1971 was preceded by lasting debates; 43 settlements were finally classified as belonging to the agglomeration zone, an area which extended 1 143 km^2 and contained about 400 000 inhabitants. However, it was obvious that the capital's sphere of attraction extended to a much larger area, since slightly more than half of the daily commuters to Budapest came from the narrowly defined agglomeration belt. Even so, the definition of the agglomeration boundary was not changed until 1997. Moreover, spatial planning co-ordination between Budapest and the settlements within the agglomeration belt, that administratively still belonged to Pest county, failed. Although a master plan was drawn up in the 1980s for the whole agglomeration, it was never implemented. Due to the lack of an overall development concept, the suburban zone evolved and changed spontaneously.[1]

Topographical and other physiographical conditions within the suburban zone were the fundamental influences in the west-east growth asymmetry and were less conspicuous in the central city. Settlements in the highly dissected landscape to the west of the Danube River, apart from a commuter rail line leading north, rely on bus service routes for their means of mass transportation. That proved to be a retarding factor in the suburban growth of this area when compared to the proliferation of growth in the eastern lowland belt.[2] On the lowland areas to the east of

[1]   In 1997, preparations for a new development plan for the broader agglomeration began.
[2]   The road network of the west bank suburbs have experienced considerable development: nevertheless, there are still some traffic bottlenecks and the dead-end suburb Nagykovácsi, though it is adjacent to Budapest, has only one single road link.

the river, suburban sprawl was faster and farther reaching along the main rail corridors than in the intermittent zones between the railway routes (PROBÁLD and RIKKINEN 1978). Asymmetrical growth is not restricted solely to the degree and rate of suburbanisation. The environmental attractions and amenities of the Buda Hills coupled with higher construction costs and a greater reliance on private vehicles resulted in a distinctive disparity in the social structure between the hilly and lowland areas. Table 4 demonstrates these differences using a few indicators and based on data provided by the last three censuses.

Table 4: Comparison of social indicators in the Buda (west) and Pest (east) suburban belt population (Buda side=100%)

Year	1970	1980	1990
Proportion between 6 and 14 years (%)	96	98	96
Proportion of the elderly (above 60 years) (%)	111	107	108
Proportion of the active population employed in agriculture (%)	114	125	148
Proportion of the active population employed in the tertiary sector (%)	90	99	87
Proportion of the active population employed in the manufac-turing industry and construction (%)	102	95	112
Proportion of persons with college or university degrees (% above 15 years)	71	69	65

Source: Hungarian census; authors' calculation.

Differences between the population's composition in the agglomeration zone's settlements can be shown by using the Hoover dissimilarity index. Among the variables listed in table 5, the population's age structure showed a marked homogeneity in the suburban belt at each observation date based on the 1970, 1980, and 1990 censuses. The highest degree of residential segregation in 1990 is characteristic of the population with higher education. Its increase since 1980 indicates a growing degree of differentiation and is comparable to the segregation observed between the central city's districts. In contrast to the elite suburban towns on the Buda side of the river, a similarly high degree of segregation appears among the agrarian population on the eastern (Pest) side of the suburban zone.

  In order to reveal finer regional distinctions in the agglomeration belt, an earlier study (PROBÁLD 1995) classified the settlements on the basis of the 1980 census data and focused on the typical features of the local labour market (i.e. the density and sectoral structure of local workplaces, enterprise ownership, and the ratio of highly qualified manpower). Except one settlement (Dunakeszi), all of the other eastern suburbs can be placed in the same category. Each suburb suffered from a shortage of local job opportunities and a high number of commuters into the central city. Moreover, a great number of poorly skilled people, some of whom used to work locally for agricultural co-operatives, which now have more industrial and service functions, also left the suburbs. Settlements on the western (Buda) side of the agglomeration and on Csepel Island, however, can be divided into five separate categories based on the above criteria. This signifies a high degree of differentiation. For the agglomeration as a whole, the relationship between the distribution of the highly educated population and the density of private enterprises in the first wave of privatisation (prior to January 1, 1992) revealed a moderate rank correlation (r_s=0.66). There was an almost function-like relationship (r_s=0.96) in the districts belonging to Budapest while the correlation (r_s=0.51) was much weaker in the suburban zone (PROBÁLD and SZEGEDI 1994).

Table 5: Index of dissimilarity among the suburban belt settlements for various indicators in the permanent population

	1970	1980	1990
Proportion between 6 and 14 years (%)	5.6	4.0	3.8
Proportion of the elderly (above 60 years) (%)	8.3	8.5	7.6
Proportion of the active population employed in agriculture (%)	19.5	17.5	16.8
Proportion of the active population employed in the tertiary sector (%)	12.9	12.6	10.7
Proportion of persons with college or university degrees (% above 15 years)	18.1	18.1	19.2

Source: Hungarian census; authors' calculation.

Table 6: Correlation coefficients between social indicators in Budapest's peripheral districts and the adjacent suburbs

	1970	1980	1990
Proportion between 6 and 14 years (%)	0.26	0.19	0.35
Proportion of the elderly (above 60 years) (%)	0.59	0.42	0.01
Proportion of the active population employed in the tertiary sector (%)	0.78	0.75	0.72
Proportion of the active population employed in the manufacturing industry and construction (%)	0.91	0.79	0.79
Proportion of the active population engaged in intellectual work (%)	0.33	0.73	0.87
Proportion of persons with college or university degrees (% above 15 years)	0.60	0.80	0.93

Source: Hungarian census; authors' calculation.

A viable assumption would be to link the asymmetrical structural pattern of the suburban ring to the sectoral imprint of the central city. An attempt to test this hypothesis has been undertaken based on the latest census figures. There are twelve districts in Budapest that have an adjoining suburban belt along its metropolitan boundary. The suburban belt was therefore classified into twelve sectors according to their proximity to adjacent metropolitan districts and to transportation links. The aggregate figures for the suburbs belonging to the same sector and data obtained for the neighbouring districts of Budapest's central city show significant correlations (see table 6). These correlations provide further evidence for the sectoral pattern of the suburban ring, and this corresponds with the regional distribution of the suburban settlement types distinguished earlier on the basis of their labour market characteristics.

3   The suburban belt during the period of transition to a market economy

During the political and economic transition from the socialist state to a market economy, the income gap between various social groups increased drastically. The differences in the country's regional development pattern, meanwhile, have been partly modified and further accentuated. A new period also began in the evolution of the suburban ring around Budapest. The 1971 official delineation of the agglomeration belt included 43 or, since the recent split of Kerepes and Kistarcsa, 44 settlements. The various publications of the Hungarian Central Statistical Office, e.g. the statistical yearbooks of Budapest and Pest county, and the series of publications on the Budapest agglomeration provide only little information about the social processes compared to the census volumes. They do not provide any information about the population's educational level and skills. Other sources, such as publications by the Finance Ministry, provide information on the changes in the population's income. This information was not available earlier.

In the 1980s, the population dynamics of Budapest and its surrounding suburban belt changed. The natural decrease of the city's population was no longer offset by massive in-migration from the countryside. The capital's population dropped from 2 059 000 to 2 016 000 by the end of the decade. This trend has accelerated since then, and by January 1998, the total number of inhabitants in Budapest was only 1 861 000. Gradually, the population loss, due to net migration, increased and by 1993, 8 000 - 10 000 inhabitants moved away from the city annually. This was further aggravated by death figures exceeding birth figures by 10 000 - 13 000 annually.

However, the overall suburban belt population continued to grow in the 1980s (from 410 000 to 413 000) and, unlike the rest of Hungary, this process accelerated in the 1990s. Although there are great differences in the growth rate among the suburbs, the agglomeration belt's total population reached 445 000 by 1997. The underlying causes may be due to a reversal of the population movement between Budapest and Pest county (mainly the agglomeration zone) since 1987 (figure 1). In the earlier development stages, the population of the agglomeration belt around Budapest increased steadily. This process was, however, unlike that in most North American and Western European cities because it was caused by a massive and continuous in-migration from rural areas. Only recently, out-migration from the central city became the major source of further suburban growth. According to the data from the mid-1990s, Pest county receives almost one and a half times as many permanent migrants from Budapest than the counties combined, and the share of the other counties has been declining. This is powerful evidence of a typical suburbanisation process instead of a counterurbanisation process.

Figure 1:   Migration between Budapest and Pest county (1980-1996)

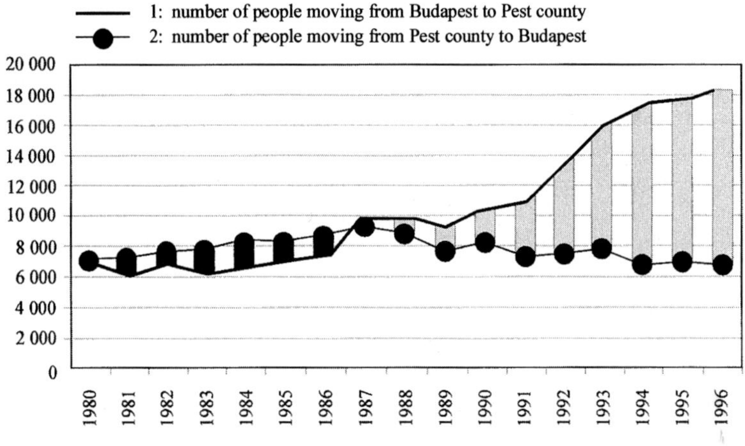

Source: NOVOTNYNÉ PLETSCHER 1998.

Two-thirds of the people moving away from Budapest into Pest county purchase or build homes in the inner suburban zone. The most popular locations in this inner belt are situated in the western sector, or to the south on Csepel Island (Budaörs, Érd, Szentendre, Budakalász, Solymár, or Szigetszentmiklós and Halásztelek). One of the finest suburbs of the Buda Hills, Budakeszi, has stopped expanding since it ran out of available construction space.

Outside the defined (inner) suburban belt, considerable housing construction activity has occurred in the north-eastern sector. This can be attributed to the demands of people leaving Budapest. The construction is concentrated in settle-ments which offer both a pleasant environment and new, modern industrial work-places (Veresegyház, Őrbottyán). In the western sector, the local council allocated cheap housing plots in Telki, thereby increasing its attractiveness for newcomers. One-third of the newly built suburban homes, most of which are single-family homes, have four or more rooms with an average total floor space of 115-120 square meters (figure 2), thus satisfying highest requirements (NOVOTNYNÉ 1998). The results of a recent survey that are later discussed in more detail clearly show that often the highly qualified, more affluent people move to the hilly Buda side of the suburban zone. Differences in the price of plots, the environmental quality, and the transportation costs influence the scale of migration.

Figure 2a: Distribution of homes (number of rooms) built by Budapest's inhabitants within the original agglomeration boundaries

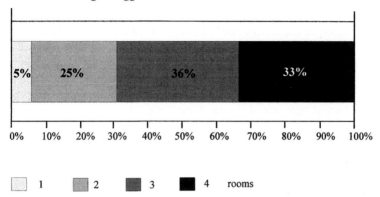

Figure 2b: Distribution of homes (floor space) built by Budapest's inhabitants in Pest county

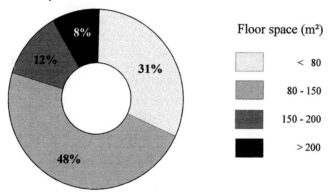

Source: NOVOTNYNÉ PLETSCHER 1998.

There is no reliable data available regarding the exact number of people commuting daily to workplaces in Budapest after 1990. The number could be about 160 000.[3] The more distant outer agglomeration zone has seen a 20% decrease in daily commuting in the past decade, this is primarily influenced by higher accessibility costs. Approximately half of the wage earners living in the suburban

---

[3]   According to the KSH (1998), the total number of commuters in Pest county exceeds 191 000 while there is an outflow of about 47 000 active wage earners who commute from Budapest to the suburban zone.

belt work in Budapest. This is despite the fact that the suburbs in the capital's vicinity, along the main routes, or along motorways have been successful in setting up investment zones and attracting capital to create new employment opportunities (e.g. wholesale centres, warehouses, office buildings and large shopping centres).

One of the most conspicuous new features in the last couple of years has been the construction of huge, American-style hypermarkets and shopping malls. Their construction began in Budapest as late as 1995 but gained rapid momentum. By January 1999, the number of shopping malls in the central city increased to thirteen, each of them with a floor space exceeding 10 000 square meters and many of them also providing entertainment facilities. They are often located in the city's former industrial belt, thereby contributing slightly to the revitalisation of the derelict areas which were abandoned when the factories closed. Several other developments, meanwhile, were placed in a greenfield area which is part of the suburban belt and also close to the motorways. The most popular suburbs for mall development are Törökbálint and Budaörs at the western end of Budapest, while a few other malls opened up in Fót in the north-east and a single mall opened at Budakalász near the northern city line. The Cora shopping centre in Törökbálint has parking for 2 500 cars, it has 20 000 square meters of usable floor space, and its investment costs reached HUF 7.2 billion. Michelfeit and Bricostore have added two specialised supermarkets worth HUF 3.6 billion in the same area. Budaörs' first Metro shopping centre was followed by the Auchan hypermarket, which opened in 1998 with a floor space of 22 500 square meters and 1 500 parking places (photos 1 and 2). The Auchan hypermarket cost HUF 8.5 billion. The OBI supermarket was completed in 1998 and IKEA and Julius Meinl started construction in 1999. Contrary to those in the central city, the suburban hypermarkets usually restrict their activities to wholesale and retail trade with much less emphasis on services to customers who come primarily from the adjacent Budapest districts.

A series of sketch maps, based on a few key indicators, depicts the strong differentiation within the agglomeration belt (figures 3-7). The high social status and affluence of people living on the western (Buda) side is most striking when contrasted with the south-eastern sector. There is only a slight regional relationship between population growth and average income or education level. This is demonstrated by correlation coefficients (0.30 and 0.26) which fall short of being significant. This may be attributed to the fact that even the less developed south-eastern suburban fringe attracts immigrants from neighbouring districts in the central city with similar social features. Increasing variation coefficients in the spatial distribution of other important variables indicate that the former social and functional disparities within the suburban belt have been further accentuated (table 7).

Photo 1: The Auchan hypermarket in Budaörs

*Photo 2:   The Auchan hypermarket in Budaörs and its main location factors: motorway
(in the foreground), parking lot, booming suburb with many family houses in a
pleasant, hilly environment (background)*

302   É. Izsák and F. Probáld

Figure 3:   Population growth in the settlements of the agglomeration belt (1990-1995)

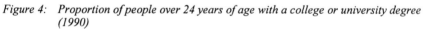

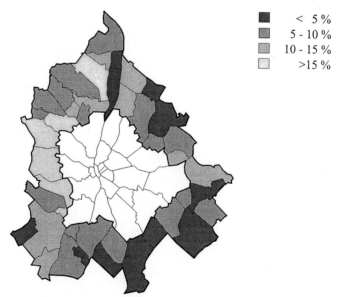

Source: Hungarian Central Statistical Office (KSH); authors' design.

Figure 4:   Proportion of people over 24 years of age with a college or university degree (1990)

Source: Hungarian Central Statistical Office (KSH); authors' design.

Figure 5:   Taxable income per capita of taxpayers in the Budapest agglomeration (1995)

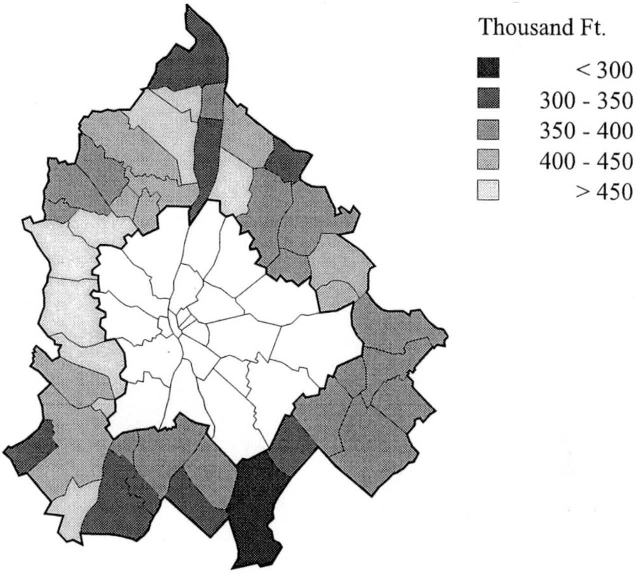

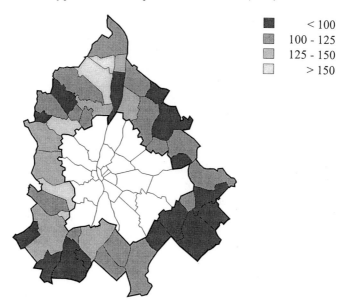

Thousand Ft.

■   < 300
■   300 - 350
▨   350 - 400
▨   400 - 450
□   > 450

Source: Ministry of Finance; authors' design.

Figure 6:   Total number of private ventures per 1 000 inhabitants (1995)

■   < 100
▨   100 - 125
▨   125 - 150
□   > 150

Source: Hungarian Central Statistical Office (KSH); authors' design.

Figure 7:   Unemployment (1995)

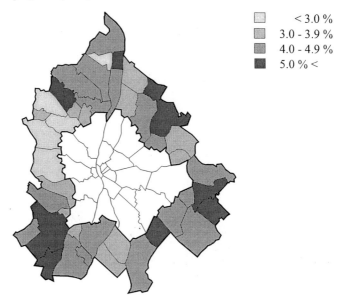

Source: Hungarian Central Statistical Office (KSH); authors' design.

Table 7:   Variation coefficients for various indicators in the Budapest agglomeration belt

	1990	1995
Natural increase (per thousand inhabitants)	0.92	2.30
Number of newly built dwelling units (per hundred inhabitants)	0.54	2.70
Length of the sewer network (meters, per hundred inhabitants)	1.23	1.83
Number of retail shops (per thousand inhabitants)	0.47	1.92
Taxable income (per capita of taxpayers)	0.12	0.15
Composite suburbanisation index	0.30	0.34

Source: Hungarian Central Statistical Office; authors' calculation.

In order to analyse the spatial differentiation processes in the period between 1990 and 1995, a composite indicator was used (called 'suburbanisation advance' or 'success index'). This composite index, which is to a certain extent limited by the availability of comparable data, contains five different variables (IZSÁK 1998). It has been developed based on figures about natural population increase, income, housing construction, the length of the sewer system and the retail trade network as related to the population size. The rank numbers summed up for each of the 44 agglomeration settlements result in a composite index with an invariable mean value and no dimension (figures 8-10). A slight increase in range and standard deviation signifies an intensification of social diversity and growing disparities within the suburban area (see also table 7).

Figure 8:   The 1990 suburbanization (success) index

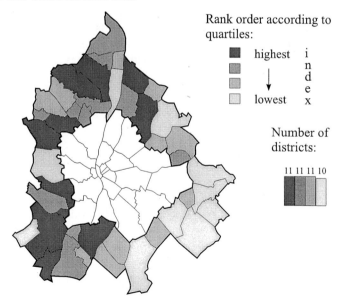

Source: Authors' calculation and design.

Figure 9:   The 1995 suburbanization index

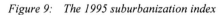

Source: Authors' calculation and design.

Figure 10:   *Settlements keeping up with development, and those lagging behind, 1990-1995* [4]

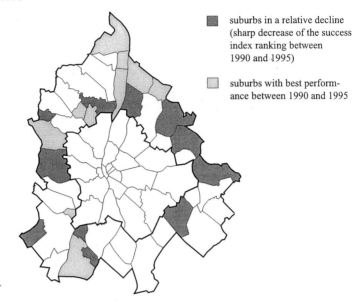

■ suburbs in a relative decline (sharp decrease of the success index ranking between 1990 and 1995)

▨ suburbs with best perform-ance between 1990 and 1995

Source: Authors' calculation and design.

Settlements that benefited from the developments between 1990 and 1995 are located in the western part of the suburban belt along the main corridors, or near the advancing development 'tentacles' close to the main transport routes. These settlements can also be found along the Danube River and on the hilly Buda area where they serve recreational purposes and are used as holiday weekend homes. The south-eastern and eastern suburban belt, found on the Pest side, is character-ised by a lower social status. It is an area lagging behind the rest. The overall structural pattern of the Budapest agglomeration is summed up in figure 11.

4   Budaörs and Gyál: a case study of different types of suburbs

This case study examines two Budapest suburbs which represent two different settlement types within the agglomeration belt according to the success index. Apart from the available data, the analysis also includes an empirical survey with

---

[4]   This map shows the settlements that experienced the greatest *changes* in the rank order. Therefore, the settlements with a stable position (e.g. at the top or bottom of the league) do not appear on this map.

emphasis on sociological interviews conducted with local inhabitants and economic actors (council leaders and entrepreneurs) and on the study of local council regulations.

Figure 11: Structural model of the Budapest agglomeration

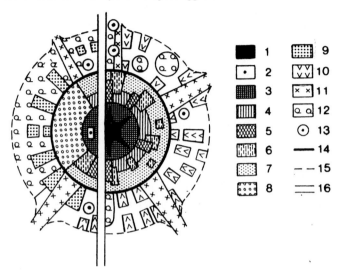

1 City (CBD)
2 historical city core (Castle Hill)
3 densely built-up inner residential belt in urgent need of renewal
4 transitory zone (industry, railways, warehouses, new shopping centers etc.)
5 industrial zone
6 housing estates of the socialist era
7 zone of former outskirts family houses
8 high prestige residential zone of the Buda Hills (villas, condominiums)
9 suburbs of high social status
10 suburbs of lower social status interspersed with agricultural land
11 development axes along the main highways and motorways
12 resort areas, leisure time housing interspersed with forest and some agricultural land
13 major satellite towns with considerable local labour market
14 border of Budapest
15 boundary of the suburban belt
16 Danube River

Source: Authors' design.

4.1 Budaörs

The town Budaörs is located in the Budaörs Basin between the Buda Hills and the Tétény Plateau. Due to its favourable geographical location, it is closely linked to the capital. Three main motorways either pass through or close by the town (M1,

M7 and M0), and it has direct and frequent bus links with Budapest and the neighbouring suburbs. This is a privileged situation in the agglomeration belt.

The recent rapid development of Budaörs was primarily due to its favourable transportation links. In the past decade the town has expanded tremendously. The built-up area grew at the expense of the outer zone as the building frenzy continued. The growth rate is outstanding even among the other settlements of the central region. Between 1980 and 1990, Budaörs experienced a 15% growth in population, most of it due to in-migration (11.7%). From 1990 to 1995, the population increased by 6% to 21 400, and surpassed 23 000 in 1998. Since 1990, the ratio of immigrants per thousand inhabitants has continued to rise in Budaörs. The influx of inhabitants occurs mostly from the nearby suburbs and from the neighbouring districts of the capital (districts XI, XII, XXII).

More than half of the newly built dwellings have four rooms or more. The average floor space in the dwellings is 128 square meters. In the period between 1990 and 1997, 950 new single family houses were built in Budaörs of which the larger plots (800-1 200 square meters) allow for small gardens. The development since 1990 incorporates wholesale, retail, warehouse and industrial areas of 74 hectares or more that have attracted investments of approximately HUF 100 billion (400 million USD). The new office buildings completed in 1996 and 1997 have a total floor space of 36 500 square meters.

Budaörs received town status in 1985. Following the Hungarian economic changes of the 1990s, Budaörs grew together with the core city, and this provided particularly favourable conditions for the emergence of an economic and trade (wholesale and retail) centre. Western European investors spotted favourable investment opportunities in the town because of its vicinity to the capital, its excellent transportation connections and the town's large areas available for construction. Consequently, the huge enterprises that located their office headquarters in this 'West Gate' suburb (Tchibo, Pannon GSM Telecommunications, Kaiser-Kraft, Matáv Telecommunications, Nokia, Honda, etc.) play a major role in its development. This is further enhanced by the favourable investment climate that was created by the local council (e.g. local taxes are held at a much lower level than in Budapest).

A new residential area in Budaörs was selected for a survey that was carried out by geography students in 1997. This area consists partly of detached condominium apartment houses and partly of rows of semi-detached, two-story, single-family houses. A sample of fifty families were randomly selected for interviews. All of them moved to Budaörs between 1990 and 1996. The answers to the most relevant questions provide insight to the social structure of recent in-migrants: The modal age group (40% of the sample) consists of the heads of family households who are between 45 and 54 years old. Among the spouses, the group between 34 and 44 years old prevails (38%). The family heads have usually attained a higher educational level (52%, university degree; 26%, college degree); this proportion is somewhat lower among the spouses (26% and 26%, respectively), but this is still about five times higher than the national average. Two-thirds of the heads of family (and almost half of the spouses) commute to work-places in Budapest. About half of the families used to live in huge apartment complexes which were part of Budapest's large-scale housing projects. As a motivating factor for moving to Budaörs, more than half of the respondents emphasised the high housing quality they could obtain in Budaörs at a much lower price than in Budapest.[5]

The results of the survey imply that Budaörs is a successful suburb with rising social status and a booming service sector that is supported by the local council. Moreover, the multinational corporations in Budaörs contribute to the tax base, and they are often willing to support local health care and cultural facilities such as the theatre or the outpatient clinic.

4.2 Gyál

Gyál is one of the youngest towns in Pest county. It was granted town status in July 1997. It is situated in the northern part of the Danube-Tisza region, and its inner core has merged with the XVIIIth district of the Budapest metropolitan area. In the 1960s and 1970s, Gyál's development could not keep pace with the rest of the suburban belt mainly because of its less favourable social structure. Primarily an agrarian village, it was inhabited by people of lower social status, and there was no industry around. Yet the settlement's population increased rapidly during the 1960s and 1970s, but people moving in from the northern and eastern parts of the Great Hungarian Plain (from Szabolcs-Szatmár-Bereg, Heves and Békés counties) were also of a lower social status. The population of Gyál mushroomed from 6 100 to 14 000 in the period between 1960 and 1970. The wave of massive in-migration created serious socialisation problems for both the new inhabitants and for those commuting to work in the city. Tension escalated over time, and by the 1980s, a crisis developed. Money for infrastructural development was in short supply, outmigration was growing, and social conditions worsened for those who stayed behind. Slight relief was brought about by the socio-economic changes of the 1990s. The outmigration stopped, and more and more people began to settle in Gyál, moving away from Budapest (predominantly from the housing projects of the neighbouring city districts). New houses were built by people from other parts of Pest county, too. Between 1990 and 1995, 46-96 houses were constructed annually while many others were enlarged and upgraded with modern amenities. The number of retail shops and catering establishments increased, and the rather poor infrastructure facilities developed, too. In 1995, the population between 18-60 years of age was 62%, and 29% of the population was less than 18 years old. However, 81% of the inhabitants were blue-collar workers and 19% were white-

---

[5]   In the survey, the apartments' usable floor space varied between 60 and 420[!] square meters.

collar workers or professionals. This is not a favourable ratio when compared to the other suburbs of Budapest. Gyál reached its development peak when it became a town at the request of the local authorities in 1997. Widening opportunities and a growing vitality of local society may have lead to the increased urbanisation.

It is sometimes difficult to determine specific causes of a suburban settlement's success based on facts and figures. The spectacular path to success that continues to happen in Budaörs is not easily replicated on the outskirts of a capital city of almost two million inhabitants. Since 1990, Gyál has followed a similarly long development path, but its point of departure was different to that of Budaörs. The local advantages of Budaörs and neighbouring suburbs with similar potential may have led to an unsurpassable good start. It is obvious that regional development aims to optimise endogenous resources which are the inherent potential of the settlements (RECHNITZER 1993). In the case of Budaörs, both the favourable external conditions (geographical location, history, cultural ties and contacts with former German inhabitants compelled to leave the country in 1946) and the internal town resources have been exploited to achieve a successful development. Endogenous potentials are much less favourable in the Gyál case and the environment is also less attractive (IZSÁK 1998). Nevertheless, the town can be proud of its recent achievements. Its decline has stopped, and this is positive, even if Gyál does not catch up with the Buda suburbs. A combined exploitation of endogenous and exogenous resources seems to be the fundamental key to success. This is demonstrated by suburban development on the Buda side which is the result of using both types of resources and has led to bright future perspectives.

5   The latest development trends and the enlargement of the Budapest agglomeration

The Regional Organisation and Development Act No. 21/1996 legalised a new type of planning after planning in Hungary had become totally disorganised during the transitional period to a market economy. A new national institutional system of middle-level regional planning was established. Inspired by preparations underway for Hungary's entry into the European Union, Hungary was divided into seven statistical planning regions, intended to be analogous with NUTS 2 regions of the EU. Budapest and Pest county together form the central region. In accordance with this new law, a government initiative suggested the formation of special development councils with less authority than those of the regional development councils. These special development councils were introduced for two specific areas, the recreational region around Lake Balaton and the Budapest agglomeration. Apart from the ministries, other organisations involved in the work of the Development Council of the Budapest Agglomeration are the local governments of Budapest and Pest county, chambers of commerce, and the six settlement development associations formed by local councils' groups, which are organised along the six radial agglomeration sectors. The role of the Council is to dissolve conflicting interests between settlements, to initiate a new comprehensive master plan for the agglomeration, and to prepare development projects and tenders.[6]

Figure 12: Changes in the official delineation of the Budapest agglomeration (1997)

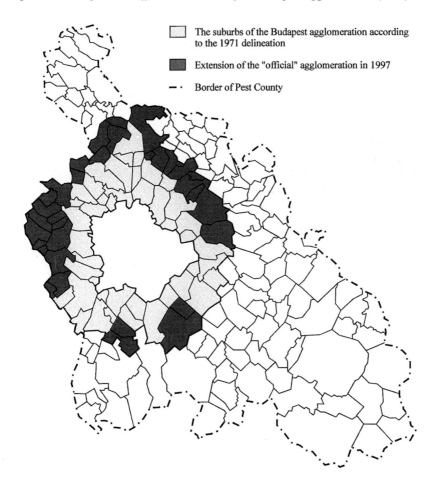

The suburbs of the Budapest agglomeration according to the 1971 delineation

Extension of the "official" agglomeration in 1997

Border of Pest County

Source: KSH (1998).

---

[6]   This Council, however, was abolished in 2000.

The 'official' delineation of the agglomeration has been modified simultaneously with the establishment of the Budapest Agglomeration's Development Council. Since 1997, the enlarged agglomeration embraces 78 suburban settlements (11 of which are towns) and the city of Budapest (figure 12). The population of this broader suburban belt rose to 618 000 on January 1, 1998 while the overall agglomeration population rose to 2 480 000. This represents one-quarter of Hungary's total population.

The agglomeration was significantly extended in a western, northern and northeastern direction. Neighbouring suburban settlements situated to the west of Budapest's have undergone spectacular development and functional reorientation. This led to the birth of new industrial and wholesale distributors and warehouse functions, and to the emergence of offices and hypermarkets (e.g. Törökbálint, Budaörs), the establishment of corporate headquarters, research institutions, and teacher retraining centres (Pilisborosjenő). In addition, the rural settlements of the Zsámbék Basin, on the eastern side of the Buda Hills, have also been influenced by the capital's sphere of attraction. The Zsámbék suburb is the most important of these settlements and many teachers and students from Zsámbék's Teachers' Training College commute out of Budapest. In 1999, Telki, another booming suburb, opened the first luxury private hospital in Hungary. The hospital serves rich patients from the Buda Hills who can afford the expensive treatments.

Figure 13: The suburbs of the inner (older) agglomeration belt

The suburban zone has also reached out along a development axis towards the north-west, thereby creating a link with the urbanised region between Esztergom and Komárom, both of which are situated along the river Danube. In the old resort suburb of Piliscsaba that is also located on this development axis, former Soviet army base barracks have been altered and reconstructed in order to accommodate the campus of the Péter Pázmány Catholic University. This institution has a close, two-way commuter link with Budapest. The closely built-up area extends as far as Visegrád to the north, which is famous for its historic monuments. This corridor along the Danube is crowded with leisure and holiday homes owned by the inhabitants of the central city and it has become fully annexed to the suburban belt. Vác, an old bishopric and major industrial town in the north-east, has been incorporated into the expanding suburban ring, though due to its local signifi-cance, its sphere of attraction reaches out beyond the Budapest agglomeration. Veresegyháza, lying to the capital's north-east, has developed tremendously in the past decade through the evolution of its privately owned pharmaceutical company, which was eventually amalgamated into an American transnational company. Gödöllő lies close to the agglomeration's new outer boundary. It has excellent transportation links (motorway, rail and commuter train) to Budapest, large amounts of foreign investment capital, and has become a favourite location for high-tech industries (pharmaceutical companies, Ganz precision engineering fac-tory, United Technologies' vehicle component manufacturing, Sony electronics production facilities). The town is the most important centre for higher agricul-tural education and agricultural training in Hungary and has its own sphere of attraction.

The growing disparities between the population processes of the core city and the surrounding suburban ring as well as the increasing contrast between the radial agglomeration sectors should be emphasised when summarising Budapest's spontaneous development trends in the past decade. The spectacular growth of the elitist Buda Hills suburbs leaves the south-eastern agglomeration sector lagging behind. The functional diversification of the suburban zone and the officially declared areal extension of the Budapest agglomeration as well as a new attempt to co-ordinate regional development policies in the whole urban region should also be mentioned. In the future, suburbanisation will probably continue. It is un-likely that a reurbanisation process will begin in the central city or that a consider-able counterurbanisation will develop in the years to come. These are some of the reasons why the internal structural changes of the agglomeration belt deserve par-ticular attention.

References

BERÉNYI, I., DÖVÉNYI, Z. (1996): Historische und aktuelle Entwicklungen der Ungarischen Siedlungsnetzes. In: Institut für Länderkunde (ed.): Städte und Städtesysteme in Mittel- und Südosteuropa. Leipzig, 104-170

BURDACK, J., HERBERT, G. (1998): Neue Entwicklungen an der Peripherie europäischer Grosstädte. Europa Regional (2), 26-44

ENYEDI, GY, SZIRMAI, V. (1992): Budapest: a Central European Capital. Belhaven Press, London

FRIEDRICHS, J. (1985): Stadtentwicklungen in West- und Osteuropa. Walter de Gruyter, Berlin

HORVÁTH, GY., ILLÉS, I. (1997): Regionális fejlődés és politika. (Regional Development and Policies). Európai Tükör, Műhelytanulmányok 16. Hungarian Central Statistical Office, Budapest

IZSÁK, É. (1998): A természeti és társadalmi környezet hatása Budapest határmenti területeinek a fejlődésére. (The Influence of the Natural and Social Environment on the Development along the City Line of Budapest). PhD értekezés. Kézirat. (PhD thesis, manuscript). ELTE TTK

KOVÁCS Z., WIESSNER, A. (1997): Prozesse und Perspektiven der Stadtentwicklung in Ostmitteleuropa. München (Münchner Geographische Hefte 76)

KSH (ed.) (1997): Az újonnan lehatárolt budapesti agglomeráció főbb adatai. (Data on the Budapest agglomeration within the extended new boundaries). Hungarian Central Statistical Office, Budapest

KSH (ed.) (1998): Közlemények a budapesti agglomerációról 8: A budapesti agglomeráció az ezredforduló küszöbén. (The Agglomeration of Budapest at the Turn of the Millennium). Hungarian Central Statistical Office, Budapest

LICHTENBERGER, E., CSÉFALVAY, Z., PAAL, M. (1994): Stadtverfall und Stadterneuerung in Budapest vor der Wende und heute. Verlag der Österreichischen Akademie der Wissenschaften, Wien

NOVOTNYNÉ PLETSCHER, H. (1998): Budapest népessége és a vándormozgalom változásának főbb területi jellemzői. (The Population of Budapest and Key Regional Features of the Changing Migration Process). Területi Statisztika 1 (38), 126-143

PROBÁLD, F. (1975): A Study of Residential Segregation in Budapest. Annales Univ. Sci. R. Eötvös IX., Budapest, 103-112

PROBÁLD, F., SZEGEDI, G. (1994): A gazdasági társaságok elterjedése a budapesti agglomerációban. (The Spread of the Private Ventures in the Budapest Agglomeration). Földrajzi Értesítő XLIII., 281-292

PROBÁLD, F. (1995): Regionale Strukturen des Arbeitsplatzangebotes in der Agglomeration von Budapest. In: MEUSBURGER, P., KLINGER, A. (eds): Vom Plan zum Markt. Physica Verlag, Heidelberg, 182-208

PROBÁLD, F. (1997): Budapest: dynamisches Zentrum Ungarns. Geographie heute (153), 32-35

RECHNITZER, J. (1993): Szétszakadás vagy felzárkózás. A térszerkezetet alakító innovációk. (Falling Apart or Catching Up. Innovations Affecting the Spatial Structure). MTA RRK, Győr, 113-124

RIKKINEN, K., PROBÁLD, F. (1978): Bevölkerungsveränderungen im Bereich von Budapest in den Jahren 1869-1966. In: KULS, W. (ed.): Probleme der Bevölkerungsgeographie. Wissenschaftliche Buchgesellschaft, Darmstadt, 150-174

SÁRFALVI, B. (1991): Neuere Tendenzen der Agglomeration von Budapest. Geographische Rundschau (12), 724-730

The Changing Hungarian Cityscape in the 1990s: A Survey of Four Sample Cities

Éva Izsák and József Nemes-Nagy

1   Introduction

There are some towns of which we do not really have any remaining impressions, and there are others which have made kind, fresh imprints on our memories, which we recall with pleasure. What are these pleasant impressions created by, and what causes the unpleasant ones? Obviously, it is the coincidence of several factors: order and disorder, cleanness or dirt, a picturesque or a dull look, outstanding pieces of architecture or tasteless buildings, simple or richly proportioned architecture, the crowdedness, peace, and vitality of streets (GERŐ 1971).

The socio-economic transition taking place at the end of the 1980s and in the early nineties can be traced in the outward appearance of the settlements, in the changing face and functions of the buildings and the whole built environment as well as in the shift of the socio-economic structure. The spatial differences in the distribution of investment show the directions of development and indicate the speed of reaction and competitiveness of each settlement. Urban decay and urban *renewal* are two key concepts in urban studies, both in the more developed western and the former socialist countries.

In Hungary, the sixties and seventies were a period of an intensive urban expansion.[1] It was in 1974 when the idea of urban *renewal* turned up for the first time in the urban policy of Budapest. At that time, this meant the complete reconstruction of a rehabilitation area determined by the leaders of the city, who treated this problem merely from a technical point of view. Therefore, the urban policy of the socialist era could be primarily characterised by the construction of new housing and institutions. There was little attention paid to the revitalisation or renovation of old districts. The recent socio-economic changes finally put an end to this kind of urban policy routine which led, however, to two major consequences: on the one hand, serious problems arise because the renewal of the inner city areas and the repair of the housing stock has been neglected for a long time; on the other hand, the fast transition to a market economy and changing ownership have made the evolution of a totally new and different urban policy possible. The aim of the former conservative practice was the preservation of functions, i.e. what used to be an apartment previous to the reconstruction had to remain an

---

[1]   In Budapest alone, 180 000 new flats were built during the first 15-year period of large scale housing estate construction which started in the mid sixties.

apartment after it too. Following 1990, four key points are to be emphasised because of their important role in breaking through former barriers:

- the major forms of large scale housing construction typical of the former system, and the building of tenement houses were no longer in practice;
- the central power has been decentralised with a simultaneous strengthening of the autonomy of local councils;
- the housing and real estate markets are controlled by market mechanisms today;
- the rehabilitation of city centres has gained special importance exerting great influence on the future of urban renewal.

These processes required a radical transformation of the Hungarian building industry.[2] The authorities commissioning constructions, and their aims and requirements have been rearranged. The new owners (local governments) formed their own privatisation strategies, and due to the differences in socio-economic structures, different models evolved in almost every settlement.

In our study, we would like to introduce a four-fold typology of the changing Hungarian cityscape. The three cities (Dunaújváros, Győr, Salgótarján), and the districts VIII and IX of Budapest represent different types of the urban development cycle, and their reactions to the process of transition showed fundamental dissimilarities. We examine the change of the built environment as a print of the political-economic transition, holding a mirror for those Hungarian settlements who have not yet found the most suitable ways to reconstruct and change their own space.

Table 1 contains some features of the four cities and their microregions. These data already indicate that from among the three cities in the countryside, Győr has the most favourable demographic and economic conditions, while Dunaújváros, and especially Salgótarján show signs of economic and demographic stagnation or even decline in the mid nineties. The survey also highlighted the positive and negative socio-economic impacts of the transition on the cities examined, and those unique features which, chiefly in the cases of Salgótarján and Dunaújváros, have led to the stiffening of the structure and architecture of the city.

In the summer of 1998, the Department of Regional Geography of the Eötvös Loránd University organised a field course. Students of geography got the task of examining the built environment in the different sample areas by walking over and mapping them. In Győr, and in the capital, the built environment and buildings were analysed on the basis of previously given categories,[3] while in Du-

---

[2]   8 500 new building companies emerged in Hungary between 1990-1993.
[3]   Categories used in the survey: Buildings in poor condition: blank, spare lot (e.g. place of demolished buildings, derelict warehouses), old buildings ready to be demolished, and old buildings in extremely poor condition, buildings in bad condition, old buildings in a moderately and slightly poor condition. Old buildings in good condition: old

naújváros and Salgótarján, status reports were produced of the inner proportions of the city structure and the processes characteristic to different districts, on the basis of experience by means of walking through the cities and making interviews. The present study summarises the most important results of this survey.

Table 1: Indicators of the sample cities

Cities	Budapest	Győr	Salgótarján	Dunaújváros
Population (thousand, 01.01.1998)	1 861	127	45	57
Population change (%, 1998/1990)	−7.7	−1.6	−6.1	−3.6
Housing stock change (%, 1998/1990)	+3.2	+4.3	+2.2	+1.8
Total number of firms (1997)	215 164	12 165	3 012	3 725
Number of employed in industry (thousand people, 1997)	91.1	22.4	6.4	12.4
Microregions of the cities				
Population change (%, 1998/1990)		−0.4	−4.4	+1.6
Unemployment (12.1997)	3.7 (city only)	4.8	12.2	6.7

Source: Hungarian Central Statistical Office (KSH).

2   Districts VIII and IX of Budapest (Józsefváros and Ferencváros)

The area of these two districts was part of the capital city at the time of its unification in 1873, so their development and history has been as long as that of Budapest. During the period of the city boom, in the last two decades of the 19th century and at the turn of the century, the built up area quickly expanded, which was manifest in the construction of three- and four-floor tenement houses with circular galleries. The residential quarters built around this time are still dominant in the area of Ferencváros and Józsefváros. Due to the urban policy described above and the neglected repair works, the poor condition of the buildings caused severe problems for Budapest already in the 1980s. Already in 1985, 30% of the dwelling stock would have required renovation, and a great deal of the problems was concentrated in the five inner districts.[4] Action has been put off, the condition

---

buildings in good condition, renovated old buildings. New buildings: this category embraces buildings erected after the Second World War.

[4]   For the rehabilitation of the inner districts in Budapest, see tables 2 and 3.

of the building stock further deteriorated. More than one third of the houses constructed before 1955 have never been renovated.

Table 2: State of the building stock in 1993

	Distr. VIII*	District IX*	Total
Number of buildings (total)	1 804	1 103	2 907 (100%)
Number of renovated buildings (1978 - 1993)	20	28	48
Percentage of renovated buildings (1973-1993)	1.1	2.5	1.6
Number of new buildings (built after 1945)	91	121	212
Percentage of new buildings (built after 1945)	5.0	10.9	7.3

Source: Authors' survey.   *Survey area of the district.

Table 3: Change in the state of the building stock between 1993 and 1998

	Distr. VIII*	District IX*	Total
Number of buildings (1993, total)	1 804	1 103	2 907 (100%)
Number of buildings demolished	20	28	48
Percentage of the total stock	1.1	2.5	1.6
Number of renovated buildings (1993-1998)	53	38	91
Percentage of renovated buildings (1993-1998)	2.9	3.4	3.1
Number of new buildings (1993-1998)	15	34	49
Percentage of new buildings (1993-1998)	0.8	3.1	1.7

Source: Authors' survey.   *Survey area of the district.

Following the political transition, the tenement houses (apartments), and the public institutions which were to serve and were founded by the given district, became the property of the district.[5] The district councils could decide freely on what they wanted to do with the new property they received. This resulted in a situation where several different 'district policies' evolved as different urban policies. The spatial dissimilarities are also reflected in the ways of reconstruction and the extent of financial support granted to the condominium apartment owners for maintenance and repair works after the privatisation of the dwellings had been almost completed.

---

[5]   At the same time, the communal buildings which served the city as a whole or were established by the city were transferred to the capital.

3   District VIII, Józsefváros

The area stretching to the east of the present *Nagykörút* (grand boulevard) was regarded as a slum zone in a survey completed in 1942. Slow rehabilitation began in the 1980s when the buildings in the worst condition were started to be demolished. Nearly a quarter of the building stock was pulled down in this area, which is also indicated by the relatively high proportion of blank plots. However, this was not followed by new construction in District VIII, and the high percentage of blank plots increases the district's depreciated appearance. Some renovation was done, however, on the blocks at the Keleti railway station, which led to an increase of the real estate prices in that area. Another early renewal attempt of the 1980s aimed to replace several pulled down blocks by a large scale housing estate between Baross and Tömő Streets, but this initiative turned out to be a financial and environmental failure. This area, however, is outside of the part of the district dealt with in the present survey.

The inner part of the district close to the City embraces also the former mansion quarter of the aristocracy and several educational and cultural institutions such as university buildings. Thus, the area between *Kiskörút and Nagykörút* (inner and grand boulevard) is in a relatively better condition. In this area, according to the survey, considerable renovation has taken place with just one completely new building erected during the last decade.

In the streets running perpendicular to the boulevards, especially in the surroundings of *Nagykörút* (grand boulevard), spacious and splendid apartments for well-off citizens and intellectuals had been built from the end of the last century.[6] The population has been completely replaced due to the neglected renovation and the deteriorating state of the housing stock. The emergence of slums and urban decay has been triggered by settling down of people of lower social status who came here partly from the newly renovated blocks of the neighbouring districts (e.g. from the rehabilitation area of District IX) and partly from the countryside. The blighted area is nowadays characterised by the highest concentration of gypsy population, criminality and prostitution. Consequently, the real estate prices in District VIII have begun to fall dramatically. The local government cannot even plan any kind of rehabilitation in the area, since it has to struggle with the increasing social problems of the population.

---

[6]   It is also indicated by the former favourable housing conditions that in these flats an elite population of higher social status resided.

4   District IX, Ferencváros

For the last decade, District IX of Budapest has been in the focus of attention in two respects: first, according to the plans which were thwarted in the last moment, part of the 1996 EXPO would have been organised in this area, second, the Middle Ferencváros rehabilitation programme was a unique attempt of its kind in Budapest in the 1990s.

4.1   Ferencváros and the cancellation of the EXPO

The world exhibition (EXPO), which was considered the greatest investment of the decade, should have been realised in 1995 and organised jointly by Budapest and Vienna. The Austrian plebiscite in 1991 decided against an EXPO in Vienna and as Budapest was left alone, the EXPO was put off to 1996, the 1100th anniversary of the Hungarian Conquest of Land. Because the programme has got in the cross-fire of more and more social and public debate,[7] the construction was repeatedly delayed, and due to political controversies finally cancelled in 1994.

The EXPO would have been held in the west bank area stretching along the Danube River (Lágymányos) in District XI. According to the plans, opposite to this EXPO area, in the eastern river-bank zone of District IX, several complementary investments would have been carried out. The quarter bordered by *Ferenc krt. – Soroksári út – Kvassay J. út* and the Danube River would have been in tight connection with the events on the other bank. Work began soon in this area, and by as early as 1993, terrain correction and demolition of all former buildings had been completed. The draft technical plan for the district had also been prepared with emphasis laid on increasing the proportion of green areas and opening up the area for private developers.

The EXPO cancellation stopped the construction in this river bank area, but a few important investments were nevertheless realised. One example is the Duna House in the southern part of *Boráros square*. There are smaller and bigger shops on the ground and first floors, and some offices and flats of luxurious quality on the floors above.[8] Further underground floors have been built in order to serve for car parking. In this area, further investments have also been made besides the Duna House, for instance, the promenade on the bank of the Danube between *Petőfi and Lágymányosi bridges.*

---

[7]   The investment would have improved primarily the infrastructure in Budapest, however, the costs would have had to be paid by the whole country.

[8]   As for the real estate prices, the price of an inward looking small apartment is Ft. 160 000 per square meter, and the bigger flats looking to the Duna cost Ft. 200 000 to 220 000 per square meter.

4.2   The rehabilitation of Middle Ferencváros

The district council has also launched a significant renewal and rehabilitation project in the area bordered by *Ferenc körút – Üllői út – Haller u. – Mester utca.* A close co-operation of the local council with a private real estate development company, involving also foreign capital by forming a joint venture, makes this investment unique in Budapest. Special emphasis has been laid on the protection of green surfaces, i.e. on the creation of parks between the separate blocks of buildings, with underground garages in order to ease parking problems in the area. Obsolete houses with apartments short of any comfort facilities have been either pulled down or modernised leaving more free space for parks and construction of new buildings with mixed functions. Thus, two zones have emerged within the renewal area:

- a densely built-up urban area primarily of closely arranged nice modern buildings with several apartments in each of them;
- an area with mixed (residential, office, touristic, recreational) functions and building density.

The City Council of Budapest contributed to the block rehabilitation project with Ft. 700 million in 1997. Therefore, while District IX formerly was thought of as an area with slums and had a rather negative image, its reputation and cityscape has improved considerably in the last 5 to 6 years. This has been the major result of the two 'action areas' mentioned before. However, population movement is further worsening the social situation of the neighbouring localities, especially that of District VIII.

The quantitative results of our field survey (table 2 and 3) clearly demonstrate, however, the still very slow pace of the renewal process. Mapping the location of renovated buildings reveals a strong concentration of them in the renewal area of District IX and in the internal part of the districts concerned, where the preservation of the architectural heritage requires substantial attention and financial support possibly provided by the establishment of new functions. Due to the spread of city centre functions, some hotels, office headquarters and quality shops appeared here already in the initial stage of renewal.

5   Győr - the western gate

The survey of the condition of the built environment included three towns in the countryside besides the two districts of Budapest. One of them is Győr, which is a traditional county centre, with great historic traditions and with an economy getting more and more significant in the last few decades. The opening of the frontiers taking place in the late 1980s heavily contributed to the increasing weight of Győr within the urban hierarchy of Hungary. The opening in the direction of

Austria was a major impulse to the city's development. Furthermore, its favourable geographical location and its advantageous position in the transportation network have to be mentioned, for the M1 motorway from Budapest to Vienna and Bratislava passes through Győr. A survey was carried out in two distinct areas in 1998: in the inner city and its surroundings as well as in Révfalu.[9]

5.1   The inner city and its surroundings

The inner city is one of the most beautiful parts of Győr today. As a result of the rehabilitation work done in the last couple of years, the old baroque-style buildings, small squares and pedestrian areas of the town centre strengthen the bourgeois image of Győr. Following the rehabilitation, considerable social restructuring and population exchange began in the last decades. The inflow and back-flow of well-off population of higher social status resembles processes of gentrification that can be found in Western Europe and North America. There are almost no new buildings here. The overwhelming majority of houses have mixed functions, there are shops, banks, restaurants on the lower floors in the frontage, and flats above them. Moving away from the core of the city, the image gradually changes. The zone of mixed functions is succeeded by the residential area, where the condition of the buildings is less favourable than of those in the centre. Further away from the city centre, the density of the built-up area gets lower, and spacious squares and wider streets become more dominant. Then the previous housing function is followed by a mixed functional zone again, where offices and shops join residential buildings, and where ten-storey apartment blocks are wedged in between.

5.2   Révfalu

This part of the city is to the north of the inner city, on the left bank of the *Mosoni-Duna*. It is bordered by the river from the west, the east and the south, which basically determined its development, and thus its present structure as well. Situated in lower and higher flood areas of the Danube River, this land was many times devastated by inundations that were usually followed by reconstruction.[10] The character of the settlement has always been that of a village, and not even its annexation to Győr (1905) could to any greater extent change this. Following the unification, the most important task was the improvement of transportation.[11] The

---

[9]   Unfortunately, there was not any survey from 1993 available (like that carried out in Budapest). Therefore, this is the first research to study the condition of the buildings and the built environment in the city.

[10] Because of the decreasing water level of Mosoni-Duna, groundwater has dropped to such a low level that it does not disturb construction.

[11] Till 1905, the only means of transport between Révfalu and Győr was the ferry boat.

extension of the infrastructure brought about the immigration of people. After the Second World War and the reconstruction period, more and more family houses were built. There are hardly any houses in the area which were built before 1945. The opening of the *Széchenyi István Technical College* in 1984 was a significant milestone in the development of this part of the town. Révfalu today is a garden-city residential quarter of Győr. The majority of its buildings are new, in good condition or renovated. One- or two-floor family houses are the most dominant, however, apartment houses of many floors are more common in areas closer to the inner city. Thus, the residential function characterises this area with the only central function represented by higher education (the college).

6   Salgótarján - captured by industrial depression

The city is the administrative centre of the smallest county in northern Hungary, Nógrád, and it was probably the greatest loser of the transition to market economy among the bigger cities. The GDP per capita is the lowest in Nógrád. It does not reach 60% of the average in Hungary. Nógrád is mainly an area of industrial depression, and after about a century of prosperity, now it is among those regions falling behind. Its prosperity was established by its coal deposits that have been exploited since the 19th century and which formed the basis of energy-intensive industrial branches (metallurgy, glass industry). The coal deposits have run out by today, the mines have been all closed down, unemployment is high, and social tensions characterise the city. A peculiar ambivalence appears in the geographical location of the city. Unlike most of the towns situated in the border zones that have partly experienced spectacular dynamism in the nineties (see essay of János Rechnitzer),[12] the Slovakian neighbourhood, for certain political and economic reasons, has exerted no considerable positive effect on the area.

However, as a peculiar paradox, all these features are reflected only partly by the image of the city. The core has been wedged into the narrow valley of a small stream (Tarján), and the only possible directions of expansion are smaller out-branching valleys and the surrounding hill sides. The city centre went through radical architectural reconstruction in the seventies; therefore, it is dominated by types of buildings which were considered the most modern ones in that period (public buildings and apartment blocks of several floors). Substantially, there is no space and possibility for changing the architecture of this area. The most typical 'change' has been the replacement of 'signboards': commercial activities and

---

[12]   Dynamism has been experienced near the western border due to accelerated reconstruction fuelled by the inflow of foreign capital, and in the southern and eastern border zones, triggered by the effects of 'grey' cross-frontier connections.

service functions are performed by the organisations, banks and shopping networks of the nineties.

While this situation provides the town centre with a relative architectural stability, the evident depression appears in the peripheries: in the industrial zone vegetating in the southern part of the city, and in the peripheral housing estates with slums. Due to a shortage of local incomes, there is no noticeably reviving place in the city, constructions by private initiatives are just isolated phenomena, and the city and its surroundings are permanently losing population and have difficulties keeping higher qualified people.

7   Dunaújváros – the open-air museum of socialist-realistic architecture

Similarly to Salgótarján, Dunaújváros,[13] which was once the greatest 'socialist city' has also been tied up by a particular heritage, yet the courses of its transformation in the nineties have been substantially different from those characteristic to Salgótarján.

The history, and life of the city has been basically determined by the Danubian Metallurgical Works (Dunaferr Vasmű). This plant is almost the only exception among the metallurgical plants of which several became victims of the economic transition, since it is one of the few industrial works of strategic importance that have survived the transition with a dominant state ownership. A fortunate circumstance contributed to its survival, because this metallurgical plant was modernised by significant capital injection in the 1980s. These investments provided a technological level that requires improvement only in the coming decade.[14] The industrial works rules the whole city in all respects (its social and political processes as well), and provides Dunaújváros with a relative economic stability.

The character of this city 'with substantially no history' (however, in Roman times, there was a significant river-crossing on the Danube here, in the eastern edge of the province Pannonia) is shaped by a mosaic of different parts of the city which are distinct in architectural character built in the 1950s to 1970s. The most peculiar elements are the blocks constructed in the initial period, in the fifties, in a Stalinist, socialist-realistic style. These places are gaining special preference today because of the historic and architectural value of some public buildings, a relatively pleasant environment (wide streets lined with trees, public spaces and squares), the traditional way of their construction (i.e. not concrete blocks), and their higher level of comfort. The new, presently evolving ideas of the city's long-

---

[13]   In the first part of the fifties, when it was built, it was called Sztálinváros.
[14]   Future modernisation, however, will most probably entail a decrease in employment.

term development make use of these features, and considers this architectural heritage a special attraction for tourists.

The city structure is stable, unlike in Salgótarján, however, here exists a special sub-centre, the 'bank quarter', where the branches of almost all the banks in Hungary can be found. This is due to the intensive financial relations of the big companies in the city, as well as to the relatively stable income position of the population. Still, characteristically, the new smaller units of commercial and service activities occupy the lower floors of the residential blocks. Thus, a picture is created in which these functions mix with the housing. Changes in population, and processes of construction on the edge of the city indicate slow but sure suburbanisation around Dunaújváros. This tendency counteracts the city's former character of being an 'industrial enclave' emerging from its agrarian environment.

The city attempts to reduce the expected unavoidable negative effects of the metallurgical technology shift by looking for new functions. A stressed element of this target is the construction of a bridge on the Danube, which is, however, rather just a local wish than a real possibility. Therefore, the short-term and long-term future of Dunaújváros may have several alternatives: relative stability can be maintained on the basis of a more diversified economic structure, yet without this, there is only the role of a 'museum of socialist realistic architecture' left for the city, which is quite unlikely to be enough for its survival.

References

GERŐ, L. (1971): Történelmi városrészek (Historical city quarters). Műszaki Könyvkiadó, Budapest

Residential Mobility during Transformation: Hungarian Cities in the 1990s

Ulrike Sailer

1   Introduction

The multidimensional transformation processes which have occurred in the fields of economic, political and social life in Hungary since the demise of socialism have been accompanied by a reform of the parameters set by state policy in the housing sector, with the abandonment of the socialist paradigm regarding housing as a social service towards the neo-liberal conception of a dwelling as an economic asset.

The comprehensive restructuring processes at the level of housing policy have included the complete withdrawal of the central state from the construction of state rentals and condominiums, the latter having been sold under very favourable conditions to households seeking a flat under the old regime. Another aspect of relevance in this context are changes in the conditions applicable to the previously highly subsidised housing loans, with state subsidies now being predominantly earmarked for new housing projects in the private sector and only a small measure of state support for housing provision for the economically disadvantaged. A further point of particular importance is the privatisation of state rentals: this sector has witnessed a virtual sell-off in view of the very favourable conditions of sale and the right-to-buy of tenants. Further aspects have been the curtailment of the quasi-ownership rights of tenants, the legalisation of rent increases and notice, triggering a considerable decline in the attraction of rentals (BALCHIN 1996; HEGEDŰS and TOSICS 1998; PICHLER-MILANOVICH 1994; SAILER-FLIEGE 1997; 1999; SMITH 1996).

In the early years of transformation, the social spectrum has widened considerably as the inequality of real socialism has given way to an increasing degree of socio-economic inequality under market economy conditions. Only small sections of the population have benefited economically from the transformation process to date (LADÁNYI and SZELÉNYI 1998). The deep economic recession, characterised by massive de-industrialisation, rampant inflation and unemployment, has resulted in shrinking incomes and falling standards of living for most of the population. Real per capita income fell to 88% between 1990-1995. The poverty ratio - defined as a proportion of the population with an income below the subsistence level - rose from 20% (1991/1992) to 33% (1993/94) (ANDORKA and SPÉDER 1995, 658/659).

The interaction between the restructuring of housing policy and economic and social upheavals has triggered drastic changes in the parameters defining the action of the players in the housing market. These changes have given rise to specific processes and problems, above all a dramatic decline in new housing construction, and this in turn has exacerbated the housing shortage inherited from the old regime. This general housing shortage has been structurally aggravated by the privatisation of state rentals and rent increases. Phenomena such as over-occupancy, rent arrears, evictions and homelessness have therefore multiplied during transformation. Since a private rental market has not yet developed, rentals in general have become very scarce in urban housing markets (BANKS et al. 1996). Decay and deficient maintenance of housing stock, a legacy from socialism past, have further deteriorated. Utility costs have climbed overproportionally as rapid price liberalisation transformed a dwelling into a very cost-intensive economic asset in the immediate post-socialist years.

This situation has been accompanied by an ongoing controversial debate on the possible effects of these multidimensional processes on residential mobility. Very low levels of residential mobility are generally assumed for the socialist era (inter alia SMITH 1996). Housing shortages and the high proportion of owner-occupied dwellings prompt some authors to assume very low rates of changes in mobility between the housing sub-markets even in the post-socialist period (cf. inter alia CHAPMAN and MURIE 1996). With reference to macro-level theories on residential mobility (cf. Zelinsky), however, the drastic upheavals in the transformation process can be expected to generate a post-socialist mobility transformation.

In this overall context, the objective of a comprehensive empirical survey was to analyse residential mobility in Hungarian cities since the demise of socialism. This survey was focused on housing estates as one of the most problematic legacy of the socialist era on account of their quantitative importance and their specific physical character.[1] Particular attention was thereby given to the extent and selectivity of mobility in the transformation process to date and the extent of residential mobility to be expected in the future.

Subsequent to a brief overview of methodology and approach, the following presents the findings of the empirical survey on mobility in Hungarian housing estates up to 1995. A further section discusses the evaluation of housing types, households' satisfaction with their housing situation, plans to move and thereby the overall question of future mobility. The article concludes with a summary and an outlook.

---

[1]   SZELÉNYI (1996) even predicts that peripheral housing estates shall be the slums of the 21st century.

2   Methodology and approach

Mobility processes in the course of transformation were examined in six Hungarian cities, selected by means of a factor and subsequent cluster analysis, in the years 1994-1996.[2] Characteristic structural features on housing supply resulting from the most recent census in Hungary in 1990, including data on tenure, housing types, quality of dwellings, dwelling and building size and age of building stock, were processed in the factor and cluster analysis. A total of six clusters were identified with cities showing considerable variances in their housing supply (figure 1). Cluster 1, which showed a very uniform housing market in 1990, with more than 90% of dwellings in single-family houses, was excluded from further investigation.

Table 1 shows a number of key values recorded in the surveyed cities to illustrate the central features of housing supply in the municipalities which comprise the various clusters. Szentendre was chosen to represent the rapidly growing suburban communities in the Budapest agglomeration (Cluster 2). In these cities with a population of mainly 15 000-25 000, 10-15% of all dwellings were state rentals in 1990, with single-family houses dominating the private sector. The cities in cluster 3, with population levels below 30 000, lie mainly in the eastern and southern Great Plain. State rentals account for 10-15% of all dwellings in these cities. Single-family houses again dominate the private sector, with a higher proportion of old building stock and a lower percentage of larger dwellings than in cluster 2. Kiskunfélegyháza was selected to represent this cluster.

Cluster 4 comprises cities which are very different in size with population figures ranging between 20 000 and 200 000. Their commonalty is the importance of the administrative function, in view of which these cities enjoyed a comparatively high inflow of state funding for the construction of flats in housing estates in socialist times. As a result, cluster 4 cities were characterised by an overproportional percentage of state rentals and a high proportion of condominiums in the private sector in 1990. Given the wide range of the population levels of these cities, Debrecen and Szombathely were chosen as the representative cities.

---

[2]   The preliminary categorisation of the housing sub-markets per size of municipalities showed that in municipalities with a population under 10 000, the percentage of privately owned dwellings already stood well above 90% in 1990, making the question of housing privatisation, another focal topic of the study, of negligible importance in these municipalities. These municipalities were therefore excluded from the factor and cluster analysis. Similarly, Budapest was not included in the survey, since the majority of the - relatively few - studies on privatisation and residential mobility in Hungary have focussed on the capital. The purpose of this empirical study was to gain an insight into post-socialist transformation processes in cities on the levels of the national urban system below the capital and in 'socialist' industrial cities.

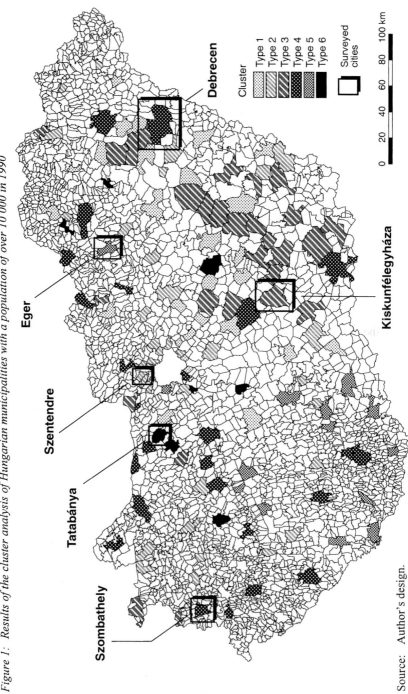

Figure 1: Results of the cluster analysis of Hungarian municipalities with a population of over 10 000 in 1990

Source: Author's design.

Table 1: Selected features of the housing stock and the population of the surveyed cities (1990)

| | Ownership types of dwellings (%) | | | |
	single-family h.	Condominiums	state rentals	Inhabitants
Debrecen	35.2	36.2	28.6	212 200
Szombathely	32.4	37.6	30.0	85 600
Tatabánya	16.4	28.6	55.0	74 300
Eger	34.4	42.2	23.4	61 900
Kiskunfélegyháza	58.7	28.5	12.8	34 200
Szentendre	55.7	25.4	18.9	19 400

Source: Census 1990, Hungarian Central Statistical Office (KSH).

Eger was chosen to represent Cluster 5. This cluster is comprised of a small number of very heterogeneous cities, their common denominator being a particularly high volume of old building stock since city expansion in the form of housing estates commenced at a date later than in the cities of the other clusters. The proportion of state rentals in cluster 5 lay in the 15-25% range in 1990.

Tatabánya represents cluster 6, consisting of typically 'socialist' industrial cities. In 1990, these cities were characterised by an overproportional percentage of state rentals - at least 50% of overall housing stock. These cities which expanded rapidly in the post-1945 period show very little variety in terms of their housing form, age of housing stock and dwelling quality.

In the period October 1994 - May 1995, comprehensive surveys involving interviews on housing privatisation and residential mobility were conducted among the residents of housing estates in the surveyed cities. The survey was based on systematic random samples with refusal rates of approx. 20%. A total of 2 018 households were surveyed in the six cities. Experts were additionally interviewed from the national ministries of finance and interior, the departments responsible for housing policy in the post-socialist period; interviews were also conducted with municipal mayors, heads of administrations and housing experts.

3   Residential mobility

3.1   Extent of residential mobility

The housing estates of the surveyed cities already showed considerable fluctuations in the period 1990-1995, when at least one change of address occurred in approx. every four to five dwellings. Depending on the year of move and the city in question, the proportion of moving households reached values of between 2%

and 8% in one year (table 2).[3] In a study conducted in Budapest, DOUGLAS (1997) shows that 21% of the surveyed households moved into their dwellings in the 1990-1995 interval, with values lying between 15% and 27% in the housing estates. A residential mobility of some 20% can therefore be regarded as characteristic of Hungarian cities in the five-year interval 1990-1995.

Table 2: Moves of households in housing estates

City	pre-1990	1990-1994	1990	1991	1992	1993	1994	N
			Rates of moving (%)					
Debrecen	73	28	2.6	4.9	6.1	5.5	8.1	346
Eger	75	25	2.9	2.9	5.1	6.1	7.7	311
Kiskunfélegyháza	76	24	3.7	2.9	5.5	4.6	6.9	348
Szombathely	80	20	3.1	2.8	3.7	4.0	6.5	353
Szentendre	79	21	4.9	3.2	4.1	5.5	3.5	344
Tatabánya	79	21	3.5	6.0	4.1	4.4	3.2	316

Source: Author's survey.

These Hungarian mobility rates can be compared with a series of corresponding values from other countries: LONG (1991) quotes mobility rates of some 50% for the first five-year interval of the 1980s in western industrialised countries where the housing market is characterised by a relatively little state interventionism and the cost of housing property is relatively low (e.g. USA, Canada, Australia), and values of 36% and 20% for Switzerland and Austria, as countries representing those with a higher degree of state regulation (restrictions on rent increases, well protected tenants' rights) and higher housing property costs.[4] LEE and STRUYK (1996) calculate mobility rates of 4.6% for seven major Russian cities and 4.5% for Moscow in 1994. According to Habitat-World Bank, annual mobility rates at the beginning of the 1990s stood at 4.4% in Budapest, 2.5% in Warsaw, 9.2% in Munich and 6% in Paris (LEE and STRUYK 1996). Studies in housing estates in other post-socialist states show very little fluctuation in the first years of transformation - despite the fact that in principle many residents plan to move (e.g. MAIER 1997; WECLAWOWICZ 1997).

These mobility rates show that in comparison with other post-socialist states, the general process of transformation in Hungary has been accompanied by a considerable mobility transformation, with patterns of mobility now approaching

---

[3]   Residential mobility is assessed by the number of the changes of address within a one-or five-year interval. Since this information is generally obtained from household surveys on the year of move, the numbers may be underestimated as additional changes of residence may have occurred prior to the move of the surveyed household in question (LONG 1991).

[4]   These figures are overall national values; higher mobility rates can be assumed for urban areas.

those observed in Western Europe. This process clearly reflects the specific parameters set by Hungarian housing policy since the demise of socialism, on the one hand, and economic and social upheavals, on the other. For the 1980s, HEGEDÜS and TOSICS (1991) and LOWE and TOSICS (1988) recorded very low mobility rates as a consequence of the prevailing housing shortages of the time, in particular in the housing estates whose residents had little opportunity to move into another housing sub-market and found themselves locked in the housing estates.

Considerable variations can be observed in the changes in address since 1990, both within the various housing estates and within the housing units of a given estate (figure 2). As more detailed statistical analyses show, residential mobility in the rental sector and in housing estates with a negative image is significantly higher than in owner-occupied units and in estates which tend to have a more positive image. Higher fluctuations can also be observed in single-room flats.[5]

The housing estates with a negative image include the more recently built peripheral neighbourhoods with a high proportion of high-rise apartment blocks and poor infrastructure whose open spaces were only provided with rudimentary green areas - sparse lawns and the odd bush or tree - as funding became scarce in the 1980s. Other areas in this category are the older housing estates close to the inner cities dating back to the early stages of industrial construction with a very compact and monotonous character, neighbourhoods often described by respondents as 'soul-less'. Housing estates with a positive image tend to be those located in the vicinity of the inner cities with a more varied architectural character, green areas and a better infrastructure.

As illustrated by the annual rates of moving, there has been a marked increase in residential mobility in the housing estates of the surveyed cities since the beginning of the transformation process (table 2)[6] - although it should be pointed out that mobility rates may have been underestimated in the immediate post-socialist period. Fluctuations increased in both the rental and owner-occupied housing stock in the 1990-1995 interval, with overall fluctuations higher in the rental sector. Nevertheless, changes in residence in the owner-occupied sector in the estates also indicate - contrary to the expectations of certain authors - that the high proportion of privately-owned units in housing estates in Hungary does not in principle imply residential stability.

---

[5]   Logistic regressions were calculated for each city on the basis of the following characteristics: year of move, tenure, dwelling and building size, period of construction and image. Only tenure and image are statistically significant for the date of move in all cities; in Debrecen and Szombathely the number of rooms is also statistically significant, it being emphasised that these samples cities included a higher number of small flats.

[6]   Mobility studies by LEE and STRUYK (1996) also record a rise in mobility in large Russian cities since the early 1990s, albeit lower than in Hungary.

Figure 2:   Moves according to tenure and housing estate image

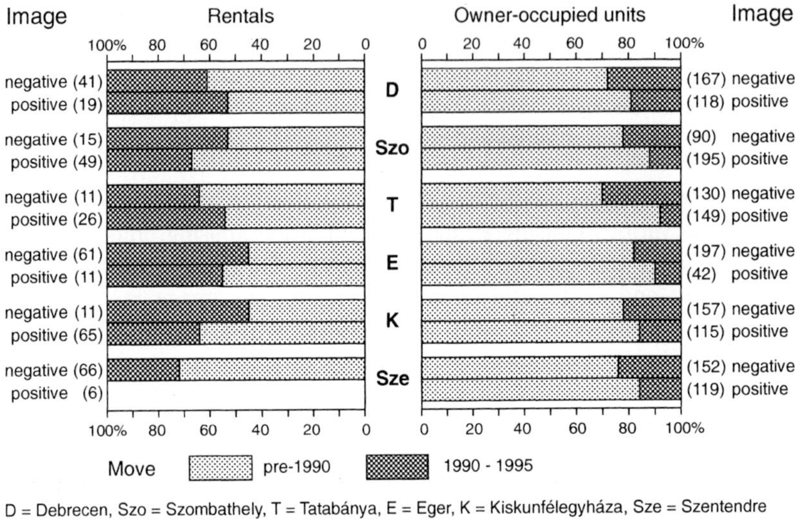

D = Debrecen, Szo = Szombathely, T = Tatabánya, E = Eger, K = Kiskunfélegyháza, Sze = Szentendre

Source: Author's survey.

This increase in mobility is not however to be observed in either the rental or in the owner-occupied sector in Szentendre and Tatabánya. In Tatabánya this can largely be explained by the fact that the urban housing market offers very little variety and households dissatisfied with their current housing situation have little opportunity of moving to a different dwelling in terms of size, form or cost. The housing market situation in Szentendre is very tight, with the migratory pressure from Budapest pushing up real estate prices in the housing sector, so that it is extremely difficult for even the better-off to move.

A major explanation for the overproportional increase in mobility in the rental sector is the fact that in all the surveyed cities - with the exception of Tatabánya and Szentendre - the reimbursement of the downpayment to tenants moving out of their state-rented units was pushed up in the first half of the 1990s.[7] Tenants were refunded up to twelve times their downpayment in Debrecen; twenty times their annual rent in Kiskunfélegyháza and 50% of the estimated market value of their dwelling in Eger and Szombathely. Experts also emphasise that due to the tight budgetary situation of the local authorities, no new municipal rented housing projects can be financed in the present context; this means that the 'buy-back' of rent-

---

[7]   The 1983 Housing Act stipulated that tenants moving out of state rentals were to be refunded three times their original downpayment; according to 1985 legislation, the level of the refund may be set by the local authorities (LOWE and TOSICS 1987).

als is the only affordable solution for municipalities wishing to increase the rental stock to cover housing emergencies.

Experts rate this 'buy-back-strategy' as relatively successful. On the one hand, it has been accepted by better-off households who have used the reimbursement from their local authority as starting capital to purchase a home of their own; on the other hand, the less well-off are attracted by the relatively high reimbursement which often triggers their decision to give up their now unaffordable full-comfort unit in a housing estate and move into less expensive units with poorer amenities, double up with other family members or move to rural areas where the cost of living is generally lower. Moreover, recent fluctuations have been kindled by the ongoing public debate on plans to scale back the quasi-ownership rights of tenants and further rent increases.

An important factor strengthening the relatively high fluctuations in owner-occupied units are the favourable repayment terms for housing loans contracted in the socialist era - prompting many housing estate property owners to pay off the outstanding balance on their loans.[8] The strategic position of these loan-free owners in the housing market has now considerably improved, so that better-off owners now have a greater opportunity to move into their preferred housing sector. In view of the overall slump in housing construction in Hungary in the 1990s, owner-occupied dwellings from the existing stock are increasingly being integrated into processes of mobility. In the 1980s, HEGEDÜS and TOSICS (1991) show that higher-status groups from housing estates primarily tended to move into the newly constructed smaller condominium complexes. It can generally be assumed that high mobility rates imply a high rate of vacant dwellings and a high level of housing construction (inter alia CLARK et al. 1986). The decline in housing construction in the 1990s has further aggravated the housing shortages inherited from the socialist era.[9] The classical filtering process via new housing stock which predominated under the old regime has largely been replaced by a filtering process in the existing stock in the post-socialist era.

As emphasised by all the experts, less well-off households have also been moving out of their privately-owned homes in housing estates where they can no longer afford the high utility costs. Like low-income tenants, these groups are

---

[8] In 1991 two alternatives were offered to those having contracted a housing loan at a fixed interest rate of 3% prior to 1989: either a rise of the fixed interest rate to 15% for the remaining term of the loan, or cancellation of one half of the outstanding amount, coupled with repayment of the remaining half at variable market interest rates. 78% of borrowers opted for the second alternative, of which 50% immediately paid back the outstanding balance (HEGEDÜS, MARK and TOSICS 1996, 88).

[9] Whereas the number of households in Hungary still exceeded the number of dwellings by 5.5% in 1990, the values recorded in the surveyed cities lay in the 2%-7% range; in view of the drastic decline in the construction of new housing and the legal and illegal conversion of dwellings into business premises, the quantitative housing shortage has exacerbated during transformation, in particular in the larger cities.

moving into the less expensive owner-occupied or rental sectors, doubling up or migrating back to rural communities. This category includes a high proportion of senior citizens selling their homes to support younger family members in solving their housing problems.

3.2   New residents and their origin

An examination of the households moving into housing estates since 1990 shows considerable variations compared to the socialist era. Distinct restructuring processes are also evident in the communities of origin of the households moving into estates since 1990 and in the housing sub-market of origin in the context of intra-urban moves.

3.2.1   The new residents

An initial approximate classification of the surveyed households according to basic demographic features shows that those moving into the housing estates in all the surveyed cities are primarily families with children and childless couples (figure 3). This is true of both rental and owner-occupied dwellings.

Figure 3:   Demographic structure of households moving into housing estates since 1990

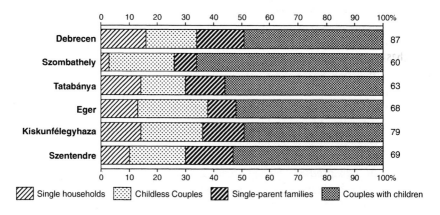

Source: Author's survey.

A more detailed classification according to further characteristics such as household income and age structure, however, reveals general variances between the households moving into rental and owner-occupied dwellings since 1990.[10] In all four household categories (figure 3), more households have moved into rentals in

---

[10]   These statements cannot be statistically corroborated on account of the low case numbers per housing sub-market per city.

the household foundation and expansion stage, with a high percentage of adults below 29 years of age in this group. In contrast, more households have moved into the owner-occupied sector in the stagnation or shrinking stage, with a high percentage of adults aged 50 plus and senior citizens in this category. Within the respective demographic household type, the economic status of the households, expressed in terms of their income, tends to be less favourable among those moving into rentals. However this is not true of senior citizen households who do not dispose of a higher regular income than their counterparts moving into rentals, despite the fact that they have moved into their own property.

Along with the differences in age structure, this demonstrates that - in contrast to the socialist era - more older households are now purchasing their own properties in housing estates. Under the old regime, new owner-occupied stock in housing estates was primarily sold to young households at favourable, highly state-subsidised conditions via the national savings bank on the basis of waiting lists.[11] Since only isolated fragments of the housing estate projects have been constructed since 1990, the stock of affordable owner-occupied units for young households has substantially diminished.[12]

Since the beginning of the transformation process, owner-occupied dwellings in housing estates up for sale have been units from the existing stock offered at market prices as private transactions with no state subsidies available. Virtually no Hungarian household can now finance a purchase of this kind from its regular income or even by taking out a housing loan, with the normal market interest rates of 20-30% being applicable since the early 1990s. This means that those interested in purchasing housing property must either already be in possession of a property - generally older households - which can be used as start-up capital for the purchase of their next dwelling - or receive generous financial support from other family members.

3.2.2   Origin of the new residents

Characteristic of the period of transformation in all cities - with the exception of Tatabánya - is a marked increase in the percentage of inter-city moves, coupled with a decline in the percentage of intra-urban migration into the housing estates (figure 4). In most of the surveyed cities, changes of residence occurring since 1990 have been involved inter-city-movers in at least every fourth dwelling. The steep rise in the percentage of inter-city movers is exemplary of the fact that the housing estates are turning the central urban gateway in many cities, a phenomenon already evident in the first years of transformation. Tatabánya is the only city in which the percentage of inter-city movers has not increased since the demise of

---

[11] State subsidies, particularly important for young households, were tied to new housing projects and not awarded for the acquisition of dwellings from the existing stock.

[12] Moreover, state subsidies for private housing since the transition has been focused on the acquisition of dwellings in new owner-occupied housing projects.

socialism; here again, intra-urban moves have to be primarily between housing estates given the lacking variety of the housing market in socialist industrial cities.

Although Debrecen also shows a marked increase in the proportion of inter-city movers since the beginning of transformation, this percentage is significantly lower than in the smaller cities in the survey - as was also the case in socialist times. This can be explained by the considerable volume of old housing stock with generally poor amenities serving as an important urban gateway for inter-city movers. The high percentage of inter-city movers recorded in Szentendre in both socialist and post-socialist times is a result of the high migratory pressure from the national capital.

Figure 4:   Percentages of inter-city moves to housing estates

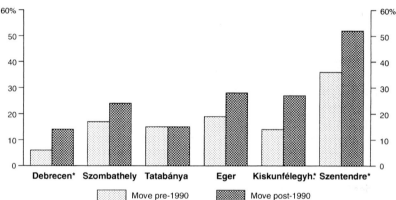

* Significantly different across inter-urban and intra-urban moves pre-1990 and post-1990 at 0.05 level using chi-square test

Source: Author's survey.

The dynamic development of the housing estates into major urban gateways, so characteristic of the post-socialist era, is a consequence of the contraction of state allocation, combined with a general loss of attraction. As under socialism, state-rentals are generally only allocated to households resident in the city for at least five years. Under the old regime, state-financed condominium projects in housing estates were generally sold to households resident in the city for several years. As a result of the rapid privatisation of state-rentals as well as the withdrawal of the state from the construction of condominiums and rentals, very few units in the housing estates remain under the state allocation system.[13]

---

[13]   As a result of the very favourable conditions of sale and tenants' right-to-buy, introduced in 1994, and with the general economic and social upheavals, the previous state-rentals have largely been sold. Whereas 18% of all dwellings in Hungary were state-rentals in 1990, this percentage had shrunk to approx. 4% by 1998.

Moreover, a portion of the state-rentals are now leased at normal market conditions to the highest bidders, with no consideration of social circumstances or aspects such as length of urban residency.[14] Illegal subletting of state-rentals and the number of privately-owned rentals dwellings in housing estates have also increased.[15] Exorbitantly high rents are charged in all these cases. Inter-city movers with very limited information on and access to other urban housing sub-markets initially set up house in these expensive rentals in the housing estates as an interim measure.

Minimal rents, a high level of comfort and the lack of alternatives in the urban housing markets made the housing estates much sought-after residential areas in socialist times. Their advantages outweighed their disadvantages - characteristics always regarded as negative, e.g. small size, monotony and the sad lack of neighbourhood quality (SZELÉNYI and KONRÁD 1974). Better-off households aspired not only towards owner-occupied dwellings but also the low-cost rentals since the quasi-ownership rights of tenants under socialism meant that a rented dwelling could be inherited, sold or exchanged. In general, moving into a rented or owner-occupied dwelling in a housing estate was a means of acquiring a value asset under very favourable conditions which could later be used as a springboard to a higher-value housing sub-market, e.g. a condominium in a small-scale multi-family house or a single-family home (HEGEDÜS and TOSICS 1983).

Price liberalisation policies have triggered an overproportional increase in utility costs since 1990,[16] especially in the case of full-comfort dwellings, the category to which most of the housing stock in the housing estates belong. Changing cost ratios have eroded the previous economic attraction of the housing estates. As a result, less well-off households now seek dwellings in less expensive housing sub-markets outside the housing estates. And even the better-off households now

---

[14]   In all the surveyed cities, state rentals are partly no longer leased as social rented housing but rented under limited lease at normal market conditions. This means a further contraction of the already small quota of state rentals available to the less well-off households; however local authority experts stress that this makes an important contribution to the funding of the urgently required maintenance and rehabilitation costs in the social rented stock in terms of cross-subsidisation.

[15]   According to the 1996 microcensus results, some 15-20% of dwellings in housing estates were occupied as rentals in the surveyed cities, of which no more than at most one half were state-rented dwellings; on account of the low case numbers, the microcensus data have only been evaluated for the cities of Debrecen, Szomabthely, Tatabánya, Eger and Kiskunfélegyháza. According to experts, a 'real' private rental sector, with the deliberate purchase of a dwelling as a form of capital investment, has not yet developed in Hungary; experts regard the emergence of private rentals as accidental, e.g. as a result of an inheritance or cohabitation leading to a dwelling becoming vacant.

[16]   Electricity costs alone climbed 231% between 1990 and 1994, heating oil 176%, hot water 333%, sewage 380% and running water 347%. Prices for food, clothing and consumer durables rose 50%-150% in the same period (HUNGARIAN CENTRAL STATISTICAL OFFICE 1995).

U. Sailer

attempt to move directly into their preferred housing sub-markets and thereby into the sector of single-family houses or condominiums in smaller multi-family houses, without the previously customary 'detour' through the housing estates. This is rendered possible since the less well-off households, faced with shrinking incomes and impoverishment, are now forced to move out of these sub-markets into cheaper dwellings.

The overall demand for dwellings in housing estates from households already resident in the city has declined considerably. In comparison with inter-city movers, long-standing urban residents generally have better sources of information on and access to the various urban housing sub-markets. This specific decline in demand has decisively facilitated the access of inter-city movers to the housing estates.

3.2.3   Intra-urban moves into the housing estates

The processes resulting from the decline in the attraction of housing estates and modified municipal allocation opportunities have left their mark on intra-urban moves, clearly distinct in the 1990s from the socialist era. In schematic form, figure 5 shows the answers given to the question "In what type of dwelling did you live before moving to this housing estate?" for the pre- and post-1990 periods, whereby only intra-urban moves have been taken into account.

In comparison with the socialist era, a substantially lower number of households with no previous dwelling have moved into condominiums in the housing estates during transformation. At least every other household moving into condominiums prior to 1990 previously had no dwelling, compared to only approx. one third of those moving in post-1990. This means that starter households are now much more dependent on the rental sector than in socialist times. Moreover, the success of housing privatisation has rendered rental flats a rare quantity in Hungarian cities, greatly exacerbating structural housing shortages in the transformation period. A substantially higher number of households have moved out of condominiums or single-family houses into the privately-owned condominiums in the housing estates during transformation. The main reasons for this phenomenon have already been discussed above.

Accordingly, a substantially higher number of households with no previous dwelling have moved into housing estate rentals since 1990.[17] Prior to 1990, approx. 25%-30% of households moving into housing estate rentals and approx. 50% of those moving post-1990 had no previous dwelling. Moreover, in view of the changing cost ratios (see above), a substantially lower number of households have moved from other rented dwellings since 1990. In comparison to the social-

---

[17]   Rented dwellings here include currently those owner-occupied dwellings which according to the resident were purchased from the state and were therefore still a rental at the time of move.

ist era, post-socialist mobility in the housing estates is therefore generally characterised by a greater degree of segmentation of the population structure according
to condominium residents, on the one hand, and those living in rentals, on the
other.

Figure 5:   Intra-urban moves into housing estates prior to and during transformation

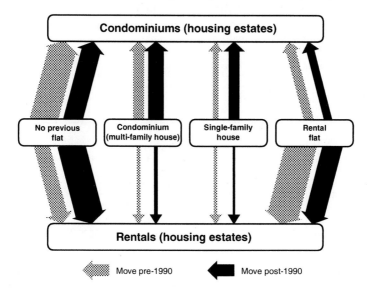

Source: Author's survey.

4   Future mobility

In order to estimate future mobility in general and, more specifically, future mobility between the urban housing sub-markets, respondents were asked to give
their personal evaluation of classic housing types, to indicate their degree of satisfaction with their current housing situation and to reveal any intentions they had
to move, with the reasons for moving. Although no direct conclusions on future
mobility can be drawn from these answers, since the current economic situation
will presumably prevent many of these households from moving into their preferred housing sub-market, an insight into these aspects nevertheless allows a
number of fundamental statements on probable trends in the extent and direction
of future mobility.

344   U. Sailer

4.1   Evaluation of housing types

Figure 6 lists the average ratings given by respondents to classic housing types in Hungarian cities on the basis of the grades used in the national education system, with 5 representing the top and 1 the bottom grade.[18] In all the surveyed cities, the only housing types rated with above-average marks are those in the privately owned sector. With an average rating of almost 5.0, single-family homes are the top preference, followed by condominiums in small multi-family houses and, thirdly, condominiums in housing estates.

Figure 6:   Evaluation of classic housing types in Hungarian cities

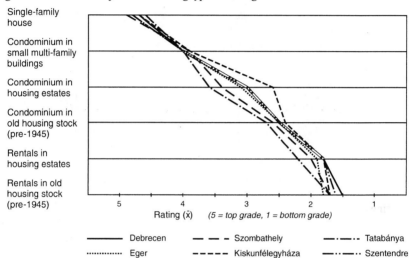

n: Debrecen 335, Szombathely 349, Tatabánya 312, Eger 308, Kiskunfélegyháza 344, Szentendre 342

Source: Author's survey.

Dwellings in old housing stock and the rental sector in general are rated very negatively in all the cities.[19] During the socialist period, rentals in housing estates tended to be given a similar rating to that of condominiums in housing estates.[20] The very poor rating currently observed clearly demonstrates the considerable downgrading of rentals as a consequence of the housing policy measures in the

---

[18]   Households were asked to rate the indicated housing forms on the basis of one of these grades.

[19]   Given the specifities of housing supply in Hungary, the term 'rentals' applies to the state-rental sector in everyday language.

[20]   LOWE and TOSCIS (1988) on this point: "For most sections of Hungarian society renting is still a desirable option. This is because rents are low, rented flats have limited capital value, and the tenancy is very secure. In physical terms there is no difference between owner-occupied and rented new state built flats."

course of the transformation process. In view of their massive state of decay and the increasing proportion of socially disadvantaged residents, dwellings in old housing stock were already rated very negatively during the socialist era. This highlights the as yet unresolved and fundamental problem of urban planning: how can the old housing stock be architecturally and socially upgraded on a comprehensive scale with such a negative perception as the point of departure? In general, the ratings given to the rentals indicate a future scenario whereby primarily lower-income households with no other alternative remain or move into these housing sub-markets, thereby exacerbating the marginalisation of these neighbourhoods. However this scenario does not apply to those dwellings in old housing stock, much sought-after even in pre-socialist times on account of their size, location and neighbourhood characteristics, and now characterised by emerging pockets of gentrification (KOVÁCS 1998).

The average ratings given to condominiums in housing estates - ranging from 2.6 to 3.6 in the various cities - reveal that housing estates have not yet become completely devalued as residential areas. Depending on the city, 20%-55% of the surveyed households give this housing form one of the top two ratings. The differences observed in the ratings for the various cities are exemplary of the specific situation in the respective urban housing markets. In Tatabánya, e.g., where the urban housing market predominantly comprises housing estates, as many as 55% of surveyed households give condominiums in housing estates ratings of 4 or 5. In contrast, in Kiskunfélegyháza, a small city with only one very compact peripheral housing estate with a negative image, this category is only given one of the top two ratings by 22% of respondents. Although housing estates with a negative image are also to be found in the other cities, these cities also have estates whose location, architectural characteristics and infrastructural amenities assure them a positive image so that households in general have a more positive perception of housing estates as a whole.

Although this statement cannot be statistically corroborated because of the low case numbers, a more differentiated analysis points to distinct demographic and socio-economic trends. An above-average number of families with children give single-family houses and condominiums in smaller multi-family houses one of the top two ratings. More childless couples and single households rate condominiums in housing estates with one of the top two positions than the other household types. It can be stated for all demographic household types that higher-income households give the top two ratings to single-family houses and condominiums in smaller multi-family houses.

It can be concluded from this analysis that - insofar as possible - the majority of current housing estate residents would move out of the housing estates and into the single-family houses sector or into the housing sub-market of condominiums in smaller multi-family houses. These include an overproportional number of families with children and above all better-off households. A higher degree of demographic and social selectivity among the remaining residents of housing estates can therefore be assumed.

4.2   Satisfaction with current housing situation

To evaluate residents' satisfaction with their housing situation, respondents were asked to qualify their satisfaction with various features of their dwelling, neighbourhood characteristics and intra-urban location.

Figure 7 shows that almost one half of all surveyed households - of which an overproportional number of households with children - are dissatisfied with the size of their dwelling. In contrast, approx. three quarters of childless couples and approx. 90% of single households are satisfied with their number of rooms and living space. The ratings given to dwelling size clearly reflect actual housing conditions: one half to two thirds of the surveyed couples with children live in a two- or one-room flat, i.e. with a living space of less than 60m².

A relatively high level of satisfaction with the quality of dwellings can be observed in all the cities. This shows that the typical classification of dwellings in housing estates as 'comfort cells' during the socialist era remains prevalent in Hungary in the early years of transformation.[21] Although this cannot be substantiated on account of the low case numbers, it can again be observed that families with children are particularly dissatisfied with the quality of their flats. Respondents mention that their tiny kitchens and bathrooms render the day-to-day organisation of a family life very difficult.

Social composition is also assessed very positively - in particular in contrast to corresponding ratings in eastern Germany (see SAILER-FLIEGE 1998). A general devaluation of the socialist housing estates on account of social composition has not yet occurred in Hungarian cities. As a result of the regressive housing allocation, many housing estates were still 'middle-class' neighbourhoods at the end of the socialist era. It is to be emphasised that in particular in housing estates with a strong social mix - in principle a positive characteristic in terms of socialist ideology - more better-off households are dissatisfied with social composition. In the surveyed cities the housing estates in question were above all the more recently built housing estates on the urban fringe with a high percentage of state rentals prior to privatisation. As a result of a wider variety of housing supply in the 1980s and the increasing allocation of rented dwellings according to social criteria, more and more working class households moved into these housing estates in the 1980s. This demonstrates that the more affluent households clearly aspire towards social distinction. With the widening of the social spectrum and an easing of the tension

---

[21]   In contrast to perceptions in eastern Germany where as a result of the rapid development of new housing projects with all modern conveniences, satisfaction with the amenities of the 'comfort cells' in the housing estates has declined considerably (SAILER-FLIEGE 1998).

in the housing market, social erosion can therefore be expected, above all in those housing estates which currently show a more heterogeneous social composition.

Ratings of neighbourhood characteristics, intra-urban location and distance-related features show no striking differences as a function of the socio-demographic structure of the households. Variations tend to relate to distance from the city centre and the architectural characteristics of the estates. In all the cities, residents of the more recently built housing estates on the urban fringe tend to be less satisfied with distance-related features.

Figure 7:   Satisfaction with selected housing features

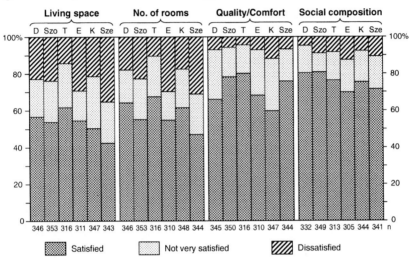

D = Debrecen, Szo = Szombathely, T = Tatabánya, E = Eger, K = Kiskunfélegyháza, Sze = Szentendre

Source: Author's survey.

4.3   Plans for moving

Considerable fluctuations in the housing estates can be expected in the future. Almost every second household in the surveyed cities plans to move. The only exception is Tatabánya where only approx. one in every three households intends to move; this again reflects the lack of alternatives in the other housing sub-markets in this city. It is also of particular interest to note that in all the cities at least every second to third household living at its current flat since 1990 also plans to move. In fact, even 30-60% of households living in their current dwelling for only one year intend to move again.

There are no fundamental differences to be observed between renting and condominium households in terms of intentions to move. Despite the high percentage of condominiums, residential stability in the housing estates is therefore not to be expected (cf. also KOVÁCS and WIEßNER 1999; BEREY 1997). It is generally assumed that renters are more ready to move and indeed move more often than owner-occupiers (e.g. QUIGLEY and WEINBERG 1977). Alongside the stronger emotional ties to a home of one's own, the transaction costs involved in moving from one rental to another are generally lower. Since there are very few rentals left in Hungarian cities, the majority of renters planning to move generally have no alternative but to buy a condominium. Such a move involves high transaction costs - whereas in the case of a household moving from one condominium to another, the costs are comparatively low since they can be offset by sale of the previous dwelling.

An overproportional number of families with children, in particular those in the higher-income bracket, intend to move out (figure 8). But childless couples and even single households, mostly senior citizens, also show a high willingness to move. Households intending to move include both the lower- and higher-income brackets, whereby the less well-off households willing to move tend to be above all single elderly households, older childless couples and to a certain extent single-parent families.

Figure 8:   Percentages of households in housing estates planning to move

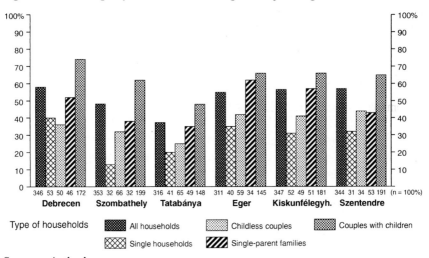

Source: Author's survey.

Every five to three household unwilling to move is in a precarious financial situation with a net income below 20 000 Forint at the time of the survey. If the wishes of even only some of those intending to move are fulfilled, the result will be an increasing concentration of low-income households in the housing estates. This poses a serious problem for social and urban development policy: social marginalisation implies not only spatial but also architectural marginalisation as many of the remaining residents will not be able to afford the high maintenance and renovation costs.

Households planning to move were also asked to name the principal reasons for moving. Their answers are a clear reflection of the fundamental problems of housing estates from the residents' point of view: the small size of dwellings, lack of quality, building deficiencies, financial reasons and a negative image of housing estates as a whole are of particular significance in all the surveyed cities (figure 9). The wish for housing property in the form of a single family house with a garden or as condominium in smaller multi-family houses is also often quoted as an important reason for moving, particularly by owner-occupiers. Financial reasons include excessively high utility costs, high rents, the desire to assist children with the sale proceeds or fear of rent increases. The negative image of housing estates is reflected in comments such as "the building dimensions are inhuman", "far too many people crowded into such a small space", "I feel so isolated", "it's so anonymous" or "not enough greenery". These features have already been identified as the essential negative elements of housing estates by SZELÉNYI and KONRÁD (1974).

In contrast to the findings of housing estate studies in Western Europe and eastern Germany, relatively few households mention social composition as an important reason for moving. The fact that the current socio-demographic structure of housing estates in Hungarian cities is still characterised by an overproportional number of middle-class households presumably has a role to play in this regard.

Recent developments in eastern Germany show that the state of the building and the quality of the flats can be decisively upgraded by comprehensive renovation measures. Although not always technically feasible, at least one specific problem of socialist housing estates - the preponderance of small two-room dwellings - can at least be mitigated by merging existing flats. However, as international experience shows, the fundamental problems of housing estates - their overall character - can only be partly alleviated by large-scale architectural improvements or by the demolition of the tallest high-rise buildings. In view of households' distinct preferences for specific housing types outside the housing estates, the implication is that even if extensive and extremely cost-intensive measures were feasible, the selective exodus and therefore the filtering down process in the housing estates can only be mitigated in the housing estates but not actually halted.

Figure 9: Main reasons for moving out of housing estates

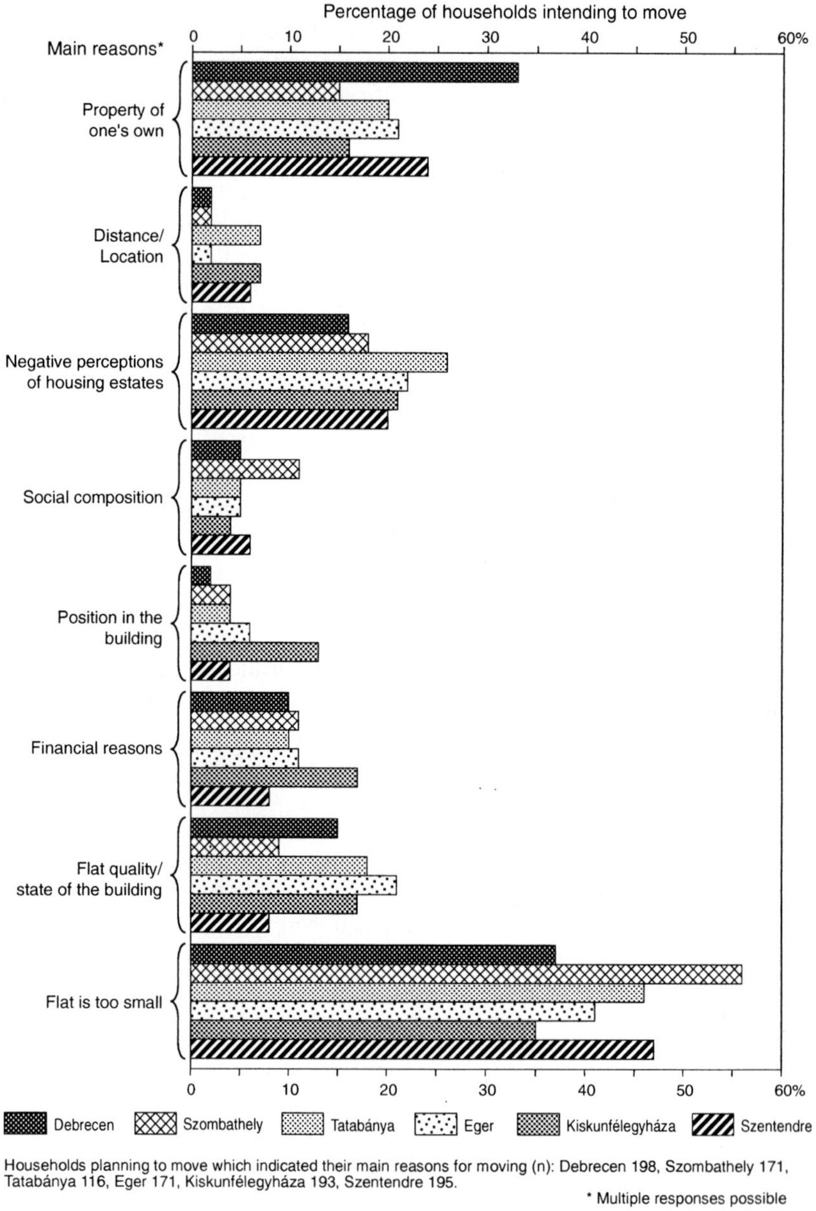
Percentage of households intending to move
0
10
20
30
40
50
60%
Main reasons*
Property of one's own
Distance/ Location
Negative perceptions of housing estates
Social composition
Position in the building
Financial reasons
Flat quality/ state of the building
Flat is too small
Debrecen
Szombathely
Tatabánya
Eger
Kiskunfélegyháza
Szentendre
Households planning to move which indicated their main reasons for moving (n): Debrecen 198, Szombathely 171, Tatabánya 116, Eger 171, Kiskunfélegyháza 193, Szentendre 195.
* Multiple responses possible

Source: Author's survey.

5 Summary and outlook

The empirical study with the objective of examining residential mobility in Hungarian cities during the transformation process focused on housing estates as the most problematic legacy of the socialist era. Comprehensive household surveys were therefore carried out in six Hungarian cities with different characteristics in terms of housing supply and interviews were conducted with experts at national and municipal levels.

The study's findings demonstrate that a considerable restructuring of residential mobility has already occurred in the comparatively short period of transformation to date. In particular in comparison with other post-socialist countries, the general transformation in Hungary has been accompanied by a considerable mobility transformation. The low mobility rates so typical of the socialist era have been replaced by a dynamic rise in mobility. This shows that the very high percentage of owner-occupied dwellings has not yet led to a fundamental residential stabilisation in Hungarian housing estates. Residential fluctuations are particularly high in the remaining rentals, in smaller apartments and in housing estates with a negative image which include the more recently built peripheral neighbourhoods and the older estates lying closer to the inner cities with a very compact and monotonous architectural character.

This dynamisation of residential mobility has occurred despite the sharp decline in housing construction and the housing shortages inherited from the socialist past. This means that the classical filtering process via new housing stock which predominated under the old regime has been largely replaced by a filtering process in the existing stock since transformation. Better-off households are now moving into more highly valued sub-markets into dwellings which have become too expensive for their previous residents in account of changing cost structures. Alongside this, low-income households are also moving out of the housing estates where they can no longer afford their full-comfort dwellings due to the dramatic rise in utility costs.

A further characteristic of post-socialist mobility in housing estates is a marked increase in the percentage of inter-city moves. In the wake of the general loss of attraction and the sharp decline in state allocation opportunities, for inter-city movers housing estates have turned into major urban gateways in the post-socialist period. Post-socialist mobility development in housing estates is also characterised by a much greater segmentation of the population according to condominium residents, on the one hand, and those living in rentals, on the other.

The results of the evaluation of classic housing types, satisfaction with current housing situation and intentions to move also show that a high level of residential mobility can be expected in the future with no fundamental differences to be observed between renting and condominium households in terms of intentions to move.

Although the wishes of all of those households intending to move will undoubtedly not be fulfilled, future mobility in the housing estates will be characterised by a considerable selectivity along demographic and socio-economic lines. The housing estates inherited from socialist times, including in particular those with lower residential value, will increasingly become 'transit points' for better-off households and families with children. The remaining households will in general tend to be those in the lower income bracket. This implies an increasing concentration of low-income households in housing estates - a factor which is highly problematic from the points of view of social and urban development policy.

Households' principal reasons for moving are a clear reflection of the fundamental problems of housing estates. As all international experience shows, the negative characteristics of housing estates which stem from their overall physical character can only partly by alleviated by large-scale architectural improvements. Although in the medium term the housing stock in the housing estates will remain indispensable for large sections of the population, especially in the larger cities, the distinct preferences for specific housing types outside the estates, intentions to move and reasons for moving imply that even if large-scale architectural improvements were feasible, it will not be possible to halt the selective exodus out of the housing estates and the concomitant filtering-down process which can already be observed to a certain extent.

Against the background of the high levels of mobility in the immediate transformation years and the generally high readiness to move, as the economic situation improves and tension in the urban housing markets eases, a further decline in the competitiveness of the housing estates can be expected. Irrespective of the future design of state policy parameters, this implies that no overall solution will be found to the fundamental problems of the housing estates inherited from the socialist past.

References

ANDORKA, R., SPÉDER, ZS. (1995): Armut in der Transformation: das Beispiel Ungarn. DIW Vierteljahrshefte zur Wirtschaftsforschung, 656-664

BALCHIN, P. (1996): Introduction to Housing in Transition. In: BALCHIN, P. (ed.): Housing policy in Europe. Routledge, London, New York, 231-243

BANKS, CH., O'LEARY, SH., RABENHORST, C. (1996): Privatized Housing and the Development of Condominiums in Central and Eastern Europe: the Cases of Poland, Hungary, Slovakia, and Romania. Review of Urban & Regional Development Studies 8, 137-155

BEREY, K. (1997): Utopia and Reality - the Example of Two Housing Estates in Budapest. In: KOVÁCS, Z., WIEßNER, R. (eds): Prozesse und Perspektiven der Stadtentwicklung in Ostmitteleuropa. L.I.S. Verlag, Passau (= Münchener Geographische Hefte 76), 231-243

CHAPMAN, M., MURIE, A. (1996): Full of Eastern Promise: Understanding Housing Privatization in Post-Socialist Countries. Review of Urban & Regional Development Studies 8, 156-170

CLAPHAM, D. (1995): Privatisation and the East European Housing Model. Urban Studies 32, 679-694

CLARK, W. A. V., DEURLOO, M. C., DIELEMANN F. M. (1986): Residential Mobility in Dutch Housing Markets. Environment and Planning A 18, 763-788

DOUGLAS, M. J. (1997): A Change of System: Housing System Transformation and Neighbourhood Change in Budapest. Netherlands Geographical Studies 222

HEGEDŰS, J., TOSICS, I. (1983): Housing Classes and Housing Policy: Some Changes in the Budapest Housing Market. International Journal of Urban and Regional Research 7, 467-494

HEGEDŰS, J. V., TOSICS, I. (1991): Gentrification in Eastern Europe - the Case of Budapest. In: WEESEP, J., MUSTERD, S. (eds): Urban Housing for the Better-off: Gentrification in Europe. Stedelijke Metwerken, Utrecht, 124-136

HEGEDŰS, J., MARK, K., TOSICS, I. (1996): Uncharted Territory: Hungarian Housing in Transition. In: STRUYK, R. J. (ed.): Economic Restructuring of the former Soviet Bloc: The Case of Housing. The Urban Institute Press, Washington, D.C., 71-138

HEGEDŰS, J., TOSICS, I. (1998): Towards New Models of the Housing System. In: ENYEDI, Gy. (ed.): Social Changes and Urban Restructuring in Central Europe. Akadémiai Kiadó, Budapest, 137-167

KOVÁCS, Z. (1998): Ghettoization or Gentrification? Post-socialist Scenarios for Budapest. Netherlands Journal of Housing and the Built Environment 13, 63-81

KOVÁCS, Z., WIEßNER, R. (1999): Stadt- und Wohnungsmarktentwicklung in Budapest. Selbstverlag Institut für Länderkunde, Leipzig (= Beiträge zur Regionalen Geographie 48)

HUNGARIAN CENTRAL STATISTICAL OFFICE (1995): Statistical Yearbook of Hungary 1994. Budapest

LADÁNYI, J., SZELÉNYI, I. (1998): Class, Ethnicity and Urban Restructuring in Postcommunist Hungary. In: ENYEDI, Gy. (ed.): Social Changes and Urban Restructuring in Central Europe. Akadémiai Kiadó, Budapest, 67-86

LEE, L., STRUYK, R. (1996): Residential Mobility in Moscow during the Transition. International Journal of Urban and Regional Research 20, 656-668

LONG, L. (1991): Residential Mobility Differences among Developed Countries. International Regional Science Review 14, 133-147

LOWE, ST., TOSICS, I. (1988): The Social Use in Market Processes in British and Hungarian Housing Policies. Housing Studies 3, 159-171

MAIER, K. (1997): Problems of Housing Estates and the Case of Prague. In: KOVÁCS, Z., WIEßNER, R. (eds): Prozesse und Perspektiven der Stadtentwicklung in Ostmitteleuropa. L.I.S. Verlag, Passau (=Münchener Geographische Hefte 76), 231-243

PICHLER-MILANOVICH, N. (1994): The Role of Housing Policy in the Transformation Process of Central-East European Cities. Urban Studies 31, 1097-1115

QUIGLEY, J. M., WEINBERG, D. H. (1977): Intra-urban Residential Mobility: A Review and Synthesis. International Regional Science Review 2, 41-66

SAILER-FLIEGE, U. (1997): Transformation of Housing Markets in East Central Europe. In: KOVÁCS, Z., WIEßNER, R. (eds): Prozesse und Perspektiven der Stadtentwicklung in Ostmitteleuropa. L.I.S. Verlag, Passau (= Münchener Geographische Hefte 76), 33-47

SAILER-FLIEGE, U. (1998): Die Suburbanisierung der Bevölkerung als Element raumstruktureller Dynamik in Mittelthüringen. Zeitschrift für Wirtschaftsgeographie 41, 97-116

SAILER-FLIEGE, U. (1999): Characteristics of Post-socialist Urban Transformation in East Central Europe. GeoJournal 49, 7-16

SMITH, D. M. (1996): The Socialist City. In: ANDRUSZ, G. et al. (eds): Cities After Socialism. Blackwell, Oxford, 70-99

SZELÉNYI, I. (1996): Cities Under Socialism and After. In: ANDRUSZ, G. et al. (eds): Cities after Socialism. Blackwell, Oxford, 286-317

SZELÉNYI, I., KONRÁD, GY. (1974): Soziologische Probleme der neuen Wohnsiedlungen. In: BALLA, B. (ed.): Vom Agrarland zur Industriegesellschaft. Band IV. Soziologie und Gesellschaft in Ungarn. Enke, Stuttgart, 98-110

WECLAWOWICZ, G. (1997): The Changing Socio-spatial Patterns in Polish Cities. In: KOVÁCS, Z., WIEßNER, R. (eds): Prozesse und Perspektiven der Stadtentwicklung in Ostmitteleuropa. L.I.S. Verlag, Passau (= Münchener Geographische Hefte 76), 75-82

Cross-Border Co-operations in Hungary in the 1990s

János Rechnitzer

Since the early 1990s, European regional policy has increasingly turned its focus towards border regions. On the one hand, national borders have hindered economic co-operations within the European Union (EU) and beyond the EU's outer boundaries; while on the other hand, goods and services have reached the border regions only in a limited way due to difficulties in accessing them. In this context, it is necessary to explore the border regions in more detail in order to highlight their problems, to make suggestions for the improvement of cross-border co-operations, and to utilise resulting joint opportunities.

This study explores the Hungarian border regions and discusses their situation before the fall of the iron curtain and during the first decade after the political change of 1989/90. It examines, how the position of these border regions changed as a consequence of new market relations, new networks and Hungary's changing spatial structure. Some characteristic features and failures of the economically successful Austro-Hungarian border region will serve as examples for positive and negative developments that evolved due to the new political and socio-economic situation.

1 Characteristics of the Hungarian border regions before 1989/90

Hungary is bordered by seven countries (Austria: 356km; Slovenia: 102km; Croatia: 355km; Yugoslavia: 164km; Romania: 453km; Ukraine: 137km; and Slovakia: 679km). 35% of the country's territory is considered as a border region,[1] and 28.2% of the population (approx. 2.7 million) live in such regions (figure 1). A significant portion of the settlements (43%) is situated close to a border. This is partly due to the fact that the majority of the village-regions are located within the border regions. When characterising the border regions by the towns' commuter belt limits, 55 of the 182 Hungarian town commuter belts are located within

---

[1] Those town neighbourhood areas (KSH small regions) are considered as border regions that adjoin the state boundary and are in direct contact with it (permanent, occasional or previously operating frontier stations).

border regions. This means that almost 30% of Hungarian towns are located near the border (table 1).

Figure 1:   Small regions alongside the border in Hungary

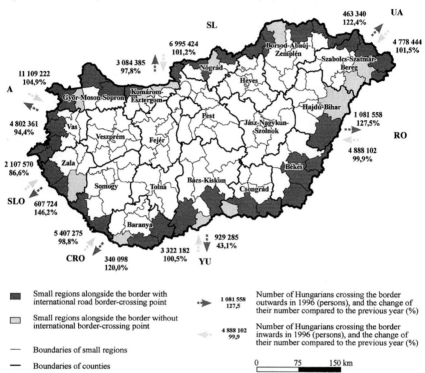

Source: RECHNITZER 1999.

In the era of state socialism, the Hungarian border regions had a very peripheral status. Compared to Hungarian central areas, however, their relative backwardness varied considerably due to historic events and changes in the Hungarian political and economic system. The centrally planned economy of state socialism, for example, made it impossible to establish and maintain contacts over the border on a local or regional level, even with 'fraternal countries' of the socialist block. There were very few contacts on the local level between the former socialist countries such as contacts between twin-cities.[2] The establishment and structure of such cross border contacts, however, followed strict rules and did not include

---

[2]   The name of housing estates in Hungarian towns is a good indicator of the contacts made, since naming the new housing estates after the twin-city or county was politically motivated.

neighbouring settlements (see figure 2, TÓTH 1996). When county-leaders decided to establish cultural contacts with a county or town situated on the other side of the border, they had to place a request with the central party and the governmental bodies. If these bodies considered the initiative acceptable, they contacted the party and governmental management of the neighbouring country through diplomacy. If the other country supported the initiative, the county's or town's contact were officially approved. After the involved counties and towns obtained approval on both sides of the border, twin-county or twin-city contacts had to be carried out according to strict procedures. However, even well established cross-border contacts between counties or towns were suspended at times since they always depended on the relation between the two countries involved.

Table 1: Importance of border regions within Hungary, 1995

Regions	Number of commuter belts	Population (thousand)	Population Rate (%)	Area (%)	Proportion of town population (%)
Border regions	55	2 969	28.2	32.2	
Austrian-Slovenian border region	7	381	3.7	3.9	56.6
West-Slovakian border region	5	348	3.3	2.2	61.4
East-Slovakian border region	11	546	5.2	6.1	45.3
Ukrainian border region	4	172	1.6	2.1	25.1
Romanian border region	14	788	7.5	7.9	58.9
Yugoslav border region	5	385	3.6	3.8	67.1
Croatian border region	8	349	3.3	6.2	38.3
Inner regions	127	7 560	71.8	67.8	
Centre (Budapest & Pest County)	16	2 840	7.8	6.1	79.9
West (Transdanubia)	46	2 123	20.2	24.2	56.2
East (Great Plain)	50	1 753	16.6	23.1	65.1
North	16	844	8.2	7.4	53.3
National average	182	10 529	100	100	63.1

Source: RUTTKAY 1995.

During state socialism, the situation of the individual border regions can be char-acterised as follows:

- The *Hungaro-Austrian border* can be considered a 'dead end' until the 1960s since personal or economic contacts did not exist at all. This situation changed in the 1970s, when tourism gradually started (shopping tourism began at the end of the decade) and generated economic contacts as well as institutional co-operations. The 1980s were characterised by increasing co-operations. Personal contacts became more dynamic, an increasing number of economic units were established, co-operation in production emerged, and the illegal influx of workforce began. When the wire fence separating the two countries was abolished in 1989, a new era of cross-border contacts between Austria and Hungary began that was characterised by the introduction of an international passport (RECHNITZER 1990).

- With 679 km, the *Hungaro-Czechoslovak border* was the longest border region in Hungary. It changed several times since 1945 due to forced relocations after World War II. Approximately 40 000 Hungarians were relocated from Slovakia by the Czechoslovak authorities, which caused more than 100 000 people to move into Hungarian territory. The remaining Hungarian population's rights were severely limited, and the first contacts over the border only began at the end of the 1950's. Based on its opportuni-ties and infrastructure, this border region can be divided into two sections: the Danube region and the East Slovak border region.

  The *Danube region* is an industrialised area and a potential transport and development zone. This section of the border was characterised by a free movement of labour which seldom occurred during state socialism (e.g. Győr textile industries, industrial zone along the Danube). Another important ele-ment of cross-border co-operation was the large Czechoslovak investment in the Gabcikovo-Nagymaros hydroelectric station, which caused a political conflict between the two countries in the 1980s, but which also became one of the most important symbols of the Hungarian transformation. Contacts between business units, such as agricultural co-operatives, occurred regularly (e.g. assistance with the harvest due to differences in ripening times, agricul-tural machinery production, mutually organised holidays), and shopping tourism (between towns along the border) as well as tourism to Balaton and Tatry became quite intense in the 1980s.

  The *East Slovak border zone* was a peripheral region that was largely agrarian although heavy industry played a predominant role in certain areas (metal-lurgy, engineering). In the past, the region fostered and even stimulated co-operation between factories. This area remained a potential zone for co-operation, mainly between the centres in Miskolc and Kosice. The other cross border co-operations between towns were substantial only in regard to shop-ping tourism.

The Hungaro-Slovak cross-border contacts, however, were institutionalised in the era of state socialism. A working committee (Working Committee for Regional Development) was established by the two countries in 1971. It prepared an interim development concept for the common frontier in 1977 (1977-1990). Some of these concepts were carried out (e.g. co-operation on water conservancy, transport management - track renewal, co-ordination of development plans, etc.) but the political and environmental objections to the hydroelectric station and dam (Gabcikovo-Nagymaros), economic hardships, and the debt crisis prevented their thorough execution (HAJDÚ 1996).

*Figure 2:   Typical possibilities for keeping contacts in the regions of the former
            COMECON countries*

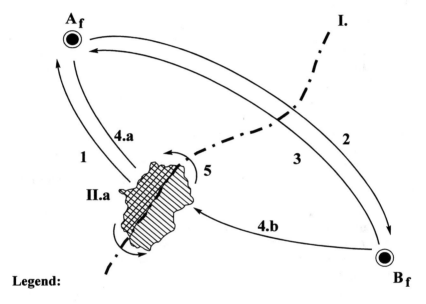

Legend:

I.   National boundaries
II.a   Border region of country 'A',- Border region of country 'B'
A_f   Capital city of country 'A'
B_f   Capital city of country 'B'
1-4.   Steps of possible contacts
5.   Normal contacts

Source: TÓTH 1996.

- Despite the economic, political and military alliance, the *Hungaro-Soviet border* was almost as closed as the Hungaro-Austrian border. This was mainly due to the fact that the Sub-Carpathian area played an outstanding strategic role, and the concentration of military forces in the region was high. The Soviet military presence, which was not aimed directly against Hungary, naturally crippled the cross-border dialogue. Co-operation was hindered by a number of factors, mostly the fear of Hungarian agitation, since 200 000 people of Hungarian ethnicity lived close to the border on the Soviet side. Only one frontier station functioned at the Záhony border, which was both a railway and a road crossing point. This station was designed to serve the larger regional centres but it also worked as a rerouting centre for the railways to change from broad to narrow tracks. A part of the shipment going towards Western Europe (predominantly Austria and West-Germany) was also rerouted here.

- Despite official propaganda ('the Romanian-Hungarian border is an area of friendship, it serves as an example for the socialist brothers'), the continuous infringement of minorities' rights and permanent hostilities hindered real co-operation along the Hungaro-Romanian border. Contacts existed on an international level, but there was a limit to everyday contacts despite the fact that the border's peripheral position could have supported co-operation. Theoretically, the territory of the Hungarians living in Romania could have been the basis for the foundation of cross-border co-operation, but in fact it became its main obstacle. The Romanian government sought to alter the ethnic character of the area rather than foster cross-border co-operation and contacts. Social relations were also limited by very severe administrative measures.

- In the fifties and sixties the *Hungaro-Yugoslavian border* was a 'dead end' since significant political opposition suspended all forms of contacts at that time. A détente began in the 1970s when economic co-operation (e.g. processing agricultural crops) became possible. During this time the Hungarian cities along the border (Nagykanizsa, Pécs, Szeged) played an increasing part in shopping tourism, and Yugoslavians also began to visit small towns along the border to go shopping. Hungarian shopping tourism targeted some Yugoslavian towns near the border. This border region, however, was not yet unified with respect to official contacts. Although a lively co-operation developed between Zala county and Slovenia, contacts were more prevalent in the Croatian section (Osijek-Pécs), and institutional co-operations were stronger in the Serbian (Vojvodinan) section (the area most populated by Hungarians).

During state socialism, proximity to the border meant a disadvantage in receiving central subsidies. When official co-operation between border regions took place, it was mainly politically driven and served propaganda purposes, and there was no significant support of personal cross border relations. The border regions were considered undeveloped, inconvenient regions until the onset of political and economic system change in 1989/90.

2   Socio-economic situation and multi-regional contacts of the Hungarian border regions in the 1990s

The political and economic transformation that began in 1989 strongly affected the spatial structure of the country and the border regions. One example of a quick adaptation to the new situation was the simplified procedure of applying for an international passport (figure 3). Most applications for passports came from the capital city, from regions at the Austrian border, the regions with major frontier stations, the tourist regions, and also from areas populated by Germans and their descendants. Mainly for economic reasons, people living in border regions quickly took advantage of free travel. People also wanted to make the best use of differences in the neighbouring country's price structure, and as a result of this, the dissolution of previous border disadvantages began.

Figure 3:   The applications for international passports, 1988

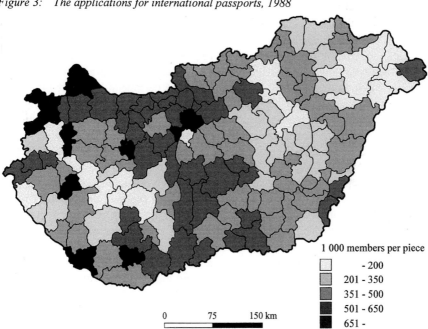

1 000 members per piece

	- 200
	201 - 350
	351 - 500
	501 - 650
	651 -

0   75   150 km

Source: RECHNITZER 1999.

2.1   Socio-economic characteristics

In spite of an overall decrease in border disadvantages through free travel, existing disparities between the different Hungarian border regions increased during the transformation period. The main reasons for this can be traced back to the centre-periphery relationship, the changes in spatial structures (RUTTKAY 1995), and the differences in the economic and social character of the regions on both sides of the frontiers (RECHNITZER 1997). The big centres and the border regions in Western Hungary belonging to the commuter belt of large cities adjusted themselves more quickly to the new situation (GOLOBICS 1995) and had better employment prospects. In the regions along the Ukrainian (formerly Soviet) border, modernisation processes did not occur. The smaller centres could not adopt the innovations of a market economy as quickly as the large centres. This allowed the structural crisis to deepen and illegal elements (underground economy, crime, smuggling people through the border) to become dominant. Altogether, however, the border regions are among the winners of the political change, since even disadvantageously positioned border regions gained some kind of development (e.g. investments, shopping and service tourism, labour, delivery and processing of goods, cultural and institutional contacts, etc.).

Table 2 shows some economic indicators for the different Hungarian border regions and compares them to the national average. The most disadvantaged border regions are those located along the Ukrainian, East-Slovak, and Romanian borders. This is indicated by their higher than average unemployment rates (17.2% in the Ukrainian border region, 18.5% in the East-Slovak region), as well as by their very low foreign capital ratios (8.0 ratio in the Ukrainian and 7.6% in the Romanian border region). Furthermore, these peripheral border regions show much lower than average per capita incomes and only moderate joint venture ratios – especially in regard to neighbouring countries. Consequently, there is almost no economic cross-border co-operation in these peripheral border regions and this suggests that the regions along these borders are unable to benefit from each other. For example, there are also no significant prize and wage differences fostering shopping tourism. In spite of ample labour resources on both sides, there are no jobs available; and the production resources (agricultural products on the Hungarian side and natural resources in the neighbouring countries) cannot be mobilised due to a lack of capital. In addition, poor transport connections, the small number and unfavourable distribution of the frontier stations, and the tedious bureaucracy involved when crossing the border make cross-border contacts difficult. Finally, unstable local currencies and high inflation rates basically prevent co-operation and investments.

In contrast, the Austrian and the west-Slovak border regions, which comprise the regions along the Danube, and which coincide with the Vienna-Győr-Budapest innovation axis, stand out in several regards such as the density of companies that are located there. The high density of companies along the Yugoslavian border indicate that the economic strength in Szeged and its surrounding area, which was established mainly in the Yugoslavian war period (1992-1996) and afterwards, was often due to business and personal wealth investments. This is also valid for the ratio of joint ventures to income positions. In general, border regions with large cities (Győr, Sopron, Szombathely, Szeged) took better advantage of the emerging market economy and modernisation than border regions without urban centres.

Table 2: Some economic indicators of the Hungarian border regions, 1995

Regions	Density of companies	Ratio of joint ventures	Income per capita	Unemploy-ment rate	Ratio of for-eign capital
Border regions					
Austrian-Slove-nian border region	8.11	24.46	123.03	7.32	30.36
West-Slovakian border region	10.72	15.03	127.02	11.08	16.71
East-Slovakian border region	4.98	8.34	90.69	18.60	17.59
Ukrainian border region	2.87	6.90	72.22	17.22	8.02
Romanian border region	7.15	7.51	92.50	14.20	7.60
Yugoslav border region	12.04	21.45	114.51	10.92	7.24
Croatian border region	4.80	15.34	93.17	13.30	16.42
Inner regions					
Centre (Budapest & Pest County)	18.66	19.03	160.91	6.99	18.84
West (Transdanubia)	8.35	13.47	114.39	10.24	10.77
East (Great Plain)	6.75	9.94	91.79	14.81	10.23
North	6.77	7.28	103.51	14.14	7.83
National average	6.88	22.98	119.00	13.00	15.84

Source: RUTTKAY 1995.

2.2   Informal and institutionalised cross-border co-operations

The border region between Hungary on the one hand and Austria and Slovenia on the other hand can be considered a success region of the transformation period. Mainly because significant foreign investment was attracted to the area, the trans-formation of the economic structure, especially privatisation and the renewal of former large-scale industries, occurred quickly. Traditionally, labour intensive, Fordist type mass production has played a dominant role in these regions, whereas small and medium-sized businesses began to appear slowly. In the late 1990s, however, the service sector is expanding rapidly, and its structure is becoming more diversified. The local and regional market is bolstered by higher incomes, and it is further strengthened by large-scale shopping tourism. Many trans-border contacts have been institutionalised, and the West/Nyugat Pannónia Euregio (figure 4) was established in 1998. However, there are still a number of opportunities for the joint development of the border region (RECHNITZER 1997).

Figure 4: Euregios with Hungarian participation

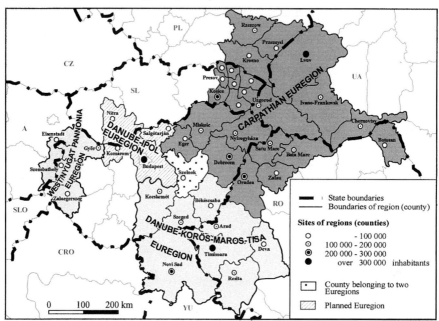

Source: RECHNITZER 1999.

The European Union subsidy that began in 1995 (Phare CBC) provided considerable impetus for co-operation in the western border regions. Organised trans-border contacts started in 1992 with the establishment of the Border Regional Council which evolved as an initiative of two Hungarian counties (Győr-Moson-Sopron and Vas) and the Austrian Bundesland of Burgenland. After the expiration of the Council's six-year-mandate (1998), the members established the West/Nyugat Pannónia Euregio (October 1998) which Zala county joined in 1999. The aim of the co-operation, in addition to deepening contacts, was to prepare for entry into the European Union. This institution aims at dealing with all questions that affect the contacts along the border. Its leading body is the Council of Euregio (with 40 members). It operates through a Secretariat (with 4 members), while working groups administer the programs. Compared to the previous situation which was characterised by informal contacts, progress has been achieved and large resources will be provided for the institution. However, problems have arisen because party politics have entered the organisation, the organisation's responsibilities are too large, and the new institution has only slowly developed to a functioning, multi-regional co-operation.

The Danube-region along the Slovak-Hungarian border successfully utilised the Vienna-Győr-Budapest innovation axis in the last third of the 1990s. The phasing-out of mining, which was the former basis of large-scale industries, resulted in a long, steady transition, while the renewal of other elements (industrial zone along the Danube) had a slow start. This border region is characterised by a gradual stabilisation and a slow rearrangement. The cross-border contacts are strongest in the sphere of private relations between Hungarians on both sides, especially close to border crossing points (Komárom, Esztergom), but as yet they have not taken an institutional form. The settlements on both sides, however, would welcome increased institutional contacts. For example, the reconstruction of the Esztergom bridge is an urgent requirement, but on a national level, this project does not have a very high priority (RECHNITZER 1999).

Along the East-Slovak border region the large-scale industrial centres of the former socialist era are in crisis, and this border region is characterised by high unemployment. The transformation of the economic structure of these centres is extremely slow; agricultural production has not been modernised; the service sector has not sufficiently developed; and the number of young people who emigrate is increasing. The elderly and an increasing number of gypsies are left behind (DANKÓ and FEKETE 1996; see also first chapter by Peter Meusburger, section 6). An economic and social renewal cannot be expected from the neighbouring countries. A more positive development in this region are the more intensive cross-border, settlement-to-settlement and interregional contacts along the Ipoly River. As a result of this, the region's mayors and some non-profit making organisations (including the Danube-Ipoly-National Park) established the Ipoly Euroregio Cross-border Interregional Co-operation in April 1999. The Ipoly Euroregio Cross-border Interregional Co-operation agreement was signed by four towns, two villages, and seven non-profit making organisations on the Hungarian side, and by an association of self-governments, four villages and six non-profit making organisations on the Slovak side. The aim of the co-operation is to strengthen the regional contacts of the Ipoly valley, to intensify economic co-operations, and to harmonise developments in an area that has 323 settlements and 450 000 inhabitants. The institutional system is composed of a governing body, secretariat, and a prospected nine working committees. The initiative for the Ipoly Euroregio has come from below and the responsibilities are clearly distributed so that this border region will probably benefit from the Ipoly Euroregio. Furthermore, Komárom-Esztergom County, Pest County, and the Nitra Region along the border are interested in establishing the Vág-Danube-Ipoly Euroregio, and the Nitra Region, that was established through an initiative of settlements, will join the Ipoly Euroregio at a later date (figure 4).

As already mentioned above, the regions along the Ukrainian and Romanian border face problems similar to those of the East-Slovak border region. They are burdened with socio-economic depression in all three countries. Even substantial amounts of subsidies from Hungary were unable to improve the socio-economic situation in the border regions between Hungary and Transylvania. The first East-European cross-border co-operation, the Kárpátok Euroregio Interregional Alliance (figure 4), was established in this region in 1993 by four countries: Hungary, Poland, Ukraine and Romania. The multinational co-operation embraces over 11 million people and a 106 000 km^2 area. This Euroregio has a permanently acting co-ordination organisation (Chief Secretariat and a foundation supporting the co-operation) that is responsible for the execution of planned activities (professional exchange programs, commercial contacts, cultural, educational and environmental protection programs). It is the first example in Eastern Europe of a former historical unit re-organising itself. However, to date, the new organisation has been unable to dissolve the economic problems and political tensions. The main problem facing the Kárpátok Euroregio Interregional Alliance is that all the regions involved can be considered as peripheries of their respective countries. A multiregional development effort would be needed in order for the region to benefit from the frontier stations, and to open new border crossing points. However, the responsible national and regional institutions are unable to pay attention to, and provide resources for, the co-operation because of differing administrative levels and competencies as well as the different priorities of the adjacent countries. Regional mobility, however, strengthens shopping tourism and markets in the larger centres. Recently, this border section has been affected the most by illegal activities such as the smuggling of cars and people by criminal gangs (SÜLI-ZAKAR 1997). It is therefore necessary to work out a joint development strategy for the Kárpátok Euroregio Interregional Alliance as soon as possible - despite significant institutional differences -, and to secure sufficient national and European Union resources for the program. With this additional funding the region would then be able to offer more opportunities for Hungary in the future.

The Romanian border region also shows characteristics of the large tri-border region; it has a typical peripheral position, although there are several centres on both sides which are becoming more and more active. On the Romanian side more significant activities can be observed in the north-eastern border region, because of the large Hungarian population living there. This is reflected in the more diversified economic contacts and the frequently occurring settlement-to-settlement and regional contacts (DANCS 1999). In the future, Szeged and Timisoara as well as Békéscsaba and Arad will probably compete with each other, and this probability might be one of the reasons for the slow development of economic contacts between these cities (LENGYEL 1996). Currently, however, there are certainly more settlement-to-settlement and regional contacts, but the small number and disadvantaged geographical location of the border crossing stations do not allow for the creation of various small co-operations. In September 1998, the Duna-

Maros-Tisza Regional Co-operation was established in this region (figure 4), but the Kosovo war prevented the organisation's operation for some time. Embracing three countries (Hungary, Romania, Yugoslavia), a 77 500 km² area and a population of nearly 6 million people, the Duna-Maros-Tisza Regional Co-operation is the first multi-regional co-operation in the Romanian border area. Its aim is to intensify cross-border contacts regarding the economy, traffic, telecommunications, environmental protection, tourism, science, culture and civil organisations. The co-operation decided that a uniform development concept should be developed and that a permanent secretariat and a common financial fund should be established.

The Yugoslavian (Serbian) border region is a continuously developing region on the fringe of the Hungarian Great Plain including Szeged, a large centre, and Baja, a medium-sized town with a strong effect on the border region, but there are a number of other small towns that could provide valuable functions. Within the region, many individual contacts have been established, but economic co-operation that developed in connection with the war in former Yugoslavia is also considerable. The operation of the border region's former 'grey-economy' was crucial for providing people with valuable experiences in the business sector (PÁL 1996). This border region could also benefit from a good accessibility through main transport routes. After the consolidation in Yugoslavia and the successful fight for autonomy in Vojvodina, there exist great opportunities for the intensification of cross border economic activity and co-operation.

The Croatian border region comprises the southern periphery of Transdanubia. On the other side of the border it faces a periphery without any urban centres. The geographic border (Dráva) does not have an adequate number of crossings, and permanent Croatian-Serbian tensions (Baranya triangle) eliminated contacts along certain border sections, despite significant, active shopping tourism elsewhere. The institutional forms of trans-border co-operation have developed very slowly, and various initiatives have often not been supported by the other side (such as the Gyurgyevác hydroelectric station on the Croatian side and the Duna-Dráva National Park on the Hungarian side). However, plans have been made to increase inter-governmental relations through the development of traffic connections such as the Rijeka-Zagreb-Budapest motorway (HAJDÚ 1996; VUICS 1996).

2.3 Future prospects

After analysing the Hungarian spatial structure, József TÓTH (1996) explored the possible future situation of co-operation along the borders. The author notes that almost 80% of Hungary's borders (1 788 km) will border a non-EU country after joining the European Union. Hungary may be the only contact point for non-EU countries such as the Ukraine, Romania, Croatia, Yugoslavia which are looking towards future European integration. A large number of Hungarian or foreign-owned enterprises with their headquarters in Hungary have already started to expand into neighbouring countries in the East (DICZHÁZY 1997), and multinational companies have established affiliations in Hungary to practice their main business activity in the Central-European region and to develop that particular market. The strengthening of contacts within the region and the development of stable economic conditions are very important for the multinational companies. The fact that nearly three million inhabitants, whose mother tongue is Hungarian and who regard themselves as Hungarian nationals, live in neighbouring countries adjacent to Hungary is also important for future cross-border co-operations. Their bond with their mother country is strong and this will play an important role in shaping these co-operations. The European Union regards Central Europe, specifically the region along the Danube and the Adriatic, as a strategic region for which development ideas have already been worked out (Central European, the Danube 1999). Altogether, these economic, political, social and transnational area development perspectives will substantially affect Hungarian cross-border co-operation, and could become a part of the national development strategy.

The western region (Austro-Slovenian-Hungarian border region) is working within an adequate organisational and institutional framework in the form of the West/Nyugat Pannonia Euregio. This is of high strategic importance, but it is clear that it is not only the formation of a uniform structure that should be promoted. The effects of the Vienna-Bratislava axis are strongly felt in the north-western regions of the country, and the Vienna-Bratislava-Győr-Sopron municipal co-operation is forming gradually and is organically merging with the Vienna-Budapest innovation axis. The formation of the Szombathely-Zalaegerszeg-Graz-Maribor municipal region is also expected to be highly successful. Euregios may provide good frameworks for co-operation, however, because of their large area, the multiple interests, and the amount of political and administrative obstacles, they are not very flexible and thus unable to co-ordinate all existing and potential contacts. Smaller organisational units (e.g. small regional, settlement-to-settlement partnerships or city alliances) may function better in the border regions for overseeing co-operations. These smaller regional units may promote a more specific and successful co-existence in the border regions than in the rather political Euregios. The western region will probably face new challenges after Hungary joins the European Union.

The northern Hungarian border region (Slovako-Hungarian border region) consists of two centres. The Danube-Ipoly area is an area where the co-operation of towns and smaller regions, that have similar capabilities on both sides of the border, may induce intensive contacts. The Miskolc-Kosice region is another area where informal contacts may provide a basis for further co-operation.

The eastern part of the Hungarian border consists of the Slovak-Ukrainian-Romanian-Hungarian border regions. Cross-border relations basically rely on city-to-city contacts, although the Carpathians' Euregio may provide a future framework for it. The northern part of the Romano-Hungarian border region gravitates towards multiregional co-operation, but it cannot develop without the co-operation of the large centres (Debrecen, Nyiregyháza, Oradea, Satu Mare). There are also large centres in the southern part of the Romano-Hungarian border region which could co-operate with each other. Szeged, Békéscsaba, Arad and Timisoara are suitable for establishing network systems for modern co-operation.

The southern border region is important for the development of relations with the Adriatic and Balkan regions. However, the present Serbian political situation does not allow the concentration of more significant resources in the Serbo-Hungarian border region, but this could change in the future. The established Duna-Tisza-Maros Euregion provides an institutional framework for future contacts. Signs of decline are visible in the Croatian border region, but the infrastructural development (transportation) in the adjacent peripheral regions may provide a foundation for co-operation. It is advisable to establish an independent institutional system within the Croatian border region.

The development of the Austro-Hungarian border region, which is explained in the following section, provides a number of lessons for the organisation of cross-border co-operations and their effects on an area's development.

3   The Austro-Hungarian border region as an example for successful cross-border co-operation

The socio-economic disparities within Hungary intensified in the 1990s, especially between the capital city and the provinces, and the western and the eastern parts of the country. Hungary's position within the European settlement network caused certain Hungarian regions to become disadvantaged and neglected while others benefited and were able to adjust themselves more quickly to the changing conditions (RECHNITZER 1993). The border position played an important role in this rearrangement, and this is true particularly for the areas along the Austro-Hungarian border.

The Austro-Hungarian border separated the two countries for many decades, and cross border contacts only began in the 1980s. The factors that had hindered co-operation started to disappear once the border opened at the beginning of the 1990s. Open travel also had an enormous effect on economic co-operation, and the growth of Austrian capital entering Hungary accelerated with the establishment of an institutional framework (RECHNITZER 1990).

The intensification of contacts was fostered by the beneficial position of the Hungarian regions. In the western regions, the economic structure adjusted itself to the conditions of the market economy more quickly because the education level of the workforce was high, its income level was better, and there was higher entrepreneurial activity. The level of infrastructure (transportation, telecommunications, social infrastructure) was also more efficient in Western Hungary. This was coupled with the fact that the regional (county) and local (settlement) economic and political players realised that by promoting the Austro-Hungarian partnership it would give impetus to the development of these regions and would diminish its problems (e.g. unemployment, income differences).

3.1   Economic contacts

The economy of the Austro-Hungarian border region is based around the development of foreign investment and foreign trade contacts. In 1994, 19.2% of the investments outside Budapest came from the two counties adjacent to the Austrian border (Győr-Moson-Sopron, Vas). 16.2% of companies selected this region as the Hungarian base for their business (table 3). This ratio is significant because it is near identical to the percentage of foreign investment in all of the Great Plain (38.7% of the territory and 28.4% of the country's population), and thus shows that the concentration of investments in these two counties is outstanding by national standards. The increase in foreign capital has been relatively even after a strong increase in 1992. In the period 1994 to 1997, the magnitude of foreign investments increased threefold in Győr-Moson-Sopron county (344.2%), and raised by nearly one and a half times in Vas county (181%). Since foreign investors favoured a 100% participation in joint ventures, the rate of 'greenfield' investments, which account for over half of the total investments, is very high. An additional one third of investments is predominantly under foreign ownership, while a share of foreign direct investment below 50% occurs only in the remaining 14-15%. Foreign investments were mainly directed towards the processing industry, a competitive area of industry, which resulted in the implementation of new manufacturing cultures, products and product lines (e.g. production of passenger car components and assembly, computer manufacturing), as well as in the modernisation of factories. Over one-fifth of all joint ventures were established in the engineering industry, and one-sixth in textile, clothing and leather goods production. Large companies with more than 300 employees dominated engineering (mainly in Győr-Moson-Sopron county). In other industrial branches, mainly to be found in Vas county, smaller company units were more typical. From 1996 onwards, foreign-owned businesses appear more in the hotel and catering business (mainly health spa tourism), in commerce (shopping centre construction in large centres such as Győr, Sopron, Szombathely), and also in branches such as banking or personal services (e.g. dentists).

The share of foreign direct investment from Austria is highest in the western border region (table 4). In Győr-Moson-Sopron county, Austrian investment is mainly in larger businesses (e.g. construction material industry, construction industry, chemical industry, food processing industry, services), while in Vas county, Austrian investment occurs both in larger economic enterprises (e.g. the chemical industry, shoe industry, food processing industry) and in numerous small and medium-sized businesses. The following relationship can be traced between the border-position and foreign investments (DÖRY and MÁTHÉ 1996):

- Important Hungarian enterprises had contacts with Austrian companies since the early 1980s (export, import, purchasing licenses).
- In the western border region, especially in Győr-Moson-Sopron county, a more developed and diversified economy operated. The Austrian capital took an active part in the privatisation of the region because of its earlier contacts. Until 1994, the share of Austrian business interest in privatised economic units was 60.7% in Győr-Moson-Sopron county, and 92.4% in Vas county.
- Austrians also invested in this region because it was easier to maintain contacts with domestic organisations, they knew Hungarians from their earlier relationships, and it was convenient to built up an Eastern European network from this region.
- The comparatively higher level of educational achievement, of foreign language knowledge, the improved infrastructure, and the presence of some multinational companies (Audi, Opel) had a significant effect on the location selection.

Table 3: Foreign-owned businesses in the counties of the Austro-Hungarian border region, 1994 and 1997

Name	Győr-Moson-Sopron		Vas	
	1994	1997	1994	1997
Number of organisations (units)	1 184	1 260	615	694
Registered capital (Billion Ft)	45.8	157.8	30.5	55.5
Foreign-owned and operated (Billion Ft)	32.5	109.6	28.1	51.6
Organisational participation (country) (%)	5.0	4.9	2.6	2.7
Organisational participation (provinces) (%)	11.0	9.8	5.7	5.4
Foreign capital participation (country) (%)	3.9	5.6	3.3	2.0
Foreign capital participation (provinces) (%)	10.3	12.7	8.9	4.5

Source: Regional Statistic Yearbook 1994, 1997.

Table 4: Austrian business interest in the counties along the Austro-Hungarian border between 1989 and 1994

Counties	Number of organisations (unit)	Participation in foreign orga- nisations (%)	Registered capital (Billion Ft)	Participation in the total registered foreign capital (%)
Győr-Moson-Sopron	485	40.9	8.84	19.7
Vas	366	59.5	4.24	13.8

Source: Data of Journal of Companies between 1989 and 1994.

The volume of foreign trade with Austria reflects the changes, or rather structural rearrangement of the economic contacts along the border. In the first half of the 1990s, the ratio of Austrian imports from Hungary varied between 23 and 28%, while its exports stabilised at 12 to 15%. Abruptly increasing German exports had jeopardised the ratio of Austrian exports. Between 1991 and 1994, the total Hungarian export entering Austria only grew in Győr-Moson-Sopron county (from 12.4 to 14.3%). However, the ratio of Austrian imports from this county decreased from 25.2 to 23.2%. In Vas county, the Austrian economy was not at all able to strengthen its position regarding the trade relations. Instead, a significant decrease in both imports and exports was registered (exports dropped from 46.3 to 26.1%; imports dropped from 47 to 27.2%). This results from the fact that, in Vas county, the Austrian contacts cannot counterbalance increasing trade relations with the German economy. However, this development does not mean that economic dependence on Austria would gradually dissolve in the counties along the border. On the contrary, they still determine the direction of the economic bases there, but, after a strong increase immediately after the borders were opened, the Austria-Hungarian trade relations have more or less stabilised in the second half of the 1990s. Austria may only be able to significantly change them again through comprehensive economic co-operation (e.g. investments, more export contacts, extended contract manufacturing systems, etc.).

3.2   Organisational frames of the co-operation

The Hungaro-Austrian Area Development and Planning Commission (MOTTB) was established at a governmental level to establish co-operation in professional areas, especially in the region along the common state frontier. Its main objectives are also to work out proposals and recommendations, and to harmonise the area's development and planning measures. This information is contained in the a work summarising the 10-year work of the commission (KOVÁCS and VÁRADI 1996): "The inter-governmental agreement that was established in 1985 is still the only bilateral contract of this nature. Its importance lies in the fact that it has opened possibilities for the cross-border co-operation of the counties' and settlements' councils and of the self-governing bodies at the border region. The contract also provides the legal and organisational framework for co-operation between the two countries in its professional circle. Its pioneering character became a pattern for similar co-operations with Austria and Slovakia."

Although the MOTTB was based on an inter-governmental agreement, regional competence is guaranteed by the fact that the representatives of Burgenland and the two neighbouring Hungarian counties (Győr-Moson-Sopron and Vas) could take part in the sub-commission. There are currently two working groups in the MOTTB: the Border Region Co-operation Working Group is engaged in co-ordinating the development plans of the border regions; the Regional Economic Working Group co-ordinates common developments (Phare CBC).

Based on the list of the participants at the commission and sub-commission sessions, it is clear that the Hungarian representatives always represent a lower level of hierarchy and political competence. This was evident at the document signing (Chancellor - Minister), and is the cause of countless problems. It is, for example, a strong limiting factor for the scope of the Hungarian counties as they do not have the Austrian level of power and competence in management and administration. The Hungarian party has to consult governmental bodies for many more decision-making questions than the Austrian.

The aim of establishing another Hungaro-Austrian co-operation that is known as the Regional Council, the Pannon Regional Council, and the Borderside Regional Council, was to create a regional parliament for contact intensification. The parties along the border established this organisation in order to increase the number of decision-making forums and to strengthen the Council's level of competence. Except at the governmental level, the same players took part in the Council's work as in the MOTTB described above. Those representatives of towns with county status were observers, and because of this situation, the Council was unsuccessful in developing any form of co-operation regarding the right of jurisdiction. The Council only made recommendations connected to the combined development of the two regions (Burgenland, Győr-Moson-Sopron and Vas Counties) to the regional and national governments. Officials and experts serve in both the MOTTB and in the Council that holds its session twice a year. Positive Council debates on the real problems of the border region were more effective in affecting change in the governmental level commission and sub-commission.

After Austria joined the European Union in 1995, the co-operation contacts described above enabled specific development co-operations to take shape. The framework was established by the 'Interreg II. - Phare CBC program' which targeted a development valued at 50 Million EURO in the border region (7.5 Billion Ft at the present exchange rate, of which 35 Million is Phare subsidy and 15 Million is a EURO national contribution). Although Zala county does not have a common border with Austria, it participates in the program because of its good political contacts, and because the Slovenian-Hungarian co-operation was still being prepared in the mid-1990s. Based on the preliminary surveys, the total value of the submissions received for the 'Interreg II. - Phare CBC program' is slightly over 20 Billion Ft. In order to keep the programs consistent, priorities were worked out by the experts of MOTTB and the Regional Council (table 5). The large (over 10 000 EURO) and small (below 10 000 EURO) programs were announced in the spring of 1997, after an extremely long phase of preparation and reconciliation. This two-year delay can be partly explained by Brussels' bureaucratic administration, by the slow formation of the Hungarian organisation, and by the limited experience of the regional (county) project managers in preparing proposals. The local office of the Phare CBC Hungarian headquarters is based in Sopron. This is unique in Hungary because all projects to date have been organised from the capital city. This situation offers a good opportunity for close contacts within the region.

Table 5:   Resources and goals of Phare CBC programs in the Hungarian Western border
regions

Programs/priorities	Breakdown		Amount of Phare subsidy (EURO)	
	Projected	Actual*	Projected	Actual*
Phare CBC Hungary-Austria program 1995-1999	100	100	35 000 000	42 000 000
Regional planning and development	10.4	3.4	3 640 000	1 429 000
Technical infrastructure development	26.7	37.4	9 345 000	15 716 00
Economy development and co-operation	39.8	38.9	13 930 000	16 319 860
Human resources development	6.2	4.5	2 170 000	1 887 000
Environmental and nature protection	11.8	9.9	4 130 000	4 153 041
Fund for small projects	5.1	3.4	1 785 000	1 445 100
Phare CBC Slovakia-Austria-Hungary program 1995-1996	-	-	3 000 000	2 250 000
Phare CBC Slovenia-Austria-Hungary program 1995-1996	-	-	3 000 000	3 000 000
Total	-	-	41 000 000	47 250 000

* Subject to change based on successful completion of project contracts.

Source: PHARE CBC PMU Sopron Office.

The Austrian representatives' opinion was vital when selecting subsidised projects. Any initiative that was competitive to the Austrian side (i.e. to the existing or planned establishments in Burgenland) was not put into the program. The Austrians have kept a watchful eye on the recent modernisation process in the three counties. If the planned establishment applies for a subsidy from the Phare CBC program and it could simultaneously interfere with the potential or existing Austrian interests, Austria has tried to prevent or slow down the project subsidies. Examples are the events surrounding the establishment of an innovation centre in Győr, and the development of health spa tourism which Austria also tried to prevent.

The basic principle in determining the direction of Phare CBC subsidies and projects has been to strengthen and broaden the contacts along the border. As a consequence, almost every socio-economic development initiative and county settlement initiative could be a part of the program. However, utilisation of these initiatives may be limited from the Austrian side since there have been many actions taken to prevent or postpone the Hungarian initiatives.

3.3   Partially legal and illegal activities

Discussing present forms of contacts and co-operation along the Austro-Hungarian border, it is also important to identify only partly legal and illegal activities. These activities cause confusion in the day-to-day contacts in economic life due to the lack of suitable legislation or inter-governmental treaties.

The first disturbing factor is the black labour market. It is not accidental that the Labour-Employment Committee of the Border Regional Council wanted to settle the question of Austrian illegal labour as it is in the interest of both sides of the border to do so, although those directly affected are satisfied with the present situation. The Committee estimates that half of the foreign workers employed in Burgenland are Hungarians (3 000 people). The actual number is higher due to the fact that 30 to 44% of the employers monitored in Burgenland have employed foreign workers illegally. Many Austrians seem to be content with illegal employment because Hungarian commuters may get a job even without a permit, and this forces wages downwards. The Austrian employment offices limit the legally obtainable permits according to settlements which also pushes illegal employment. And since many farmers in the Burgenland do not live exclusively from their work in agriculture but also have jobs in Vienna, illegal employment helps to keep the family business going. The severity of the problem is shown by the fact that Austrian trade unions have opened advisory offices in Sopron and Szombathely so that Hungarians seeking employment in Austria can become familiar with the employees' rights and therefore know who to contact should their rights become infringed. There are approximately 12 000 to 15 000 daily cross border commuters, and this causes problems on the Hungarian side, since it is becoming more difficult to find a carpenter, cook, waiter or plumber in Sopron, and the local specialists ask for much higher wages and service fees.

In the commuter business, seasonal work is common. The neighbourhoods along Neusiedler Lake, for example, are filled with Hungarian conversation during the autumn grape harvest. Year after year, the same teams return for the harvest with the same farmers. The employment authorities of both countries try to control this through an agreement that allows employment in Austria up to a certain quota for Hungarian citizens with a permanent residence in one of the two counties along the border. This quota is determined annually by the Austro-Hungarian Joint Committee. Trainee employment is allowed for an additional 300 persons between the ages of 18 and 30. Thus, after nineteen trades have been mutually acknowledged, the employment conditions have gradually improved.

However, the next problematic factor is property purchase and cross border ownership. This is caused by the fact that the 'Austrian' landlord has appeared in both counties and this has triggered different reactions: approval, support or fierce anger towards private persons, mayors and the managers of the agricultural federation. Naturally, the appearance of foreign buyers/owners in the property market significantly influences the market structure. Their actions and presence in residential real estate is in most cases legal, but their presence in the land property market is questionable. Foreign possession of arable land has provoked a fierce reaction from Hungarian agrarian federations because Hungarian arable land does not yet have a solid internal market.

Foreign land ownership is legal only for former Hungarian citizens who were compensated for expropriation in socialist times. The ratio of foreign citizens who received land as compensation for expropriation may be 3 to 5% of all people who received compensation in these Hungarian counties. The circumstances of land property obtained by foreigners through other ways are uncertain. It is clear, however, that lawyer's offices had an active role in executing the purchase of compensation coupons, concluding pocket-contracts, and making bids on commission. In this system, it was possible that the person entitled for compensation sold its property rights before a transaction took place. Unfortunately, there is no information available on the magnitude of this method of becoming a land owner. However, this kind of illegal transactions appears to have considerably contributed to an increase of foreign land ownership in Hungary. It only came to light in the course of lawsuits against the field inventories: the piece of land sold at bid was already with the sixth owner.

The farming lease system is fairly widespread in the Austro-Hungarian border region. This method of land utilisation is supported by the Hungarian agrarian federation because it brings agriculture, machines and new experiences to Hungarian farmers. The Austrian farmers grow hay for their animals and sunflowers on the leased land. According to the present legal provisions, these two products can be brought into Austria without paying duty. The Austrian landowners can lease Hungarian land cheaper than at home, but they are in competition with agricultural co-operatives and businesses. The leasing fee in the counties along the border is 15 kg wheat/gold crown, but the Austrian landowner has to pay twice that amount. Experts estimate that 10 to 15% of the arable land in the two Hungarian counties along the Austro-Hungarian border could be utilised by Austrian landowners as the ratios have grown in the past years. However, the high demand in the land leasing market in Győr-Moson-Sopron and Vas county has increased the costs for the Hungarian producers.

In contrast to the problems that are associated with foreign land ownership in Győr-Moson-Sopron and Vas county, the appearance of Austrian property owners in settlements can be regarded as positive for the settlement's development since Austrian house-owner often renovate old houses, take care of them, and thus start a renewal process in Hungarian villages. A number of examples show that Austrian property purchases stopped the decline of certain settlements.

3.4   Factors hindering co-operation and outlook

The governor of Burgenland, Mr Stix, who is at the same time national deputy chairman of the Austrian Social Democratic Party, and one of the confidants of the previous Austrian Chancellor, provoked a small 'revolution along the border' with a statement in November 1996. He considered Hungary's ambition for a quick entry into the EU as a false move because the country was not yet ready for membership, and the Hungarian EU membership would be detrimental to the Burgenland. Leading to the free movement of employees and businesses between the two regions, Mr. Stix argued that the moderate Hungarian wage and production costs would gradually squeeze the Burgenland companies and employees out of the market.

This statement was interpreted by several people in several ways, but it obviously triggered a new development in Austro-Hungarian cross-border co-operation. Border control has increased over the last two years on the Austrian side, and more statements have appeared on the implementation and observance of the Schengen Agreement, and on the growing expenses associated with guarding the eastern Austrian borders. Information shared at conferences and in professional journals indicates that the number of emigrants to Austria will grow when Hungary joins the European Union, that this will affect the Austrian economy, and that the eastern extension of the European Union will also affect the competitiveness of Austrian industries and services. Less research has been completed concerning the land purchases of the Austrian landlords in the neighbouring countries, on the expansion of Austrian capital, on the cross-border shopping and service tourism, on the lower wages offered for Hungarian guest-workers, and on employment tensions that could arise in times of labour shortage.

In the context of EU integration, the Austro-Hungarian border region has become an issue of national politics. It is likely that during the period when Hungary prepares to join the EU, it will be a constantly monitored region, a 'barometer' for the level of arguments on both sides. The role model character of this border region is further strengthened by the fact that the Phare CBC program has been extended to other border regions of Hungary. Thus, the EU-bureaucracy and politicians can use the border region to test how much the Hungarian regional organisations are able to use the subsidies in a sensible way, as well as to plan, apply and implement the related objectives.

A number of other factors influence co-operation along the Austro-Hungarian border (RECHNITZER 1996). One factor is that a peripheral Austrian region meets a central region on the Hungarian side. Several others essays in this volume have shown that the economies of Győr-Moson-Sopron and Vas county adjusted more quickly to the market economy conditions than those in other Hungarian areas, and that privatisation occurred without major problems, mainly due to an high influx of foreign direct investment. Also enterprise activity is stronger, the level of educational achievement is higher, and the population is more diversified in the north-western part of Hungary than elsewhere. Finally, the living conditions are better, and there is a large consumer base with higher incomes. Altogether, the population of the two counties, that comprises of almost 700 000 people, forms lively local and regional markets. The same is not true of Burgenland, since the population is small (270 000 inhabitants), the economic structure is one-sided (agrarian production), the level of education is relatively low, and there are no urban centres. The Burgenland's industrial base is weak, and it is strongly dependent on Vienna and other Austrian economic-industrial centres (e.g. Graz). Although the Burgenland possesses significant development resources as a primary target area of EU subsidies, some leading politicians seem not to be interested in cross-border co-operation, but rather, as indicated by the examples given in this text, in hindering joint opportunities. However, based on the existing co-operations, it should be worth to adhere to the long-term area development concept that has been worked out in both countries. This concept suggests to initiate the development of an economic environment that would simultaneously foster the extension and internal renewal of the existing structures by developing the production infrastructure, training and education systems, and by encouraging rural development.

Vienna is another element in the restructuring of the Austro-Hungarian border region since it has a strong effect on this area. The Austrian capital will probably influence the future position of Győr-Moson-Sopron and Vas county much stronger than the Burgenland. It will take some more time to develop intensive contacts and co-operations with Vienna, however, as a result of the 'Interreg II. - Phare CBC program,' some Viennese interests have already turned towards the Austro-Hungarian border region, and thus provide good future prospects in regard to joint development programs.

Finally, one more problem of cross-border co-operation along the Austro-Hungarian border should be addressed. It refers to the communication and co-ordination within the *Hungarian* border region in question. As outlined above, three Hungarian counties are involved in Austro-Hungarian cross-border co-operation, namely Győr-Moson-Sopron, Vas and Zala county, although the latter does not directly border Austria. Each of these three counties has particular interests, and tries to integrate them into bilateral agreements with Austria. The formation of common interests on the Hungarian side is even more difficult because of the counties' five cities with equal self-government independence (Győr, Sopron, Szombathely, Nagykanizsa, Zalaegerszeg).

A serious problem in regard to effective cross-border co-operation results from the fact that the Regional Development Council of West Transdanubia, which was established by the three counties involved in Phare CBC,[3] is only formed on an organisational level. The functions and financial resources of the planning and development regions are not clear, and there is no distinct separation between the Council's development tasks, and the counties' competencies. This may be the result of limited co-operation between Győr-Moson-Sopron and Vas county, as

---

[3]   Act No. XXI. of 1996 made it possible to organise planning-development regions on the basis of the counties' voluntary association.

well as between Vas and Zala, while there has as yet been no regional co-operation at all between Zala and Győr-Moson-Sopron. Therefore, each county, and city, continues to follow its own particular interests. The northern part, which comprises of Győr-Moson-Sopron county, strongly gravitates towards Vienna, and is interested in forming the Vienna-Bratislava-Győr-Sopron municipal region (or association) in the near future, although the details of such a potential organisation have still to be discussed. In this part of Hungary, the development resources are concentrated in the *regional* authorities which therefore primarily organise *regional* cross-border co-operations. In the border region's southern areas, however, co-operation between small regions dominates. This is expressed in several settlement-to-settlement contacts along the border. The Graz-Szombathely-Zalaegerszeg-Maribor municipal co-operation could also develop, but, in contrast to the Vienna-Bratislava-Győr-Sopron municipal region, this potential municipal co-operation would be composed of centres with nearly identical capabilities. Altogether, the quite different structures and opportunities in the three counties Győr-Moson-Sopron, Vas, and Zala have not yet been sufficiently recognised by the Regional Development Council of West Transdanubia and thus require further attention. Furthermore, the Hungarian part of the Austro-Hungarian border region decreases the potential for cross-border co-operation with Austria, if the involved actors do not agree on some common interests.

As mentioned above, the effect of the Austrian partners on the Hungarian part of the Austro-Hungarian border region and its economy will probably not grow in the future but stabilise on the present level, or even slightly decrease. According to a survey of foreign enterprises in Hungary (DŐRY and MÁTHE 1996), a significant number of Austrian enterprises appeared in Győr-Moson-Sopron and Vas county before and after 1989/90. Most were multinational companies that invest also in other Central and Eastern European countries. Aiming at an international network of affiliations, they also plan to establish plants in the Hungarian regions along the Ukrainian, Romanian and East-Slovak borders. According to the survey's findings, however, the companies' belief is that Hungarian privatisation is finished, and that there are no more economic units in the western border region that are worth buying out. Also the dynamics of greenfield investments have considerably decreased in the two counties in question. There are now business parks in all of the centres (e.g. Győr, Szombathely, Sopron), and the settlement of companies specialising in mass and labour-intensive production even caused a labour shortage in some of these places. Regarding future investments of Austrian companies, however, the businesses that were included in the survey seem rather to refocus on defending their local (Austrian) position. Since Austria's EU-membership triggered fierce competition within Austrian markets, Austrian-owned companies in the north-western Hungarian border region prefer, at least in the near future, to concentrate their resources and developments in their home country rather than to reinvest in Hungary.

Finally, it should be stressed that day-to-day life shapes the cross-border relations within the Austro-Hungarian border region. The region's attraction is considerable within Hungary. There are increasingly more people moving to the region from other parts of the country and increasingly more new businesses being established. Individual and institutional actors do not only appreciate the proximity of the Austrian market and the attractive opportunities offered by the neighbouring country, but also the vivid character of the local and regional economies. The functions of the centres are becoming more diversified due to an expansion of contacts and business networks. Sopron, Szombathely, Kőszeg, Mosonmagyaróvár and Győr are similarly-sized centres in the Austro-Hungarian border region that offer more business opportunities in regard to catering, trade or numerous other services than the neighbouring, less populated Austrian towns. In day-to-day life, both Hungarian counties are strongly connected to the Austrian areas, and the contacts with Hungary are also important for the population living in the adjoining Austrian settlements. In regard to everyday life, therefore, the Austro-Hungarian region can be regarded as a well functioning Euregion, but, as the latter part of this essay has shown, there are still several problems to be solved at the level of institutionalised socio-economic co-operation, especially in regard to the Burgenland, but also within the Hungarian part of the Austro-Hungarian border region.

References

DANCS, L. (1999): Határon átnyúló kapcsolatok Északkelet-Alföldön. (Cross-Border Connections in North-East Great Plain) In: BARANYI, B. (ed.): Észkkelet-Magyarország és a határok. (North-East Hungary and the Borders). Manuscript. MTA RKK Research Institute in the Great Plain, Research Group of Debrecen, Debrecen, 329-371

DANKÓ, L., G. FEKETE, É. (1996): Az országhatáron átnyúló gazdasági kapcsolatok és ezek kölcsönhatása a területi fejlődésre Északkelet-Magyarországon. (Cross-Border Economic Connections and Their Effects on the Spatial Development in North-East Hungary). Manuscript. MTA RKK Department of North Hungary, Miskolc

DICZHÁZY, B. (1997): Javaslatok a hazai vállalkozások külföldi terjeszkedésének elősegítésére. (Proposals for the Promotion of Foreign Expansion of Domestic Enterprises). Valóság (Reality) 2, 93-106

DÖRY, T., MÁTHÉ, M. (1996): Az osztrák-magyar gazdasági kapcsolatok értékelése. (Evaluation of the Austrain-Hungarian Economic Connections) MTA RKK NYUTI, Győr

GOLOBICS, P. (1995): A határ menti térségek városainak szerepe a regionális együttműködésben. (The role of towns in cross-border areas in regional co-operation). Pécs (= Proceeding of JPTE Department of General Social Geography and Urbanistics 3)

GORZELAK, G. (1998): Regional and Local Potential for Transformation Infrastructure Poland. European Institute for Regional and Local Development, Warsaw (= Regional and Local Studies 14)

HAJDÚ, Z. (1996): A magyar-horvát határ menti együttműködés dilemmái. (Dilemmas of Hungarian-Croatian Cross-Border Co-operation). In: PÁL, Á., SZÓNOKYNÉ ANCSIN, G. (eds): Határon innen - határon túl. (From and Beyond Borders). International Geographical Scientific Conference, JPTE, Szeged, 306-313

KOVÁCS, K., VÁRADI, M. (1996): Karöltve: A regionális együttműködés esélyei a Bécs-Pozsony-Győr háromszögben (Go Pari Passu With: The Chances of Regional Co-operation in the Vienna-Bratislava-Győr Triangle) Műhely (Workroom) 3, 1-11

LENGYEL, I. (1996): Határtalan lehetőségek? Békés megye gazdasága és határ menti kapcsolatai. (Borderless Possibilities? Economy and Cross-Border Connections of Békés County). Manuscript. Kőrösi Csoma Sándor College, Institute of Economics, Békéscsaba-Gyula

PÁL, Á. (1996): Dél-alföldi határ menti települések társadalom-gazdaságföldrajzi jelentősége. (Social-Geographical Economic Importance of Border Settlements in the South Great Plain) In: PÁL, Á., SZÓNOKYNÉ ANCSIN, G. (eds): Határon innen - határon túl. (From and Beyond Borders). International Geographical Scientific Conference, JPTE, Szeged, 181-191

RECHNITZER, J. (ed.) (1990): A nyitott határ. (The Open Border: The Innovation-oriented Development of Economic and Intellectual Resources in the Austro-Hungarian Border Regions). MTA RKK NYUTI, Győr

RECHNITZER, J. (1993): Szétszakadás vagy felzárkózás? A térszerkezetet alakító innovációk. (Parting or Closing Up? Innovations Forming the Spatial Structure). MTA RKK, Győr

RECHNITZER, J. (1996): Az osztrák magyar határ menti térségek együttműködésének új dimenziói, egy potenciális eurorégió körvonalai. (New Dimension of Austrian-Hungarian Cross-Border Co-operations, the Sketch of a Potential Euroregion). Manuscript. MTA RKK NYUTI, Győr

RECHNITZER, J. (1997): Eurorégió vázlatok az osztrák-magyar-szlovák határ menti térségben. (Outlines of an Euroregion in the Austro-Hungarian-Slovakian Border Region). Tér és Társadalom (Space and Society) 2, 29-58

RECHNITZER, J. (1999): Bécs-Pozsony-Győr innovációs tengely és a magyar területfejlesztési koncepció. (The Vienna-Bratislava-Győr Innovation Axis and the Hungarian Spatial Development Concept). Manuscript. MTA RKK NYUTI, Győr

RECHNITZER, J. (1999): Határ menti együttműködések Európában és Magyarországon. (Cross-border Co-operations in Europe and in Hungary). In: Nárai,

M., Rechnitzer, J. (eds): Elválaszt és összeköt a határ. (The Dividing and Connecting Border). MTA RKK, Győr and Pécs, 9-73

RUTTKAY, É. (1995): Határok, határ mentiség, regionális politika. (Borders, Border Situation, Regional Policy). Comitatus (December), 23-35

SÜLI-ZAKAR, I. (1997): A Kárpátok Eurorégió szerepe a határon átnyúló kapcsolatok erősítésében. (The Role of the Carpathians Euroregion in the Intensification of Cross-border Connections). Comitatus (June), 30-44

TÓTH, J. (1996): A Kárpát-medence és a nemzetközi regionális együttműködés. (The Carpathian Basin and the International Regional Co-operation). In: PÁL, Á., SZÓNOKYNÉ ANCSIN, G. (eds): Határon innen - határon túl. (From and Beyond Borders). International Geographical Scientific Conference, JPTE, Szeged, 22-27

VUICS, T. (1996): A határ menti együttműködés lehetőségei Baranya megyében. (The Possibilities of Cross-border Co-operations in Baranya County). In: PÁL, Á., SZÓNOKYNÉ ANCSIN, G. (eds): Határon innen - határon túl. (From and Beyond Borders). International Geographical Scientific Conference, JPTE, Szeged, 299-306

Contributions to Economics

Druck: Strauss Offsetdruck, Mörlenbach
Verarbeitung: Schäffer, Grünstadt